CLIMATE DYNAMICS IN HORTICULTURAL SCIENCE

VOLUME 1

Principles and Applications

CLIMATE DYNAMICS IN HORTICULTURAL SCIENCE

VOLUME 1

Principles and Applications

Edited by

**M. L. Choudhary, PhD, V. B. Patel, PhD,
Mohammed Wasim Siddiqui, PhD,
and Syed Sheraz Mahdi, PhD**

Apple Academic Press Inc. | Apple Academic Press Inc.
3333 Mistwell Crescent | 9 Spinnaker Way
Oakville, ON L6L 0A2 | Waretown, NJ 08758
Canada | USA

©2015 by Apple Academic Press, Inc.

First issued in paperback 2021

Exclusive worldwide distribution by CRC Press, a member of Taylor & Francis Group
No claim to original U.S. Government works

ISBN 13: 978-1-77463-097-6 (pbk)
ISBN 13: 978-1-77188-031-2 (hbk)

Library of Congress Control Number: 2014957360

Library and Archives Canada Cataloguing in Publication

Climate dynamics in horticultural science.

Includes bibliographical references and index.
Contents: Volume 1. Principles and applications/edited by M.L. Choudhary, PhD, V.B. Patel, PhD, Mohammed Wasim Siddiqui, PhD, and Syed Sheraz Mahdi, PhD -- Volume 2. Impact, adaptation, and mitigation / edited by M.L. Choudhary, PhD, V.B. Patel, PhD, Mohammed Wasim Siddiqui, PhD, and R.B. Verma, PhD
ISBN 978-1-77188-031-2 (v. 1 : bound).--ISBN 978-1-77188-070-1 (v. 2 : bound).--ISBN 978-1-77188-071-8 (set)
1. Horticulture. 2. Horticultural crops--Climatic factors. 3. Climatic changes. I. Choudhary, M. L., 1953-, editor

---

SB318.C42 2014   635   C2014-907285-6

Apple Academic Press also publishes its books in a variety of electronic formats. Some content that appears in print may not be available in electronic format. For information about Apple Academic Press products, visit our website at **www.appleacademicpress.com** and the CRC Press website at **www.crcpress.com**

CLIMATE DYNAMICS IN HORTICULTURAL SCIENCE

VOLUME 1

Principles and Applications

Edited by

**M. L. Choudhary, PhD, V. B. Patel, PhD,
Mohammed Wasim Siddiqui, PhD,
and Syed Sheraz Mahdi, PhD**

APPLE
ACADEMIC
PRESS

Apple Academic Press Inc.	Apple Academic Press Inc.
3333 Mistwell Crescent	9 Spinnaker Way
Oakville, ON L6L 0A2	Waretown, NJ 08758
Canada	USA

ISBN 13: 978-1-77463-097-6 (pbk)
ISBN 13: 978-1-77188-031-2 (hbk)

Library of Congress Control Number: 2014957360

Library and Archives Canada Cataloguing in Publication

Climate dynamics in horticultural science.

Includes bibliographical references and index.
Contents: Volume 1. Principles and applications/edited by M.L. Choudhary, PhD, V.B. Patel, PhD, Mohammed Wasim Siddiqui, PhD, and Syed Sheraz Mahdi, PhD -- Volume 2. Impact, adaptation, and mitigation / edited by M.L. Choudhary, PhD, V.B. Patel, PhD, Mohammed Wasim Siddiqui, PhD, and R.B. Verma, PhD
ISBN 978-1-77188-031-2 (v. 1 : bound).--ISBN 978-1-77188-070-1 (v. 2 : bound).--ISBN 978-1-77188-071-8 (set)
1. Horticulture. 2. Horticultural crops--Climatic factors. 3. Climatic changes. I. Choudhary, M. L., 1953-, editor

| SB318.C42 2014 | 635 | C2014-907285-6 |

Apple Academic Press also publishes its books in a variety of electronic formats. Some content that appears in print may not be available in electronic format. For information about Apple Academic Press products, visit our website at **www.appleacademicpress.com** and the CRC Press website at **www.crcpress.com**

ABOUT THE EDITORS

Dr. M. L. Choudhary

M. L. Choudhary, PhD, is currently Vice Chancellor of Bihar Agricultural University in Sabour, Bhagalpur, Bihar, India. He received a master's degree in horticulture from Banaras Hindu University, Varanasi, Utter Pradash, India, and his PhD in the USA. The Government of India deputed him for an advanced study on Hi-Tech Horticulture and Precision Farming in Israel and Chile. In a career spanning 30 years, he has held several executive positions, including Horticulture Commissioner, Ministry of Agriculture, India; Chairman, Coconut Development Board, Kochi (Kerala); Ministry of Agriculture, India; National Project Director, FAO; Visiting Scientist at Rutgers University, New Jersey, USA; Head of the Department of Ornamental Crops, IIHR, Bangalore; and Head of the Division in Floriculture at IARI, New Delhi. As Horticulture Commissioner of the Ministry of Agriculture, India, he has conceived, conceptualized, and implemented flagship programs such as the National Horticulture Mission and the National Bamboo Mission and Micro-irrigation.

Apart from his professional career, he was also Chairman and Member Secretary of various committees constituted by the Government of India as well as state governments. He has guided 18 PhD students and 21 MSc students in the field of horticulture and has published 124 book chapters, 19 books, and 109 research papers of national and international repute. Dr. Choudhary has represented India at various international forums in the capacity of Chairman and Member. He was also the Chairman of the Codex Committee of the Scientific Committee for Organic Standard of Ministry of Commerce. He has been conferred with 15 awards from various scientific and nonscientific organizations and government committees for his outstanding contribution in the field of horticulture/ floriculture. He has also been awarded the Fellowship of Horticultural Society of India. Dr. Choudhary in his 30 years of academic, research, and administrative career has visited more than 15 countries to participate in various professional meetings.

Dr. V. B. Patel

V. B. Patel, PhD, is presently working as University Professor and Chairman at Bihar Agricultural University, Sabour, Bhagalpur, Bihar. He has worked on developing leaf nutrient guides, nutrient management strategies, use of AMF for biohardening and stress tolerance, and a survey of indigenous germplasm of fruit crops. He developed the Leaf Sampling Technique and Standards as well as made fertilizer recommendations for several fruits. He has developed nutrient management through organic means for high-density planted mango.

Dr. Patel has organized eight national/ international seminars and workshops as convener, associate convener, or core team member, including the Indian Horticulture Congresses, the International Seminar on Precision Farming and Plasticulture, the National Seminar on Organic Farming, Seminar on Hitech Horticulture, and National Seminar on Climate Change and Indian Horticulture. He has guided three MSc and one PhD students. He earned his PhD from the Indian Agricultural Research Institute, New Delhi.

Dr. Patel has received a number of national awards and recognitions for his work in the field of horticultural research and development, including being named a Fellow of The Horticultural Society of India, New Delhi; Associate, National Academy of Agricultural Sciences (NAAS), New Delhi, India; Agricultural Leadership Award (2012) by Agricultural Today, Centre for Agricultural and Rural Development, New Delhi; Hari Om Ashram Trust Award (2007); Lal Bahadur Shastri Young Scientist Award, ICAR, New Delhi (2009); Young Scientist Award, NAAS (2005-06); Yuva Vigyanic Samman from Council of Science and Technology, Govt. Uttar Pradesh (2005); and AAAS Junior Award (2005) from the Indian Society for Plant Physiology, New Delhi. He has also received best research paper/ poster paper presentation awards from different organizations. He has published 41 research papers, 16 books and bulletins, and many popular articles, among other publications.

Dr. Mohammed Wasim Siddiqui

Mohammed Wasim Siddiqui, PhD, is an Assistant Professor/Scientist in the Department of Food Science and Technology, Bihar Agricultural University, India, and author or co-author of 21 peer-reviewed journal papers, 12 book chapters, and 16 conference papers. Recently, Dr. Siddiqui has initiated an international peer-reviewed journal, *Journal of Postharvest*

Technology. He has been serving as an editorial board member of several journals.

Dr. Siddiqui acquired a BSc (Agriculture) degree from Jawaharlal Nehru Krishi Vishwavidyalaya, India. He received MSc (Horticulture) and PhD (Horticulture) degrees from Bidhan Chandra Krishi Viswavidyalaya, Mohanpur, Nadia, India, with specialization in the Postharvest Technology. He was awarded the Maulana Azad National Fellowship Award from the University Grants Commission, New Delhi, India. He has received several grants from various funding agencies to carry out his research works during his academic career.

Dr. S. Sheraz Mahdi

S. Sheraz Mahdi, PhD, is working as Assistant Professor (Agronomy) at the Bihar Agricultural University, Sabour, Bhagalpur, Bihar, with specialization in crop modeling. He worked as a Research Associate at the Indian Meteorological Department (IMD), Srinagar, Kashmir. He took his PhD (Agronomy) degree from Sher-e-Kashmir University of Agricultural Sciences and Technology of Kashmir (SKUAST-K) and is the recipient of the Young Scientist Award from the Department of Science and Technology, New Delhi. He has also worked as an independent researcher in the cold arid region of Kargil Ladkh for about four years in collaboration with the Department of Agriculture, KVK-Kargil and the Himalyan Ecological and Conservation Research Development Foundation (HECRF). Dr. Mahdi has also published more than 25 research papers and some popular articles, covering aspects of advanced agronomy in particular.

CONTENTS

LIST OF CONTRIBUTORS

Tarun Adak
Central Institute for Subtropical Horticulture (ICAR), Rehmankhera, P.O. Kakori, India

M. S. Ahmad
Department of Food Science and Technology, Bihar Agricultural University, Sabour, Bhagalpur, Bihar, India

Nazeer Ahmed
Central Institute of Temperate Horticulture, Old Air Field, Rangreth–190007, Srinagar, J&K, India

Shirin Akhtar
Department of Horticulture (Vegetable and Floriculture) Bihar Agricultural University, Sabour, Bhagalpur, India

M. Anandaraj
Indian Institute of Spices Research, Kozhikode, 673 012, Kerala, India

S. J. Ankegowda
Indian Institute of Spices Research, Kozhikode, 673 012, Kerala, India

Ram Asrey
Division of Post Harvest Technology, Indian Agricultural Research Institute, New Delhi 110012, India, E-mail: ramu_211@yahoo.com

Kalyan Barman
Department of Horticulture (Fruit and Fruit Technology), Bihar Agricultural University, Sabour, Bhagalpur 813 210, Bihar, India

A. Chatterjee
Molecular Biology Section, Centre of Advanced Study in Botany, Banaras Hindu University, Varanasi–221 005 (U.P.), India

B. Chempakam
Indian Institute of Spices Research, Kozhikode 673 012, Kerala, India

W. S. Dhillon
Punjab Horticultural Post-Harvest Technology Centre, PAU Campus, Ludhiana - 141 004, India

P. P. S. Gill
Department of Fruit Science, PAU, Ludhiana, India

Bharat Sing Hada
Division of Floriculture and Landscaping Indian Agricultural Research Institute, New Delhi, India

Pranab Hazra
Department of Vegetable Crops Faculty of Horticulture Bidhan Chandra Krishi Viswavidyalaya, Mohanpur, Nadia, India

T. Janakiram
Division of Floriculture and Landscaping Indian Agricultural Research Institute, New Delhi, India

K. K. Jindal
Dr Y.S. Parmar University of Horticulture and Forestry, Nauni Solan–173 230, India

Pritam Kalia
Division of Vegetable Science Indian Agricultural Research Institute, New Delhi, India

K. Kandiannan
Indian Institute of Spices Research, Kozhikode 673 012, Kerala, India

K.S. Krishnamurthy
Indian Institute of Spices Research, Kozhikode 673 012, Kerala, India

Amit Kumar
Division of Fruit Science, Sher-e-Kashmir Univ. of Agri. Sci. & Tech., Kashmir, Shalimar, India
Email: khokherak@rediffmail.com

M. S. Ladaniya
Principal Scientist National Research Centre for Citrus, P.B. No. 464, Shankar Nagar P.O., Nagpur–440010, India

Prativa Lakhotia
Division of Floriculture and Landscaping Indian Agricultural Research Institute, New Delhi, India

S. Lal
Central Institute of Temperate Horticulture, Old Air Field, Rangreth–190007, Srinagar, J&K, India

S. Sheraz Mahdi
Department of Agronomy, Bihar Agricultural University, Sabour–810 213, Bhagalpur (Bihar), India

Abhishek Naik
Department of Vegetable Crops Faculty of Horticulture Bidhan Chandra Krishi Viswavidyalaya, Mohanpur, Nadia, India

H. R. Naik
Division of Food Technology, Sher-e-Kashmir University of Agricultural Sciences and Technology of Kashmir Shalimar Campus–191121 (J &K), India

B. K. Pandey
Central Institute for Subtropical Horticulture (ICAR), Rehmankhera, P.O. Kakori, India

Sapna Panwar
Division of Floriculture and Landscaping Indian Agricultural Research Institute, New Delhi, India

V. B. Patel
Department of Horticulture (Fruit and Fruit Technology) Bihar Agricultural University, Sabour, Bhagalpur (813210) Bihar, India

K. V. Prasad
Division of Floriculture and Landscaping Indian Agricultural Research Institute, New Delhi, India. Email: kvprasad66@gmail.com

H. Ravishankar
Central Institute for Subtropical Horticulture (ICAR), Rehmankhera, P.O. Kakori, India

P. K. Ray
Horticulture Department, Rajendra Agricultural University, Pusa–848 125, India

T. R. Rupa
Directorate of Cashew Research Puttur – 574 202, Karnataka, India. E-mail: dircajures@gmail.com, dircajures@yahoo.com

Kamesh Salam
World Bamboo Organization (WBO) and South Asia Bamboo Foundation (SAbF)

B. R. Salvi
RFRS, Vengurle, Maharashtra, India

P. L. Saroj
Director, Principal Scientist (Soil Science) E-mail: dircajures@gmail.com, dircajures@yahoo.com

Nirmal Sharma
Division of Fruit Science, Sher-e-Kashmir Univ. of Agri. Sci. & Tech., Jammu, Chatha, India

Mohammed Wasim Siddiqui
Department of Food Science and Technology, Bihar Agricultural University, Sabour, Bhagalpur, Bihar, India

A. K. Singh
Central Institute for Subtropical Horticulture (ICAR), Rehmankhera, P.O. Kakori, India

Anil K. Singh
Department of Horticulture, Institute of Agricultural Sciences, Banaras Hindu University, Varanasi–221 005, India

Balraj Singh
NRC on Seed Spices, Tabiji, Ajmer–305206, Rajasthan, India

V. K. Singh
Central Institute for Subtropical Horticulture (ICAR), Rehmankhera, P.O. Kakori, India

Anjana Sisodia
Department of Horticulture, Institute of Agricultural Sciences, Banaras Hindu University, Varanasi–221 005, India

S. S. Solankey
Department of Horticulture (Vegetables and Floriculture), Bihar Agricultural College, Bihar Agricultural University, Sabour (Bhagalpur) – 813 210, Bihar, India

R. K. Solanki
NRC on Seed Spices, Tabiji, Ajmer–305206, Rajasthan, India

R. K. Yadav
Division of Vegetable Science Indian Agricultural Research Institute, New Delhi, India

LIST OF ABBREVIATIONS

ABA	Abscissic Acid
ASP	Amnesic Shellfish Poisoning
AZP	Azaspiracid Shellfish Poisoning
BMPTC	Building Materials and Technology Promotion Council
BVOCs	Biogenic Volatile Organic Compounds
BVP	Bacteria, Viruses, and Parasitic Protozoa
CBF	C-Binding Factor
CFCs	Chlorofluorocarbons
CI	Chilling Injury
CMS	Cell Membrane Stability
CMT	Cell Membrane Thermo Stability
CSRB	Cashew Stem and Root Borers
CWSI	Crop Water Stress Index
DSP	Diarrheic Shellfish Poisoning
EC	Electrical Conductivity
ECU	Effective Chill Units
ESP	Exchangeable Sodium Percentage
ET	Evapotranspiration
FAO	Food and Agriculture Organization
FBD	Fruit Bud Differentiation
FCF	Fungal Culture Filtrate
GDHOC	Growing Degree Hours Celsius
GDP	Gross Domestic Product
GHGs	Green House Gases
GLOFs	Glacial Lake Outburst Floods
GRAS	Generally Recognized As Safe
GWP	Global Warming Potential
HCFCs	Hydrochlorofluorocarbons
HDP	High Density Planting
HSGs	Heat Shock Granules
HSPs	Heat Shock Proteins
ICAR	Indian Council of Agricultural Research

ICFRE	Indian Council of Forestry Research and Education
ICT	Information and Communication Technologies
INM	Integrated Nutrient Management
IPCC	Intergovernmental Panel on Climate Change
IPM	Integrated Pest Management
IWRM	Integrated Water Resource Management
LAR	Leaf Area Ratio
LEA	Late Embryogenesis Abundant
MAS	Marker-Assisted Selection
MSL	Mean Sea Level
NBDA	Nagaland Bamboo Development Agency
NEH	North Eastern Hilly
NO	Nitric Oxide
NPP	Net Primary Productivity
NSP	Neurotoxic Shellfish Poisoning
OA	Osmotic Adjustment
OTC	Open Top Chamber
PR	Pathogenesis-Related
PSP	Paralytic Shellfish Poisoning
PUT	Putrescine
QTL	Quantitative Trait Locus
RCTs	Resource Conserving Technologies
REC	Relative Electrical Conductivity
RH	Relative Humidity
RWC	Relative Water Content
SA	Salicylic Acid
SD	Short-Day
SLA	Specific Leaf Area
SNP	Sodium Nitroprusside
SPD	Spermidine
SPM	Spermine
TMB	Tea Mosquito Bug
TYLCV	Transfer Tomato Leaf Curl Virus
UNCED	UN Conference on Environment and Development
UNFCCC	United Nations Framework Convention on Climate Change
WHO	World Health Organization
WUE	Water Use Efficiency
YP	Yield Potential

LIST OF SYMBOLS

A	photosynthesis
E	transpiration
N	nitrogen
P	phosphorus
AN	carbon assimilation
gm	mesophyll conductance
gs	solute potential
Si	silicon
Ta	air temperature
T_c	canopy
CO	carbon monoxide
F0	base fluorescence
CH_4	methane
CO_2	carbon dioxide
N_2O	nitrous oxide
SF_6	sulphur hexafluoride
SO_2	sulphur dioxide
$^1O^{2-}$	singlet oxygen
O_{2-}	superoxide anions
O_3	ozone
H_2O_2	hydrogen peroxide
b_1	biochemical limitation
K_1	leaf diffusive conductance
Ψ_{leaf}	leaf water potential
Ψ_p	turgor potential
Ψ_s	rate of change in solute potential

PREFACE

Climate change and increased climate variability in terms of rise in temperature, shifting rainfall pattern, and incidences of extreme weather events like severe drought, devastating flood, etc., have now started posing a real threat to the production of agricultural/horticultural crops and natural resource management in the several countries. The rate of climate change and warming expected over the next century is anticipated to be more than what has occurred during the past 10,000 years. Predictions clearly indicate that the impact of the climate change is going to be more intensified in coming years. Climate change is already affecting—and is likely to increase—invasive species, pests, and disease vectors, all adversely affecting agri-horticultural crops productivity. Climate change makes agriculture more difficult to sustain with higher frequencies of floods, droughts, heat weaves, erratic rainfall, increased salinity, as well as a rise in sea level. These phenomena have direct impact on crop yield, quality, and consequently food security. Attempts to develop pioneering solutions to adapt climate change and address its own needs are required urgently.

Horticulture has emerged as the most dynamic and indispensable part of agriculture, offering a wide range of choices to farmers for crop diversification and much-needed nutrition for people. The changing of dietary habits of people with an improved standard of living has increased the demand for horticultural products. Horticultural crops are sometimes more vulnerable to the changed scenario of climatic aberrations. The dependence of crops on natural resources is sometime more pronounced on cereals or pulses, and so is their reaction to the extreme variations in seasonal temperature. Increase in productivity of horticultural crops is essentially required to cope with the demand of the increasing populace with diminishing per capita arable land and water, with degrading soil resources, and with expanding biotic stresses, and efforts to mitigate the adverse impact of climate change on productivity deserve attention of crop scientists/growers. Meeting the goal of enhanced production of nutritious fruits and vegetables in altered climatic conditions is achievable, but the strategies that enhance tolerance of crops to such variability in the growing environment of this region are yet to be formulated.

This book, *Climate Dynamics in Horticultural Science: Volume 1: Principles and Application*, deals with the basic concepts of climate change and its effects on horticultural crops. Different effects of climate change and their possible mitigation strategies are discussed. This book—along with its companion volume, *Volume 2: Impact, Adaptation, and Mitigation*—will be a standard reference work for policymakers who are charged with developing programs and legislation. Moreover, researchers will also understand the implications of climate change to provide a sound basis for decisions on adaptive strategies that enhance the agri-horti sector's ability to manage climate risks and take advantage of opportunities.

— **M. L. Choudhary, PhD, V. B. Patel, PhD,**
Mohammed Wasim Siddiqui, PhD, and
Syed Sheraz Mahdi, PhD

CHAPTER 1

HI-TECH HORTICULTURE AND CLIMATE CHANGE

P. K. RAY

Horticulture Department, Rajendra Agricultural University, Pusa - 848 125, India; E-mail: pkray@gmail.com

CONTENTS

1.1   INTRODUCTION

Horticulture crops, in general, are more knowledge and capital-intensive than staple crops. Today horticulture in India is a more vibrant and dynamic sector than ever before. It contributes nearly 30% of the agricultural GDP. Annual production of 74.8 million tons of fruits, 146.5 million tons of vegetables and 1.03 million tons of loose flowers (NHB, 2011) have to be increased substantially to cope with increasing demand of these commodities due to increasing population and expanding domestic and external markets. Short-term growth and long-term viability of any sector are critically dependent on access to technical knowledge, the ability to adapt that knowledge to local conditions and the flexibility to develop new production systems as market conditions change. Successful production of a horticultural crop depends on understanding of various factors affecting plant growth, fruiting, and manipulation of these factors for higher productivity and improved quality cultural activities.

In the past three decades aspects like plant density, planting time, manuring, irrigation, weed management, intercropping, training and pruning, disease and insect pests management have attracted attention of the scientists and a great deal of work has been done on standardizing these practices for higher productivity per unit area and possible reduction in the cost of production. In the recent past, the growth of horticulture sector has been far better than the overall growth of the agriculture sector. However, expanding the scale of horticultural production is often complicated by substantial problems. In the last 20–25 years, research and development scenario in horticulture has gained momentum with an impressive public and private support. The research institutes have developed a large number of technologies to improve the productivity and quality of fruits, vegetables and flowers. Some of these technologies require high degree of instrumentation and involves specific skills and accuracy to perform them. They are generally considered under hi-tech horticulture.

1.2   CLIMATE CHANGE

There are growing evidences to show that climate change has already affected agricultural productivity and will put increasing pressure on agriculture in the coming decades. Record breaking extreme weather events in the recent past different parts of the world offered a glimpse of the challenges climate change would bring. Analysis of recorded climatic datasets clearly indicates that there has been a 0.3 °C to 0.6 °C warming of earth surface since the late nineteenth century. The average global temperature has increased by 0.8 °C in the past 100 years and is expected to rise by 1.8 °C to 4.0 °C by the year 2100. For Indian region (South Asia), the Intergovernmental Panel on Climate Change (IPCC) has predicted 0.5 to 1.2 °C rise in temperature by 2020, 0.88 to 3.16 °C by 2050 and 1.56 to 5.44 °C by 2080, depending on the scenario of future development. The atmospheric warming will also be associated with changes with rainfall patterns, increased frequency of extreme events of drought, frost and flooding. Since the late 1970s, there have been increases in the percentage of the globe experiencing extreme drought or extreme moisture surplus.

The Intergovernmental Panel on Climate Change (IPCC) predicts that by 2050, mean temperatures around the planet may rise by between 2 and 5 °C or more and atmospheric CO_2 concentration are likely to be >550 ppm (cf. 380 ppm at present). Tropical and semitropical climates in particular are expected to experience dramatic increases in temperatures, as well as more variable rainfall (Jarvis et al., 2010). Of serious concern is the fact that most of the world's low-income families dependent on agriculture live in vulnerable areas, namely in Asia and Africa. Farmers having small land holdings in India will need to adapt to higher temperatures and shifting precipitation patterns. In addition, climate variability will likely cut into global food production, exacerbating the existing problems of poverty, food insecurity, and malnutrition. Furthermore, the greenhouse gas emissions are once again rising rapidly, making the climate change challenge to food security much greater.

In general, alterations in our climate are governed by a complex system of atmospheric and oceanic processes and their interactions. In the context of crop production, relevant atmospheric processes consist of losses in beneficial stratospheric ozone (O_3) concentration and increasing concentrations of the surface-layer trace gases, including atmospheric carbon dioxide (CO_2), methane (CH_4), nitrous oxide (N_2O) and sulfur dioxide (SO_2). Surface level O_3; SO_2; and CO_2 have direct impacts on crops, while CO_2, CH_4 and N_2O are critical in altering air temperature.

Particular attention is paid to likely changes on extreme events and sea level alterations. It is reported with high to very high confidence that in the 1990–2100 periods most extreme events will increase in intensity or frequency, or both. The published reports on the subject predict higher maximum temperatures and a greater number of hot days, higher minimum temperatures and fewer cold days, reduced diurnal temperature ranges, more intense precipitation events, increased risk of drought in summer periods, increases in peak wind intensities of cyclones, and increases in mean and peak precipitation intensities of tropical cyclones. On top of that, sea level is predicted to increase by 0.09–0.20 m.

1.3   CONSEQUENCES

The major changes in the earth's atmosphere are the concentrations of CO_2, which have increased by about 25% since the beginning of the industrial revolution. The CO_2 concentration has increased from preindustrial level of about 280 ppm to 393 ppm in 2010. Carbon dioxide enhances photosynthesis and depresses plant respiration; these effects are expected to increase plant growth as well as affecting various other processes. However, a number of plants physiological processes are also affected by changes in temperature, ozone, ultraviolet radiation, nutrients and water, all of which are variable factors often associated with climatic change.

Weather is the most important cause of year-to-year variability in crop production, even in high-yield and high technology environments. There has been considerable concern in recent years about possibility of climatic changes and their impact on the crop productivity. Today the entire world is suffering from global warming and its consequent climate change. Its impact on productivity and quality of crops has been documented fairly well. Since a crop could be defined as a biological system tailored to give certain products, the product output and quality is bound to vary with change in the growing environment.

Crop productivity will not only be affected by changes in climatically related abiotic stresses (i.e., increasing temperatures, decreasing water availability, increasing salinity and inundation) and biotic stresses (such as increases in pests and diseases), but also changes in the atmospheric concentration of carbon dioxide, acid deposition and ground level ozone. Hence, a key challenge is to assess how crops will respond to simultaneous changes to the full range of biotic and abiotic stresses. Responding to these challenges will require advances in crop research and the adoption of appropriate technologies.

1.3.1 IMPACT ON FRUIT CROPS

The recurrent developmental events of phenology and seasonality in vegetative flushing or bud differentiation distinguish trees from annual or agricultural crops. The fact that trees live over multiple growing seasons implies that every year there is a considerable renewal cost of some organs (leaves and fine roots), and that trees are more responsive or susceptible to climatic changes. When horticultural productivity is the goal, the allocation of resources toward reproductive processes must be maximized. However, the tree must also preserve its growth potential for future years; thus, a delicate C balance must be maintained between vegetative and reproductive needs. In spring, stored sugars and nutrients support actively growing shoots and inflorescences. Competition occurs between vegetative and reproductive meristems, and the fruit is growing essentially on the currently produced photosynthates. Fruits represent a major C sink in tree crops. The relationship between fruit load and photosynthetic activity (Palmer et al., 1997), as well as the effects of several climatic variables, has been intensively studied (Buwalda and Lenz, 1995; Wibbe et al., 1993). Among all tree crops, cultivated fruit trees are the ones most adequately supplied with water and nutrients; thus, there should be few, if any, constraints to a positive CO_2 response (Janssen et al., 2000).

Climatic change effects are not caused by a single factor (e.g., elevated CO_2), but originate from complex interactions among various factors such as atmospheric CO_2, air temperature, nutrient supply, tropospheric ozone level, UV-B radiation, drought frequency, etc. A reduction of stomatal conductance under elevated CO_2 might have a significant effect on water transport in trees, since the latter is roughly proportional to stomatal conductance. Hydraulic conductivity was reported to decrease with elevated CO_2 (Tognetti et al., 1996) but this effect is very species-specific. A decrease in stomatal conductance in response to CO_2 enrichment is commonly observed in many crops. However, under elevated CO_2, the increase in WUE is usually greater than the reduction of stomatal conductance, especially under drought conditions. Among all plant organs, fine tree roots generally show the greatest response to elevated CO_2. In addition to increases in fine-root density, trees may enhance their nutrient uptake capacity through alterations in root morphology and architecture. Trees grown under elevated CO_2 initiate more lateral root primordial, leading to increased root branching and a more thorough exploration of the soil. In addition to changes in fine-root density, morphology and structure, alterations in root functioning are also frequently observed in the changed environments. However, it has been a general observation that the long-lived plants have more time to acclimate to changing environmental conditions than the short-lived organisms. On a time scale, this acclimation might occur in the order of several years. The acclimation process might be influenced by seasonal changes in environmental conditions.

1.3.2 IMPACT ON VEGETABLES

Increasing CO_2 will enhance photosynthesis and improve water-use efficiency, thus increasing yield in most vegetable crops. Relative benefits from increased CO_2 can often be maintained with modest water and N deficiency, but yield benefits on an absolute basis are reduced when water or N limit growth. The impact of increasing temperatures is more difficult to predict. Seed germination will probably be improved for most vegetables, as will vegetative growth in regions where mean daily temperatures during the growing season remain under 25 °C, assuming adequate water is available. Reproductive growth is extremely vulnerable to periods of heat stress in many important vegetable fruiting crops, such as tomato, pepper, bean and sweet corn, and yield reductions will probably occur unless production is shifted to cooler portions of the year or to cooler production regions. This vulnerability results from the shortened duration of grain, storage tissue, or fruit filling and from failure of various reproductive events, especially the production and release of viable pollen. Processing crops, which are sometimes direct-seeded and are more frequently grown in cool-summer areas, are more likely than fresh-market crops to benefit from higher temperatures. In general, crops with a high harvest index, high sink demand, indeterminate growth and long growth seasons are considered most likely to respond positively to the combination of higher CO_2 and temperature. Relatively few crops have been studied, however and cultivars within a crop often differ in their responses, thus making generalizations difficult.

In many crops, high temperatures may decrease quality parameters, such as size, soluble solids and tenderness. For fresh-market vegetable producers, even minor quality flaws can make their crops completely unsaleable in some markets. Reduced or more irregular precipitation will also decrease vegetable yields and quality, although soluble solids and specific weight may increase in some crops. Leafy greens and most cole crops are generally considered to be cool-season crops, so heat stress during the growing season would be detrimental to these species. High-temperature effects on lettuce and spinach and low-temperature effects on cole crops include induction of flowering and elongation of the seed stalk. Perennial crops also require an overwinter cool period. Thus, planting dates, production areas and cultivars may need to be adjusted if temperatures change.

1.4   HI-TECH HORTICULTURE

Hi-tech horticulture is now widely employed for the profitable commercial production of horticultural products. In general, hi-tech horticultural practices include practices that demand high level of precision for application of inputs and management of the crop right from sowing to harvesting. Examples of hi-tech horticulture are biotechnological tools used for characterization and developing new superior strains, cryopreservation, micro propagation, greenhouse or protected cultivation, hydroponics, drip and sprinkler irrigation, fertigation, integrated pest management (IPM), integrated nutrient management (INM), molecular diagnostics, HDP mechanization for harvesting, grading, packing and storage of fruits, vegetables and flowers, and quality management of value-added horticultural products throughout the entire cold chain.

Precision farming of fruits and vegetables is also sometimes synonymous to hi-tech horticulture as it focuses on the very latest techniques and innovations with respect to production of the crops. It implies that the grower knows precisely how to steer his production process to achieve optimal yield and quality of the concerned crop. By combining minimal input with maximal output, without wasting resources, he not only promotes environmental well-being, but also increases his profitability. In this article, discussion is limited to important technologies that are helpful in providing opportunities for reducing the impact of climate change in future, as climate change has been predicted to result in disruption of many farming or crop production systems. The Food and Agriculture Organization (FAO) predicts a 15–20% fall in global agricultural production by 2080. Consequently, developing suitable strategies to mitigate the impact of climate change is one of the biggest challenges for plant scientists today.

1.5 BIOTECHNOLOGY

A rapidly changing climate will require rapid development of new varieties of fruits, vegetables and flowers that are capable to withstand vagaries of the weather. As an adjunct of the well-established conventional approaches of plant breeding, biotechnology can enhance the speed, flexibility and efficiency of developing new hardy cultivars. Important contribution of plant biotechnology would be to improve stress tolerance traits in fruits, vegetables and other crops of commercial value in horticulture. Molecular markers are used to provide greater focus, accuracy and speed in crop breeding programs. In general, the opportunities for enhancing crop performance under stress conditions lie in identifying key traits that require enhancement, stress-relieving candidate genes and the appropriate plant stage of development where enhancement should occur. This process of gene identification and gene expression patterns associated with quantitative genetic traits in crops is becoming more attractive and promising as the scientific community develops and gains access to large-scale genomics resources such as EST (expressed sequence tag) databases, high-throughput gene expression profiling technologies and genomic mapping information. In recent past stress tolerance of plants has been improved through gene transfer, or at least identifies cases in which genetic engineering of plants has shed some light on the mechanisms by which stress tolerance or resistance is conferred.

Further, to improve the ability of crops to cope with existing and new stresses, it is imperative to develop a basic understanding of the mechanisms and processes by which plants respond to biotic and abiotic stresses. Many of the major abiotic stresses arise because of a common biochemical phenomenon, efforts to improve tolerance to one abiotic stress have potential to confer tolerance to other abiotic stresses. Research stations working on horticultural crops are maintaining a large number of germplasms of their mandate crops. They can help in breeding strains resistant to biotic and abiotic stresses. Cryopreservation i.e. storage of tissues, for example, meristem, embryo or recalcitrant seeds, etc. in liquid nitrogen at 196 °C has been adopted for storage of germplasm and to ensures safety of rare or endangered plant spp. This type of storage arrests all metabolic activities resulting in storage of material for a longer time.

In the past two decades, more emphasis has been given to modifying crops with stress tolerance-conferring genes. These efforts have demonstrated enhanced abiotic stress tolerance to drought, salinity, temperature, soil pH, and drought stress, and have provided validation to the concept of enhancing abiotic stress tolerance in crop plants via biotechnology. Biotechnology is thus providing new solutions to old enigmatic problems. We are discovering and understanding how to optimize metabolic pathways and physiological processes in plants to increase crop yield potential and crop production under challenging agricultural environments. Efforts are also under way to improve crop quality and nutritional value for human consumption as well as animal feed.

Climate change is expected to have major effects on population thresholds of microorganisms and disease vectors. The dynamics affecting host-pathogen interactions lead to the selection of new pathotypes or pathogens. They also determine the emergence of new diseases and pests. Increases in yield per unit of area will continue to depend largely on more efficient control of (biotic) stresses rather than on an increase in yield potential. Integrated crop management is therefore the basis for sustainable agriculture. The range of options for adapting to the changes increases with technological advances. Breeding for pest and disease resistance using molecular tools is critical and will remain an essential part of crop improvement programs in horticultural crops.

It is now apparent that the recent advancements in biotechnology are providing the research communities with new tools such as genomics and proteomics that will allow researchers to discover new genes and understand their function in higher numbers, and with greater speed and more precision. The knowledge gained from these technologies has helped bringing more opportunities and tools to solve agricultural problems that once were hard to approach or understand. Several transgenic varieties of horticultural crops showing resistance to biotic and abiotic stresses are available today (Kaur and Bansal, 2008; Mou and Scorza, 2011). However, molecular biology tools will never replace the input and role of crop breeders in improving agronomic traits, but these tools will enable them to be more responsive in both time and breadth of environmentally sensitive traits to meet agricultural market needs and opportunities. Biotechnology tools combined with conventional breeding should position us to be able to take greater care of the production environment and allow us to achieve adequate food production and security for the growing world population.

1.6   MICROPROPAGATION

Fast, large-scale vegetative multiplication of plants by application of tissue culture technique is known as micropropagation. It is used as an accelerated form of clonal propagation and usually conducted in growth chambers. The propagation is completed in several stages under controlled conditions and through this technique large quantity of disease free quality planting materials can be supplied within stipulated time. HDP of crops has necessitated faster multiplication of plants through this technique. Each step of the process can be manipulated by control of the tissue culture environment. Many people consider it as a biotechnology component as it is helpful in faster multiplication of a newly bred, superior cultivar or regeneration of genotypes conserved or kept under cryopreservation. In a situation, when large scale crop devastation occurs due to climatic catastrophes this can be a powerful tool for producing large number of plants of a particular clone or clones. Since most of the operations are performed under room conditions under sterile conditions, the impact of climate change on efficiency of production or multiplication is lesser in magnitude. However, hardening of TC plants at later stages suffers from weather aberrations. Although hardening is also done under polyhouse or nethouse conditions, storms/hurricanes can tear the coverings of these structures and cause substantial damage.

It is expected that more than 100 tissue culture units are operational in public and private sectors in our country, out of which 25% have gone for commercial production of plants. Although all these tissue culture units are having high production capacity, there is a cause for concern because the target crops are the same in all. Many institutions are working on developing protocols in different crops. However, today well-tested protocols are available only in case of banana, papaya, strawberry, orchids and anthuriums. Protocols are also available but not yet commercially adopted in respect of apple, citrus, black pepper, potato, ginger, rose and grape. Bulk of tissue-cultured material being exported is primarily of ornamental plants for which the domestic demand is limited.

1.7   MICRO-IRRIGATION/FERTIGATION

With apparent change in climatic conditions growth rate of plants is expected to be faster due to increased plant temperature, which reduces the window of opportunity for photosynthesis since the life cycle is truncated, whileboth heat and drought stress may inhibit growth directly at the metabolic level. This ultimately is responsible for lower productivity of crops. As frequency of drought like situations is increasing in many areas, efficient irrigation technology like drip or sprinkler irrigation can play a major role in improving productivity of crops not only in drought-prone areas but also in areas receiving normal rains. Drought and salinity are the major and most widespread environmental stresses that substantially constrain crop productivity in bothnonirrigated and irrigated agriculture. Irrigation management practices used to increase crop outputs exacerbate the detrimental impact of salinity in agriculture. In the recent years, the more emphasis has been laid on microirrigation techniques. In majority of the crops, yields for micro irrigation systems were better than surface method of irrigation. Micro irrigation system was found to result in 30 to 70% water savings as compared to traditional irrigation methods (flooding/furrow or check basin) in various orchard crops and vegetables along with 10 to 60% increases in yield. Mulching with drip further enhanced the crop yield to the tune of 10–20% and controlled weeds up to 30–90% (Rajput and Patel, 2008).

Fertilizer solutions can be injected into the irrigation system (fertigation) in commercial crop plantings to avoid stress and poor growth of plants. The water (and some of the fertilizers present) can then be recycled by pumping it back out of the holding tank or pond, after some of the impurities (sand and silt) have settled out. Recycled water has actually been shown to improve plant growth. In experiments with more than 100 species of ornamentals grown in 2.8 L containers, the mean relative growth of plants irrigated with continuously recycled water was 103% over that of the control. Another way to reduce runoff is to use pulse irrigation. In this system, instead of applying one heavy watering daily, a small amount of water is applied five or six times during the day. Very little water escapes from the container or runs off from the field. The production advantage to this is that less fertilizer has to be applied, because there is less leaching. Plant growth may be more effectively maximized by reducing moisture stress than by increasing fertilizer concentration. A fertilizer concentration ranging between 50 and 200 mg/L of nitrogen gives good results for potted plants. Sufficient evidences are available to show that water stress might limit growth more frequently than does limited nutrition under container production of flowering and foliage plants.

1.8   HIGH DENSITY PLANTING

HDP of perennial fruit trees like apple, pear, peach, mango, guava or litchi is considered as one of the technological advancements of recent past as it can lead to cost-output and fruit quality enhancements. The main factors here are intra and inter-row spacing, training system, and canopy and hedgerow profiles. This is perhaps the most important, irrevocable and decisive set of criteria, for these decisions are made before planting and often determine yield and the orchard's overall economic viability. The primary reasons for the adoption of high-density orchard systems have been earlier cropping and higher yields, which translate to higher production efficiency, better utilization of land, and a higher return of investment. Trees in high-density orchards may be free-standing, staked, or supported by a trellis. This is a function of the training system, the tree species the rootstock and cultivar selected, and the goals of the enterprise. Intensive orchards require a greater outlay of capital, labor and managerial skills, especially during establishment. The need for greater investment is a function of the larger number of trees and tree supports and will be especially significant if a wire supported training system is proposed.

The concept of HDP is becoming widespread all over the country, although there are limits to it, and varying constraints linked to given circumstances when the goal is to produce high-quality fruit at competitive prices. It is an approach dictate by two key factors: shorter life-span of the orchard to quickly depreciate start-up outlays (by years 4 to 8 of orchard life) by planting early cropping cultivars, and to establish a management system that will reduce per-unit production costs. Once established, high-density orchards with smaller trees require less labor per unit of fruit produced than low-density orchards. Smaller trees and readily accessible canopies are easier to harvest, and the need for using and transporting ladders is minimize or eliminated. The ability to harvest most of the fruit from ground level is also valuable in pick-your-own operations where the absence of ladders reduces concerns of liability. Pruning is less labor-intensive in many systems, and the trees are easier to manage, provided plant growth is regulated by early and regular cropping.

Pest control in high-density orchards is facilitated because tree canopies are smaller and in many systems (especially trellised ones) are not very deep. This allows enhanced spray penetration into the canopy and reduces the need for large orchard spray equipment. Studies on spray deposits have shown that the coverage of spray materials on the leaf surfaces of trellised and nontrellised high-density trees is better than on standard trees. This, however, varies with the training system. For example, horizontal canopies have reduced spray penetration compared to vertical ones and consequently may have higher insect and disease damage. One of the greatest advantages of dwarf trees is that the interior shaded areas of the trees is greatly reduced on each tree, and with many more trees per hectare, light penetration and orchard efficiency are dramatically increased.

1.9   INTEGRATED NUTRIENT MANAGEMENT

Modern nutrient management practices rely on fine-tuning the application of nutrients of satisfy specific needs of different tree organs at times most beneficial from the standpoint of tree productivity and fruit quality. An improved understanding of how tree nutrient reserves are built up and mobilized leads to fertilizer practices that optimize yield and fruit quality while minimizing excessive vegetative growth. The use of different rootstocks with various abilities to acquire nutrients from the soil is being explored to solve tree nutritional problems via genetic means rather than fertilizer manipulations alone. A better understanding of the genetic control of plant nutrient uptake and translocation on a molecular level will open new frontiers for further improving the efficiency of mineral nutrient acquisition and utilization with the use of less fertilizer. All these modern approaches to plant nutrition are aimed at minimizing or eliminating the environmental pollution that can potentially result from the use of fertilizers. In wake of the changing climatic scenario fertilizer practices will increasingly be assessed by their overall impact on yield, quality of the produce, soil health, and environment.

Accomplishments of the enhanced rate of productivity require that soil quality be either enhanced or at least sustained at the present level. The soil fertility can be managed in complete harmony with sustainable production of the crops by careful analysis of soil and plant tissues. Sophisticated instruments are available to do these analyzes rapidly and draw suitable inferences so that type of fertilizers and their quantity can be decided accurately as per the need of the situation.

Land use pattern and soil management practices during crop production over a period have large impact on soil health and whether the soil functions as a source or as a sink for carbon. Intensive cultivation has been depleting more nutrients than what was added externally through fertilizers. It is therefore not surprising that farmers have to apply more fertilizers to get the same yields. Micronutrients such as sulfur, zinc, manganese, etc. are generally not replaced externally and so have compounded the problem of nutrient deficiency. Integrated use of organics (organic manures and bio-fertilizers) and chemical fertilizers has been found to be promising in not only maintaining and sustaining higher productivity but also in providing stability in crop production.

Integrated nutrient management holds great promise in maintaining sustainability of soil health and its fertility. Use of vermicompost and bio-fertilizers has been found highly effective in sustaining productivity of the soil. Similarly, use of a consortium of biofertilizers has helped in reducing the recommended fertilizer dose in many horticultural crops. More than 50% micronutrient fertilizers could be saved by mixing them with just 1.0 to 1.2t BGS (Bio-Gas Slurry)/ha. Returning crop residues to soil over long periods has great bearing on its productivity and physicochemical and microbial properties. Crop residues at 50% of straw produced take care of micronutrient nutrition to crops in the long run. Similarly, use of green manure often takes care of micronutrients deficiency in the soil. Under changing climate when there is widespread deficiency of nutrients in the soil, integrated approach for nutrient management would definitely be more sustainable and rewarding.

1.10   INTEGRATED PEST MANAGEMENT

Fruits and vegetables suffer more from residual toxicity of chemicals in their flesh than cereals and pulses, as they are highly perishable and consumed soon after harvest. A plethora of insect pests and diseases frequently invade them. It has been reported that the climatic aberrations can alter pest dynamics of a place. Changing patterns of drought and heat, as well as elevated CO_2 are likely to be accompanied by a change in the whole spectrum of biotic stresses. For most of the commercial crops, more tropical environments are also associated with greater numbers of foliar pests and diseases. Therefore, climate change would be likely to result in increased risk of epidemics (Legrève and Duveiller, 2010). In dry regions, root diseases such as nematode infestation are also problematic since they further reduce the plant's ability to extract scarce water.

Climate change is likely to alter the distribution and severity of soil borne diseases affecting both intensive and low-input agricultural production systems. Naturally occur ring disease suppressive soils have been documented in a variety of cropping systems, and in many instances, the biological attributes contributing to suppressiveness have been identified. While these studies have often yielded an understanding of operative mechanisms leading to the suppressive state, significant difficulty has been realized in the transfer of this knowledge into the development of effective field-level disease control practices. Early efforts focused on the inundative application of individual or mixtures of microbial strains recovered from these systems, and known to function in specific soil suppressiveness. However, the introduction of biological agents into nonnative soil ecosystems typically fails to yield commercially viable or consistent levels of disease control. Of late, greater emphasis has been placed on manipulation of the cropping system to manage resident beneficial rhizosphere microorganisms as a means to suppress soil borne plant pathogens. One such strategy is the cropping of specific plant species or genotypes, or the application of soil amendments with the goal of selectively enhancing disease suppressive microbial communities.

Three essential components are required simultaneously for a disease to occur: a virulent pathogen, a susceptible host and a favorable environment for multiplication often referred to as the 'disease triangle.' Climate change, as well as sometimes fulfilling the last link of that triangle, can also drive evolutionary change in pathogen populations by forcing changes in reproductive behavior (Gregory et al., 2009). One of the most challenging aspects of adapting crops to climate change will be to maintain their genetic resistance to pests and diseases, including herbivorous insects, arthropods, nematodes, fungi, bacteria and viruses. Breeding for host resistance will continue to have a pivotal role in offsetting the adverse impact of climate. Rising temperatures and variations in humidity affect the diversity and responsiveness of agricultural pests and diseases and are likely to lead to new and perhaps unpredictable epidemiologies (Gregory et al., 2009). Factors driving new outbreaks include extraordinary climatic events and trends in temperature selecting pathogens and their natural enemies towards new critical thresholds for inoculums survival. Disease cycle components such as survival, infection, colonization processes and latency period, in addition to production and dispersal of inoculum, are all affected.

Integrated management for controlling potential new disease and pest epidemics are need of the hour. The management methods used for a particular insect or disease problem will depend on the insect species or pathogen involved the extent of the problem, and a variety of other factors specific to the situation and local regulations. Nevertheless, the efficient IPM practice will depend on a thorough knowledge of the pest life cycle, environmental conditions, cultural practices, and minimizing host plant abiotic stresses. A stressed plant grown under drought, saline or water-logged situation is much more susceptible to pest problems.

The need for more expensive control practices of a pest would be required if the problem is permitted to spread. Total elimination of a pest is not always feasible nor is it biologically desirable, if the process is environmentally damaging or leads to new, more resistant pests and eliminates beneficial fungi and insects. IPM uses as many management (control) methods as possible in a systematic program of suppressing pests to a commercially acceptable level, which is a more ecologically sound system.

InIPM better-targeted control with less chemical usage occurs because of the integration of additional biological and cultural management measures. Cultural control begins with the preplant treatment of soil mixes to suppress pathogens and pests. Other cultural control techniques include sanitizing of soil or growing media, not allowing any drought stress, providing good water drainage to reduce the potential of *Phytophthora* root rot and other damping-off organisms, reducing humidity to control *Botrytis*, minimizing the spread of pathogens by quickly disposing of diseased plants from the field. Biological control measures include predator insects and mites; beneficial nematodes and beneficial fungi and bacteria.

Bioinsecticides or biofungicides are preventative, rather than curative, and must be applied or incorporated before disease onset to work properly. Recently, neem (*Azadirachta indica*) products have been found highly effective among several botanicals used for pest control. Pheromone traps are being used to manage fruit flies in mango. Likewise, the beneficial fungus *Gliocladium virens* is an alternative to the chemical fungicide Benomyl. As higher plants have evolved, so have beneficial below-ground organisms interacting with the plant root system (the plant rhizosphere). Examples of this include symbiotic nitrogenfixing bacteria, which are important for leguminous plants, and selected nematodes that control fungal gnats. It is wellknown that beneficial mycorrhizal fungi (which naturally colonize the root systems of most major horticulture plants) can increase plant disease resistance, and helps alleviate plant stress by enhancing the host plant water and nutrient uptake. Mycorrhizae can also benefit propagation of cuttings, seedlings, and transplanting of liner plants. *Tricoderma virde* and *Tricoderma harzianum* have been frequently used in recent years to manage Fusarium wilt in crops. These species have plant growth-enhancing effects, independent of their biocontrol of root pathogens.

These beneficial microorganisms suppress fungal root pathogens by antibiosis (production of antibiotic chemicals), by parasitism (direct attack and killing of pathogen hyphae or spores), or by competing with the pathogen for space or nutrients, sometimes by producing chemicals such as siderophores, which bind nutrients (such as iron) needed by the pathogen for its disease-causing activities. The inhibitory capacity of these biocontrol antagonists increases in the presence of mycorrhizal fungi, and in the absence of plant pathogens there is a stimulation of plant growth by bacterial antagonists; somehow these bacteria stimulate plant growth, but the mechanism is not well known.

1.11   PROTECTED CULTIVATION

The name *'protected cultivation'* involves a series of techniques for the modification of the natural environment of plants, which totally or partially alter the microclimate conditions, with the aim of improving their productive performance. Among the protected cultivation techniques, it is worth noting that mulches, direct covers, net houses, low and high tunnels, glass and polyhouses and microirrigation, mist system, etc., play major roles. The main objectives of protected cultivation are, among others, to protect the crops from harmful temperatures, wind, rain, hail, snow, insect pests, diseases and predators, and creating a favorable microclimate that allows for the improvement in their productivity, product quality and early maturity.

Protected cultivation is thus ideal for the areas, which face frequent fluctuations in weather conditions or experience extreme events like prolonged cold or heat wave or severe droughts or unprecedented heavy rains. Walk-in tunnels are suitable to raise off-season nursery and off-season vegetable cultivation. Insect proof net houses can be used for virus-free cultivation of tomato, chili, sweet pepper and other vegetables particularly during the rainy season. Green houses can be used for high quality vegetable cultivation (Ghosh, 2009).

Greenhouses provide better control over environmental factors affecting plant growth. If environment is controlled, crops can be produced to specific market dates and the quality maintained by eliminating many of the variations and hazards associated with weather and biotic pests. Temperature can be regulated with varying degree of precision, damage from wind and rain is avoided; injury from plant diseases and insects is reduced to a great extent. Growing media, moisture content, and fertility levels can be adjusted to meet plant requirements. However, the precision with which the environment is regulated is determined by the ability of the grower to manage the greenhouses equipments and controls. A profitable operation demands maximum and efficient utilization of greenhouse space. Crop must be mature when demand is greatest. To meet this demand, complete control of the greenhouse environment is essential.

Climate change in the form of temperature rise, a distinct change in the rainfall pattern has already worried farmers in areas where crop production is monsoon dependent. It has also raised questions over the profitability of growing high value crops like roses, gerbera, anthurium, capsicum, tomato, cucumber, etc. under open field conditions. This year droughts in the initial phases and heavy rainfall towards the end disturbed the farming schedule in parts of our country. This has raised the scope of protected farming in our country. Fortunately, some farmers and enthusiasts are already reaping benefits of the system. Protected farming provides distinct advantages with respect to quality, productivity and better market price of the produce. Vegetable and flower growers can increase their income by cultivation of vegetables/flowers during the off-season. The vegetables produced during the normal season generally do not get good returns due to their large availability or glut in the markets. Considering the volatile climate, protected farming is now becoming more relevant to quality production of high value crops, for example, vegetables, flowers and other ornamental crops.

1.11.1 TEMPERATURE CONTROL

A basic reason for using a greenhouse is to control the temperatures at which plants are grown. During the winter months, night temperatures can be maintained at any level with a well-designed heating system and overhead or perforated tube ventilation. Good control is possible in the spring and fall for temperatures equal to or higher than ambient levels. In the summer, however, night temperatures will be higher than outdoor temperatures until the radiant energy absorbed by plants, benches, and walks is dissipated. Heat transfer can be accelerated with exhaust fans. Evaporative cooling, however, is not very effective at night and it raises the relative humidity to levels that favor disease development, particularly Botrytis. Summer night temperatures above ambient levels can be maintained easily if the heating system is in operation.

Day temperature control is an entirely different situation. Although greenhouses are constructed to regulate temperatures, they actually interfere with temperature control. The problem is mainly one of keeping the greenhouse cool. When outdoor temperature are low, it is relatively easy to maintain day temperatures within desired limits. Heat can be added through the heating system or it can be removed by overhead or perforated tube ventilation. As seasonal temperatures increase, however, precise control of day temperatures becomes more difficult. It generally requires forced ventilation or evaporative cooling. In the summer months, acceptable control of day temperatures requires ventilation with roof and side ventilators, use of curtains for roof shading, ventilation with exhaust fans, fan and pad for evaporative cooling, misting and the operation of an evaporative cooling system. Several methods for cooling greenhouses are available. They depend on convection or forced air movement with or without the evaporation of water.

1.12   CONCLUSIONS

Predictions on future climatic conditions indicate possibilities of increases in temperature from 1 to 3 °C by 2050 combined with some complex spatially explicit changes in rainfall. However, there remains high uncertainty in predictions of extreme events, especially hurricanes. Consequently, climate change is likely to invoke substantial changes to production of horticultural crop in a region and the severity with which biotic and abiotic stresses will affect the productivity of these crops. Since climate change can be expected to have varying effects in different areas on the expression of drought, salinity, water logging and pest infestation, the mitigation strategies also vary according to the prevailing situations. While there will be increased irrigation under drought conditions, urgent measures are required for irrigation in drought areas and drainage of water from localities getting excessive rain or flood causing waterlogged situations. New advance technologies can be helpful in offsetting the negative effects of climate change.

Biotechnology can enhance the speed, flexibility and efficiency of developing new cultivars showing resistance to biotic and abiotic stresses arisingdue toimpact of the climate change. Cryopreservation can be adopted for long-term storage of genetic materials. This can eliminate the chances of losing the genotype due to bad weather from the field gene bank. Through micropropagation technique large quantity of high quality, disease free planting materials can be prepared within a stipulated time. The concept of HDP is becoming widespread all over the country. The primary reasons for its adoption have been earlier cropping and higher yields, which translate to higher production efficiency, better utilization of land, and a higher return of investment. This system offers better opportunities than the traditional planting approach. Under changing climatic scenario, drought and salinity are the major and most widespread environmental stresses that substantially constrain crop productivity. Drip irrigation and fertigation systems have been found highly effective and beneficial with a yield gain of 10 to 60% and water saving from 30 to 70% compared to traditional irrigation methods.

Rising temperatures and variations in humidity not only affect the responsiveness of crops to various pests and diseases but are likely to have distinct effects on population thresholds of the insect pests, disease causing pathogens and both beneficial as well as harmful soil microbes. INM and IPM are better solutions for sustainable higher production of crops even in altered climatic conditions. Protected cultivation can be a suitable option to protect the crops from harmful temperatures, wind, rain, hail, snow, insect pests, diseases and predators in the areas, which face frequent fluctuations in weather conditions or experience extreme events like prolonged cold or excessive heat. Advancement in technologies has done fairly well in tackling various abiotic and biotic stress related problems. Tomorrow's horticultural industry will be dominated far more than today's by crop improvement, production, and protection technologies. Research generates technologies and technology generates a series of benefits to the growers. Thus, tomorrow's growers would be far more well versed in tackling the climatic vagaries only when more research is carried out on stress management of fruits vegetables and other horticultural crops.

KEYWORDS

- **Climate Change Consequences**
- **Hi-Tech Horticulture**
- **High Density Planting**
- **Integrated Nutrient Management**
- **Integrated Pest Management**
- **Protected Cultivation**

REFERENCES

Buwalda, J. G. & Lenz, F. (1995). Water use by European pear trees growing in drainage lysimeters. J. Hort. Sci., 70, 531–540.

Ghosh, A. (2009). Greenhouse Technology The future concept of horticulture, Kalyani Publishers, Ludhiana, India. 223.

Gregory, P. J., Johnson, S. N., Newton, A. C., & Ingram, J. S. I. (2009). Integrating pests and pathogens into the climate change food security debate. J. Exp. Bot., 60, 2827–2838.

Janssens, I. A., Mousseau, M., & Ceulemans, R. (2000). Crop Ecosystem Responses to Climatic Change: Tree Crops Climate Change and Global Crop Productivity, (Reddy, K.R. and Hodges, H. F. Eds.), CABI, Wallingford, Oxon OX10 8DE, UK, 245–270.

Jarvis, A., Ramirez, J., Anderson, B., Leibing, C., & Aggarwal, P. (2010). Scenarios of Climate Change within the Context of Agriculture Climate Change and Crop Production, (Reynolds, M. P. Ed.), CABI Wallingford, Oxfords hire, Cambridge, UK, 9–37.

Kaur, P., & Bansal, K. C. (2008). Transgenic in Horticulture Status, Opportunities and Limitations; Recent Initiatives in Horticulture, The Horticultural Society of India, New Delhi, 511–520.

Kumar, S. N. & Aggarwal, P. K. (2011). Climate Change and Future Strategies Horticulture to Agribusiness (Chadha, K. L., Singh, A. K. & Patel, V. B. (Eds.)), Westville Publishing House, New Delhi, 1–15.

Legreve, A., & Duveiller, E. (2010). Preventing Potential Disease and Pest Epidemics under a Changing Climate. Climate Change and Crop Production, (Reynolds, M. P., Ed.), CABI, Wallingford, Cambridge, UK, 50–70.

Mou, Beiquan, & Scorza. (2011). Transgenic horticultural crops challenges and opportunities, CRC Press, Boca Raton, USA 342.

NHM. (2011). Indian Horticulture Database 2011 (Kumar, B., Mistry, N. C., Singh, B. & Gandhi, C. P. (Eds.)), National Horticulture Board, Gurgaon, Haryana, 280.

Palmer, J. W., Giuliani, R., & Adams, H. M. (1997). Effect of crop load on fruiting and leaf photosynthesis of 'Braeburn M26' apple trees. *Tree Physiology*, 17, 741–746.

Peet, M. M., & Wolf, D. W. (2000). Crop Ecosystem Responses to Climatic Change Vegetable Crops Climate Change and Global Crop Productivity, (Reddy, K. R. & Hodges, H. F. (Eds.)), CABI, Wallingford, Oxon OX10 8DE, UK, 213–244.

Peet, M. M., Sato, S. & Gardner, R. (1998). Comparing heat stress on male-fertile and male-sterile tomatoes. *Plant, Cell and Environment*, 21, 225–231.

Rajput, T. B. S., & Patel, N. (2008). Micro irrigation and fertigation Development, Opportunities and Constraints, Recent Initiatives in Horticulture, The Horticultural Society of India, New Delhi, 337–350.

Ramkrishna, Y. S., Rao, G. G. S. N., Rao, G. S. & Vijay Kumar, P. (2006). Climate Change Environment and agriculture (Chadha, K. L. & Swaminathan, M. S. Eds.), Malhotra Publishing House, New Delhi, 1–30.

Reddy, K. R. & Hodges, H. F. (2000). Climate Change and Global Crop Productivity an Overview; Climate Change and Global Crop Productivity, (Reddy, K. R. & Hodges, H. F. (Eds.)), CABI, Wallingford, Oxon OX10 8DE, UK, 1–6.

Reynolds, M. P., Hays, D. & Chapman, S. (2010). Breeding for Adaptation to Heat and Drought Stress Climate Change and Crop Production, (Reynolds, M. P. (Ed.)), CABI, Wallingford, Oxfords hire, Cambridge, UK, 71–91.

Singh, B & Hassan, M. (2008). Protected Cultivation Status, Gaps and Prospects; Recent Initiatives in Horticulture. The Horticultural Society of India, New Delhi, 502–510.

Swarup, A. (2006). Fertilizer use; Environment and agriculture (Chadha, K. L. & Swaminathan, M. S. (Eds.)), Malhotra Publishing House, New Delhi, 172–191.

Tognetti, R., Giovannelli, A., Longobucco, A., Miglietta, F. & Raschi, A. (1996). Water relations of oak species growing in the natural CO_2 spring of Rapolano (central Italy*). Annales des Sciences Forestières,* 53, 475–485.

Whitford, R., Gilbert, M. & Langridge, P. (2010). Biotechnology in Agriculture Climate Change and Crop Production, (Reynolds, M. P. (Ed.)), CABI, Wallingford, Cambridge, UK, 219–244.

Wibbe, M. L., Blanke, M. M. & Lenz, F. (1993). Effect of fruiting on carbon budgets of apple tree canopies. *Trees*, 8, 56–60.

CHAPTER 2

CLIMATE CHANGE AND FRUIT PRODUCTION

W. S. DHILLON[1] and P. P. S. GILL[2]

[1]Punjab Horticultural Post- Harvest Technology Centre, PAU Campus, Ludhiana - 141 004, India; E-mail: wasakhasingh@yahoo.com

[2]Department of Fruit Science, PAU, Ludhiana, India.

CONTENTS

2.1 INTRODUCTION

In India, diverse horticultural crops are grown due to the availability of different climactic zones ranging from extreme temperate (chilgoza, pecannut, walnut) to tropical crops (banana, cashewnut, avocado). Hence, after China, India is the second largest country in terms of area and production of fruits in the world. The impact of horticulture industry in India is visible as it contributes to about 29.5% of Agriculture GDP from 13.5% of area, and also contributes substantially to the earning from total agricultural exports. Horticultural crops need high investment due to long juvenile period. Apart from investment the growers should have technical knowhow to minimize the risk involved in this venture.

The long-term climatic variations and sudden fluctuation in weather parameters raised doubts for investment in fruit industry. In order to sustain farm income and provide nutritional security, it becomes imperative to understand the possible impacts of climate change on various horticultural crops. The ever increase in greenhouse gasses emission resulted warming of climate by 0.74 °C over the last 100 years (Sthaye et al., 2006; Ghude et al., 2009). However, it is uncertain to comment on future, but according to IPCC 2001 the increase in global temperature and other weather events will continue with higher frequency in twenty-first centurydue to emission of greenhouse gases. The phenology of horticultural plants is greatly affected by maximum, minimum and mean temperatures. Every fruit crop has specific requirements for temperature for growth, flowering, fruit development and maturity. The fruits plants are being grown in area where it is best adapted to prevailing climates. Increased global temperature would require demand for more high temperature adaptable varieties. The effect of increasing air temperatures is the prolongation of the growing season (Chmielewski and Rötzer, 2001; Menzel and Fabian, 1999) and the modification of the phenological phases of individual plants, i.e. the sequence of the developmental stages. The climatic change will affect suitability and adaptability of current cultivars by altering the growing period. The broad effects can be summarized as:

- Shift in growing location of fruits particularly temperate crops;
- Effect on chilling requirement;
- Flowering and fruit set of fruit crops through heat stress, bee activity, etc.;

- Distribution of insect-pests and diseases and thus increased cost of inputs;
- Elevated requirements of irrigation; and
- Higher incidence of physiological disorders of fruits.

Most of the temperate fruit crops like apple, pear, peach, almond, cherry, plum, etc. have specific chilling requirements for breaking of dormancy (Samish and Lavee, 1982). The insufficient chillingdue to increased temperature may leads to development of one or more associated symptoms including delayed foliation, reduced fruit set, increased buttoning and reduced fruit quality. The dormancy symptoms may persist when winter is neither long nor not cold adequately to break the dormancy and such situation is referred as 'Delayed foliation' when temperate fruits do not complete their rest period because of mild winter conditions (Black, 1952; Ruck, 1975). Hauagge and Cummins (1991) reported that in addition to chill unit's accumulation, there are important interactions among cultivars and environmental factors that are responsible for terminating bud dormancy. In hill regions, plants grown with a winter dormancy requirement, an active growing period in summer, plant phenology is mainly driven by temperature. After the dormancy break the development of plants strongly depends on air temperature. With higher temperature the biochemical reactions are accelerated up to a threshold where enzyme systems are destroyed and cells die (Chmielewski et al., 2004). Over last 20 years, the mean temperature elevated between 1.45 °C and 2.32 °Cwhich is adversely affecting vernalization of temperate fruit crops (Ahmed et al., 2011). The northern hilly region of India showed a general rise in temperature by approximately 0.65 °C to 2.3 °C North-western Himalayas about 0.5 °Csince last century (Bhutiyani and Kale, 2002). A shift in area of these crops has already been witnessed in temperate zones of India. Higher temperature also increased demand for irrigation water, which is eventually affecting the productivity of fruit crops especially being cultivated in rain-fed regions.

The warmer temperature during dormant season induce early flowering in plants,which become vulnerable to spring frost. In Himachal Pradesh, the earlier plantation was primarily focused on varieties of delicious group. During various stages of growing season, it has been noticed that these varieties are adversely affected with changing climatic patterns (Bhatia, 2010). Above normal temperatures during flowering season affected fruit setting due todrying of flowers. Similarly, in fruits grown under subtropical conditions, climatic shifts is impacting fruit bud differentiation pattern, which affects time of flowering and fruit production (Ravishanker and Rajan, 2011). Lower temperatures could result in early flowering and poor fruit set due to floral abnormalities.

On the other hand, higher temperatures during flowering reduce the effective time for pollination by pollinators, poor pollinator activities and desiccation of pollen. The Tommy Atkins mango trees flowered within 10 weeks when held at day/night temperatures of 18/10 °C, whereas trees held at 30/25 °C produced vegetative growth and did not flower (Nunez Elisea et al., 1933). Higher temperature during microsporogenesis of pollen decreased pollen viability from 85 to 60 percent with increase in temperature from 33 °C to 36 °C (Issarakraisila et al., 1993). While studying the effect of higher temperature on twig drying in fruited and nonbearing four varieties of mango, Reddy and Singh (2011) observed that drying of shoots was more in fruited plants when compared to un fruited ones. Among different varieties, Dashehari (77.28%) showed maximum twig drying followed by Baneshan (32.0%) and Rajapari (33.93%), while lowest twig drying was recorded in Langra (19.05%). The extent of leaf scorching in plants of Baneshan and Rajapari was higher than other varieties. In banana, with increase in temperature of 1–2 °C beyond 25–30 °C resulted an increased leaf production and thereby reducing crop duration and increased production (Chaddha and Kumar, 2011).

2.2 HIGH TEMPERATURE

Most of the plant processes related to growth and yield are high temperature dependent and there can be an optimum temperature range for maximum yield. The optimum growth corresponds to the optimum temperature for the photosynthesis and other metabolic reactions. Plant growth and development is speeded up in annual crops, but, in perennial crops increase of several degrees of temperature than optimum could reduce photosynthesis thereby affecting productivity through reduced growth period. The particular crop that is currently being grown in areas where it is usually best-adapted to the prevailing climate may not be tomorrow. A significant increase in growing season temperatures will require shifts to new varieties that are more heat tolerant and have a higher optimum temperature for photosynthesis. Under prevailing unlimited conditions, development of such varieties is not an easy task. In extreme cases the well-suited fruits crops presently may have to be replaced with newer crops in the new environments. In areas where current temperature is below optimal for specific crops, there will be a benefit, but in areas where plants are near the top of their optimal range, yields will decrease. Even a minor climate shift of 1–2 °C could have a substantial impact on the geographic range of these crops. As fruit crops are perennial, moving production areas is not so easy.

2.3   COLD TEMPERATURE

The occurrence of frost due to change in climate is a major threat to fruits grown under the subtropical conditions of North-western India. Occurrence of cold wave in from west to eastern India during 2002–2003 caused severe damage to fruit plantations (Samra et al., 2003). The level of damage was much higher in mango (40–100%) and litchi (30–80%). Similarly, other evergreen fruits like guava, Kinnow and Ber were also affected, resulting in reduced fruit size with poor fruit quality. Among various mango varieties, the mortality was highest in Dashehari, followed by Amrapali and the Langra was least affected in Northern states of India. In Uttaranchal, two years old mango plantation recorded 80 percent mortality while the severity of damage was lesser with age of trees. The extent of injury in different fruit plants was in order of mango>papaya>banana>litchi>pomegranate>Indian goose berry. The cold wave affect to fruit trees was greater in low-lying areas where cold air settled and remained for a longer time on ground. Contrarily to response of evergreen plants, the deciduous fruit plants likepeach, plum, apple and cherry gave higher yield due to extended chilling hours in 2003.

Similarly, in 2007–2008, the temperature dipped down and frost occurred in many parts of Punjab due to 'Western disturbances' that caused freeze and chilling damage to various fruit plants particularly the mango (Gill and Singh, 2012). The damage was significantly high in frost prone belts of submontane parts of Punjab where most of the mango plantations were damaged severely. The magnitude of extent of injury varied from 50–100 percent irrespective of the age of fruit trees. Symptoms of cold injury damage included drying of shoots up to 2–3 meters from tips and entire upper tree plant foliage was killed with bark cracking and stripping. Furthermore, the old and young grafted mango trees were more vulnerable to frost than seedling plants. The extent of frost damage on various mango growing cultivars/strains at PAU Fruit Research Station, Gangian (Dasuya) in Hoshiarpur district of Punjab was recorded during January–February, 2008. The plants growing adjoining to the boundary or blocks covered with windbreak (jamun, eucalyptus trees) recorded minimum foliage damage. Mango trees planted without boundary plantation or away from the windbreak or shelterbelt had recorded maximum foliage damage. This situation resulted complete fruit yield loss in 2008 and the following year due to damage of terminal shoot with frost injury.

In banana, only a half leaf is produced if temperature falls below 15 °C and leaf emergence stopped below 9–10 °C (Turner, 2003). In subtropical regions the bunch emergence during winter season show uneven ripening and deformed fruits. Chilling symptoms are turning of leaf lamina to yellow in color, midrib may show brown areas which are water soaked underneath. The symptoms on older leaf resembles to that of potash deficiency. If temperature falls below freezing, bunches become water soaked, blacken and die (Ravi and Mustaffa, 2011). High temperatures reduce both the duration of flushing and the interval between flushes. Low temperature also affects the rate of reproductive development with panicles emerging earlier but taking longer time to reach anthesis. In litchi, average temperature of 18 °C is associated with higher numbers of female flowers and temperature of 23 °C decreases number of hermaphrodite flowers. It has been observed that the areas with winter temperature maximum of above 25 °C are not well suited for litchi cultivation.

2.4   HIGH CARBON DIOXIDE LEVEL

One of the possible beneficial effects of climatic shift for the plants may be obtained from elevated level of carbon dioxide in the atmosphere. In general, the plants that are raised under enhanced carbon dioxide show increased rate of net photosynthesis. This increased photosynthesis ultimately result in higher dry matter production and consequently increased yields of the plants (Kimball, 1983; Cure, 1985). But, the enhanced rate of photosynthesis may lead to enhance respiration rate thereby requiring more water to draw from the soil, resulting water stress under water deficient conditions. In a study, Robinson et al. (1998) reported that during the initial period of growth of sour orange plants, the bark, limbs, and fine roots growing in an atmosphere with 700 ppm of CO_2 showed rates of growth more than 170% higher than those at 400 ppm. When the trees grown older and are matured, this CO_2 induced enhancement lowered to approximately 100%. However, 127% increase in orange production was recorded from the trees under 700 ppm CO_2 exposure. Experiments have also shown that stomata conductance of C3 plants is initially reduced with elevated CO_2, leading to reduced transpiration and increased conservation of water (higher carbon assimilation per water lost); but, such responses are short- lived in some species (Bunce, 1992). Increased CO_2 has also been reported to ameliorate the negative effects of drought stress (Tolley and Strain, 1985).

2.5   FRUIT QUALITY

Temperature is an important factor that affects anthocyanin biosynthesis in plants. The expression of anthocyanin biosynthetic genes has been in-duced by low temperature. However, it has been repressed by high tem-perature in various plants such as apple (Ubi*et al.*, 2006), grape (Yamane*et al.*, 2006) and red oranges (Lo Piero*et al.*, 2005). In grapes, high tem-peratures above 115 °F cause thick skin of berries. Similarly, higher night temperatures reduce anthocyanin accumulation in berry skin and the de-crease in anthocyanin accumulation under high temperature results from factors such as anthocyanin degradation as well as the inhibition of mRNA transcription of the anthocyanin biosynthetic genes (Mori et al., 2005). In ripening apples, anthocyanin's are apparently induced at low temperatures (<10 °C) (Curry, 1997) while synthesis takes place under high irradiation at mild temperatures (20 °C to 27 °C) in detached, mature apples (Curry, 1997; Reay, 1999).

High temperatures results in poor red peel color development of apples (Wand et al., 2002, 2005). Felicetti and Schrader (2008) reported three types of sun-burn: a disorder that occurred due to heat and or light stress. The first type is sun-burn necrosis and is heat-induced when the tempera-ture of fruit surface reaches 126 °F for 10 min. It induces thermal death of cells in the peel followed by a necrotic (dark brown or black) spot. The second type of sun-burn is sun-burn browning and it is the most common type, which results in a yellow, brown, or dark tan spot on the sun exposed side of the fruit. Third type of sun-burn is induced when fruits are sud-denly exposed to full sun light. These apples have been shaded earlier and are not acclimated to sun exposure. The apples with sun-burn necrosis had higher relative electrical conductivity (REC) than REC for fruits with no sun-burn or with sun-burn browning (Schrader et al., 2001). Increased REC in fruits with sun-burn necrosis showed that membrane integrity was damaged, allowing electrolytes to leak freely.

In view of the drastic changes in climates which will beimpacting fruit crop production seriously there is urgent need to train the work force including horticultural researchers, extension workers and more importantly the fruit growers on climate change issues. The attention should be focused on studying the impact of climate change on growth, development, fruit yield and quality of fruit crops. Farmers in developing countries will be least able to adapt to climate change because of a relatively weak agricultural research base, poor availability of inputs and inadequate capital for investing in fruits farming. Adapting to climate change will be costly as farming cost includes increased use of water, fertilizers and pesticides to maximize harvest under the higher CO_2 regime. In case of fruit crops, the investment should be made for new plantations, and to create appropriate storage facilities for the produce. Also the research should be shifted towards evolving new fruit varieties or introduce new crops which could adapt to the changing climatic conditions in better way.

KEYWORDS

- **Climate Change**
- **Elivated CO2**
- **Fruit Production**
- **Fruit Quality**
- **High Temperature**
- **Postharvest Quality**

REFERENCES

Ahmed, N., Lal, S., Das, B., & Mir, J. I. (2011). Impact of climate change on temperate fruit crops. In: *Impact of Climate Change on Fruit Crops* (Eds. Dhillon, W. S., & Aulakh, P. S). Narendra Publishing House, New Delhi, 141–150.

Bhatia, H. S. (2010). Evaluation of new apple cultivars under changing climate in Kullu Valley of Himachal Pradesh. *Proc. National Seminar on Impact of Climate Change on Fruit Crops* (ICCFC–2010) at PAU Ludhiana, from 6–8 October, 2010, 1–6.

Bhutiyani, M. R., & Kale, V. S. (2002). Climate change in the last century Are the Himalaya warming. *Sapper* 13, 37–46.

Black, M. W. (1952). The problem of prolonged rest in deciduous trees. *Proc. 13th Int. Hort Congress*, held at London, 2, 1122–1131.

Bunce, J. (1992). Stomatal conductance, photosynthesis and respiration of temperate deciduous tree seedlings grown out doors at an elevated concentration of carbon dioxide. *Plant Cell Environ.*15, 541–549.

Chadda, K. L., & Kumar, S. N. (2011). Climate change impacts on production of horticultural crops In: *Impact of Climate Change on Fruit Crops* (Eds. Dhillon, W. S., & Aulakh, P. S.) 3–9.

Chmielewski, F. M., & Rötzer, T. (2001). Response of tree penology to climate change across Europe.*Agric. Forest Meteorol.*108, 101–112.

Curry, E. A. (1997). Temperatures for optimal anthocyanin accumulation in apple skin. *J. Hort. Sci.* 72, 723–729.

Felicetti, D. A. & Schrader, L. E. (2008a). Photo oxidative sun-burn of apples Characterization of a third type of apple sun- burn. *Int. J. Fruit Sci.* 8(3/4), 160–172.

Frank, M., Chmielewski, F. M., Müller, A., & Bruns, E. (2004). Climate changes and trends in penology of fruit trees and field crops in Germany 1961–2000.*Agricultural Forest Meteorology* 121, 69–78.

Gill, P. P. S., & Singh, N. P. (2012). Decline of mango diversity in sub-montane and *Kandi* zone of Punjab. An overview. *Indian J. Ecology* 39(2), 313–315.

Hauagge, R., & Cummins, J. N. (1991). Phenotypic variations of length and bud dormancy in apple cultivars and related Malus species. *J Amer. Soc. Hort. Sci.* 116(1), 100–106.

Issarakraisila, M., Considine, J. A., & Turner, D. W. (1993). Effects of temperature on pollen viability in mango cv. Kensington.*Acta Horticulturae* 341, 112–124.

Lo Piero, A. R., Puglisi, I., Rapisarda, P., & Petrone, G. (2005). Anthocyanins accumulation and related gene expression in red orange fruit induced by low temperature storage. *J Agricultural Food Chemistry* 53, 9083–9088.

Menzel, A., & Fabian, P. (1999). Growing season extended in Europe. *Nature* 397, 659.

Nunez Elisea, R., Davenport, T. L., & Calderia, M. L. (1993). Bud initiation and morphogenesis in 'Tommy Atkins' mango as affected by temperature and triazole growth retardants. *Acta Horticulturae* 341, 192–198.

Ravi, I., & Mustaffa, M. M. (2011). Impact of climate change on growth and development of banana. In: *Impact of Climate Change on Fruit Crops* (Eds. Dhillon, W. S., & Aulakh, P. S.) 87–93.

Ravishanker, H., & Rajan, S. (2011). Possible impact of climate change on mango and guava productivity. In: *Impact of Climate Change on Fruit Crops* (Eds. Dhillon, W. S., & Aulakh, P. S.) 151–156.

Reay, P. F. (1999). The role of low temperature in the development of the red blush on apple fruit (Granny Smith). *Scientia Hort.* 79, 113–119.

Reddy, Y. N., & Singh, O. (2011). Role of heat shock proteins in adaptivity of plants to higher temperatures and alleviation of heat shock in mango. In *Impact of Climate Change on Fruit Crops* (Eds. Dhillon, W. S., & Aulakh, P. S.) 57–64.

Robinson, A. B., Baliunas, S. L., Soon, W., & Robinson, Z. W. (1998). Environmental effects of increased atmospheric carbon dioxide. Oregon Institute of Science and Medicine, Cave Junction, Oregon, US. Email: infor@osim.org.

Ruck, H. C. (1975). Deciduous fruit tree cultivars for tropical and sub tropical regions. *Hort. Rev* 3, Common Wealth Burr. Hort. and Plantation Crops, East Malling, UK.

Samish, R. M., & Lavees, S. (1982). The chilling requirement of fruit trees. In: *Proc. XVI Int. Hort. Cong*, held at Brussels 5, 372–388.

Samra, J. S., Singh, G., & Rama Krishna, Y. S. (2003). Cold wave of 2002–2003 *Impact on Agriculture*. Nat. Res. Mgt. Div. ICAR, KrishiBhavan, New Delhi.

Schrader, L. E., Sun, J., Felicetti, D., Seo, J. H., Jedlow, L., & Zhang, J. (2003). Stress induced disorders Effects on apple fruit quality. http://postharvest.tfrec.wsu.edu/PC2004E.pdf.

Schrader, L. E., Zhang, J., & Duplaga, W. K. (2001). Two types of sun- burn in apple caused by high fruit surface (peel) temperature. *Plant Health Prog.* 10, 1094.

Tolley, L. C., & Strain, B. R. (1985). Effects of CO_2 enrichment and water stress on gas exchange of *Liquidambar styraciflua* and *Pinustaeda* seedlings grown under different irradiance levels. *Oecologia* 65, 166–172.

Ubi, B. W., Honda, C., Bessho, H., Kondo, S., Wada, M., Kobayashi. S., & Moriguchi, T. (2006). Expression analysis of anthocyanin biosynthetic genes in apple skin effect of UV–B and temperature.*Plant Science* 170, 571–578.

Soon, W., Baliunas, S. L., Robinson, A. B., & Robinson, Z. W. (1999). Environmental effects of increased atmospheric carbon dioxide. *Climate Research* 13, 149–164.

Wand, S. J. E., Steyn, W. J., Mdluli, M. J., Marais, S. J. S., & Jacobs, G. (2002). Over tree evaporative cooling for fruit quality enhancement. *South Afr. Fr. J.* 2, 18–21.

Wand, S. J. E., Steyn, W. J., Mdluli, M. J., Marais, S. J. S., & Jacobs, G. (2005). Use of evaporative cooling to improve 'Rosemarie' and 'Forelle' pear fruit blush colour and quality. *Acta Horticulturae* 671, 103–111.

Yamane, T., Jeong, S. T., Goto-Yamamoto, N., Koshita, Y., & Kobayashi, S. (2006). Effects of temperature on anthocyanin biosynthesis in grape berry skins. *American J. Enology Viticulture* 57, 54–59.

CHAPTER 3

CLIMATE CHANGE: IMPACT ON PRODUCTIVITY AND QUALITY OF TEMPERATE FRUITS AND ITS MITIGATION STRATEGIES

NAZEER AHMED and S. LAL

Central Institute of Temperate Horticulture Old Air Field, Rangreth-190007, Srinagar, J&K, India; E-mail: cith@nic.in

CONTENTS

3.1 INTRODUCTION

The North Western Himalayas have a unique and fragile ecosystem, where people are heavily dependent on their natural environment for their sustainance and livelihood and draws about 60% of Gross Domestic Product (GDP) from agri-horticultural system. The climate in this region temperate type mainly characterized by extreme cool winters and mild summers. It offers tremendous opportunity to produce high quality horticulture crops like apple, pear, peach, plum, almond, apricot, walnut *and* off-season vegetable, and ornamental crops. These crops covers an area of more than 500 thousand hectares and produces fruits of approximately 31 lakh tons. After independence there has been seen marked growth in area and production of these crops but on the other hand productivity has left far behind as compared to advanced countries. The low productivity is mainly attributed to several factors including environmental, physiological and biological. But over the years, environmental changes playing a significant role like occurrence of erratic rain and snowfall, droughts increase in temperature, etc. resulting in fast receding of glaciers. A significant change in climate at global and national level is certainly impacting horticulture and affecting our fruit production and quality. But understanding of impact of climate change on perennial horticultural production system and the potential effects on fruit quality has drawn a little attention of researchers. The current and future changes in climate patterns and presence of higher concentrations of anthropogenic gases in the atmosphere can have dramatic effects on yield, flavor and nutritional quality of fruit. These effects are likely to affect both growers and consumers. The projected increase in surface-air temperature can have both positive and negative effects on eating quality of fruit. The yield and quality may be inferior in fruit produced under water-deficit or high rainfall conditions. The staggered flowering due to inadequate chilling in temperate fruit crops can lead to low fruit set, non-uniform fruit quality and wider harvest windows. Fruit coloration may be severely affected because the biosynthesis of coloring pigments is strongly influenced by the temperature. The nutritional value and antioxidant potential of fruit may be affected due to decrease in skin pigments. The severity of certain physiological disorders in fruits is likely to increase, contributing economic losses to the growers. The projected increase in the incidence of insect-pests and diseases would further affect the fruit quality and consequently low pack-outs for farmers.

3.2   IMPACT OF CLIMATE CHANGE ON TEMPERATE FRUITS (A CASE STUDY)

The IMD monitoring reveals that temperatures are increasing in both Jammu region and Kashmir valley, with significant increase in maximum temperature of 0.05°C per year. The average mean temperature in Kashmir has risen by 1.45°C in last 28 years while in Jammu region, it has increased by the rise is 2.32°C.

As a result, of rise in temperature and decline in rainfall, the apricot and cherries are fast disappearing from some areas of Kashmir Valley. Due to general rise temperature and less availability of water, the yield and quality of apples in valley and mid temperate region of Jammu are fast deteriorating. Over the last few years, there has been distinct slow growth in production and productivity in rain-fed Kashmir's Karewas areas. Due to unusual hailstorms and windstorms in summer fruits like cherry, apple, plum, peach and apricot are getting damaged heavily.

In recent years there marked change in the pattern of snowfall in Kashmir, which is effecting all the pome and stone fruits (Fig. 3.1). It has been observed that the snowfall and flowering in some years is coinciding leading to great loss in quantity and quality.

FIGURE 3.1   Occurrence of late snowfall-causes heavy damage to almonds-coincided with full bloom.

Due to shortage of water for agronomic crops like rice shift has been recorded from agronomic crops to temperate fruits and nut in J&K as fruit crops are more remunerative as compared to agronomic crops (Fig. 3.2).

FIGURE 3.2   This formerly paddy land has been converted into an orchard in Khan Sahib Karewa due to water shortage. Similarly, thousands of Canals of irrigated paddy land have been converted into dry land.

In Himachal Pradesh, the study examines the impact of climate change in recent years on apple indicated its cultivation shifting towards higher altitudes. It is evident from the data that temperature in apple growing regions of Himachal Pradesh showed increasing trends whereas precipitation decreased over years (Table 3.1). The chill units calculated showed decreasing trends of chill units up to 2400 msl from Bajaura in Kullu at 1221 msl to Sarbo in Kinnaur at 2400 msl. The Dhundi station situated at 2700 msl showed increasing trend of chill unit at the rate of 25.0 CUs per year. The increasing trends of chill unit at 2700 msl suggested that area is becoming suitable for apple cultivation in higher altitude. These findings have also been supported by the farmers' perceptions, which clearly reflected that apple cultivation is expanding to higher altitude in Lahaul and Spitti. The average land use per farm in Lahual and Spitti showed more than 2% shift towards apple cultivation but it showed reverse trend in other apple growing regions. The income of the farmers increased more than 10% in Lahual and Spitti whereas it showed a decrease of more than 27% in Kullu and Shimla districts from fruits in recent decade compared to 1995. The secondary data on area under apple cultivation also compounded statement that apple cultivation is expanding in Lahaul and Spitti in recent decade (Singh et al., 2005).

In Uttarakhand the area under apple cultivation has drastically been reduced. It might be due to less rainfall and higher temperature in winter. It is causing major problem of chilling requirement, which is very important to meet for higher and quality production of apple.

TABLE 3.1 Changes in Temperature and Precipitation vis-a-vis Apple Acreage in Himalayan States During 1980–2010

State	Period	Avg. Annual Temp. (°C)	Rise in Temperature (°C) (1980–2008)	Precipitation (mm)	Area (ha)	Approx. new area covered under higher elevations
J & K	1980–85	13.01		726	63.09	-
Ladakh	1986–90	13.58		817	66.85	-
	1991–95	13.12		784	71.33	-
	1996–2000	13.91	1.45–2.32	585	82.18	6510
	2001–2005	14.46	0.5–1	682	96.34	8496
	2006–2008	13.32	(S) (W)	763	138.19	25,110
	2001–2002	-		Reduced	0.609	-
	2009–2010	-		(1973–2008)	0.836	227
Uttara-khand	1980–85	12.40		1394	-	-
	1986–90	11.45		1430	-	-
	1991–95	13.69	1.51	1104	52.70	-
	1996–2000	13.90		1067	51.80	-
	2001–2005	13.84		935	55.98	4180
	2006–2009	13.91		1245	31.66	−24,320

TABLE 3.1   *(Continued)*

State	Period	Avg. Annual Temp. (°C)	Rise in Temperature (°C) (1980–2008)	Precipitation (mm)	Area (ha)	Approx. new area covered under higher elevations
H.P.	1980–1985	13.03		1323	46.80	-
Solan/ Kangra	1991–2000	13.77		1270	83.20	36,400
	2001–2007	14.40		1023	90.20	7000
Mandi/ Chamba/ Sirmaur	1980–1985	-		-	953	−404
	2005–2007	-		-	549	16,804
Shimla/ Kullu	1980–1985	-	1.37	-	12368	20,516
	2005–2007	-		-	29172	6433
Kinaur/ Lahul-Spiti	1980–1985	-		-	30975	
	2005–2007	-		-	51491	
	1980–1985	-		-	2532	
	2005–2007	-		-	8965	

Source: IMD, Srinagar.

3.2.1   EFFECTS TEMPERATURE ON FRUIT QUALITY

Each plant species has its own characteristic response to temperature. The most biological activity low almost to zero below 5 °C. At still lower temperatures cell functions may be impaired and the plant damaged. But in recent years observed that higher average temperatures are major concerns, which lead to earlier dates of bloom and maturity, and greater fruit size in temperate fruit crops. These effects are likely to be small over the next 50 years.

3.2.2   EFFECTS ON SPROUTING

The impact of temperature change is most in apple and almond where trees sprout 2–3 weeks early but normally apples trees sprout in mid April. As a result last few years about 70% of trees began to open their buds in mid

March. At the end of March it can definitely become very cold again. At this time most trees have their buds open are very susceptible to frost damage.

3.2.3   EFFECT ON FRUIT SIZE

Warmer temperatures are very beneficial for some newer varieties. But it is a disadvantage though for the apple cv. *Cox Orange*; because the warmth produces too many cells per fruit and this results in fruit that is too big, too soft and goes off quickly. Fruit size is generally smaller, where low temperate occurs during spring, which affects cell division in the post-bloom period.

3.2.4   EFFECT ON FRUIT COLOR

In Kashmir valley, the failure of apples to change into their specific red shades, or an increase of apples with sunburn. The deep red color is a result of low temperatures during the night in autumn, just before harvesting. If the temperatures are not low enough, most apples fail to turn into their specific red shades. For many apples their red color is a trademark of quality but Ladakh province becomes potential area for apple cultivation due to climate change (Fig. 3.3).

FIGURE 3.3   Apple Fruit Quality at Ladakh.

3.2.5   EFFECT ON CHILLING REQUIREMENTS

Most deciduous fruit trees need sufficient accumulated chilling, or vernalization to break winter dormancy Inadequate chillingdue to enhanced greenhouse warming may result in prolonged dormancy, leading to reduced fruit quality and yield. The low warming scenario is less than 1 °C is unlikely to affect the vernalization of high-chill fruit (Apple, walnut, apricot almond, cherry varieties), and if warming scenario exceeds 1.5 °C and would significantly increase the risk of prolonged dormancy for both stone-fruit and pome-fruit. Japanese plums have a CR of between 500–800, while European cultivars have a CR in excess of 1,000 h. Periods of mild weather can upset the accumulated CR requiring further periods of cold weather to achieve sufficient hours. Mild winters may result in delayed or irregular flowering, reduced fruit set and an extended flowering period. The CR is a major concern in the marginal temperate area of North Western Himalayas where fruit trees with a low CR have to be grown. Therefore in some areas, it is impossible to grow cherries. For example, sweet cherry requires the accumulation of 1000 chill units at 3.8 °C in order to complete or 'break' dormancy. If chilling is inadequate, the development and/or the later expansion of leaf and flower buds may be impaired. Problems have already been experienced with poor cropping of blackcurrant after mild winters and the same might happen with raspberry, apple and other fruits as winter temperatures continue to increase.

3.2.6   EFFECT ON POLLINATION

Climate change during the past eight years has played a critical role in apple pollination failure. There are rains during the flowering season, which affect pollination by wind and insects. Low temperatures also adversely affect fruit set in apple. At present these Delicious group of cultivars mainly Starking Delicious and Red Delicious constitute nearly 80% of apple trees. Although India ranks 10thin world production of apple, yet the decreasing trend in productivity of its orchards, in the last decade due to changing climate scenario has caused a serious concern to the fruit growers and planners of the country.

More than 70% of orchards have less than 20% pollinizer proportion, whereas a minimum of 30–33% is required in our agroclimatic conditions for good fruit set. Moreover there is lack of diversity in pollinizing cultivars as mainly Golden Delicious and Red Gold are being predominantly used which have attained biennial bearing tendency and their bloom seldom coincides with the flowering period of Delicious cultivars. The population of natural pollinators has gone down due to indiscriminate use of pesticides and deterioration in ecosystem. Managed bee pollination is very limited and available bee hives during bloom hardly meet 2–3% of the demand. All these factors have lead to poor fruit setting of Delicious (Kjøhl et al., 2011).

3.2.7   EFFECT ON PRE-COOLING

Fruit crops are generally precooled after harvest and before packing operations. Cooling techniques have been used to remove field heat from fresh produce, based on the principle that shelf-life is extended 2 to 3 fold for each 10 °C decrease in pulp temperature. Rapid cooling optimizes this process by cooling the product to the lowest safe storage temperature within hours of harvest. By reducing the respiration rate and enzyme activity, produce quality is extended as evidenced by slower ripening/senescence, maintenance of firmness, inhibition of pathogenic microbial growth and minimal water loss (Talbot and Chau, 2002). Rapid cooling methods such as forced-air cooling, hydrocooling and vacuum cooling demand considerable amounts of energy. Therefore, it is anticipated that under warmer climatic conditions, fruit and vegetable crops will be harvested with higher pulp temperatures, which will demand more energy for proper cooling and raise product prices.

3.2.8   FRUIT RIPENING

High temperatures on fruit surface caused by prolonged exposure to sunlight hasten ripening and other associated events. One of the classical examples is that of grapes, where berries exposed to direct sunlight ripened faster than those ripened in shaded areas within the canopy. For fruits exposed to direct sunlight, pulp temperatures reached 35 °C and required 1.5 days longer to ripen than those than grew in the shade (Woolf et al., 1999). Cell wall enzyme activity (cellulose and polygalacturonase) was negatively correlated with fruit firmness, indicating that sun exposure, that is, higher temperatures during growth and development, can delay ripening. However this delay did not occur via a direct effect on the enzymes associated with cell wall degradation. In apples, treatments of 38 and 40 °C for 2–6 days did not have marked effects on respiration, although ethylene production was reduced.

3.2.9   QUALITY PARAMETERS

Flavor is affected by high temperatures. Apple fruits exposed to direct sunlight had a higher sugar content compared to those fruits grown on shaded sides increase in 10 °C increase in growth temperature caused a 50% reduction in tartaric acid content and malic acid synthesis is more sensitive to high temperature exposure during growth than was the synthesis of tartaric acid. Fruit firmness is also affected by high temperature conditions during growth. Changes in cell wall composition, cell number, and cell turgor properties were postulated as being associated with the observed phenomenon.

3.2.10   ANTIOXIDANT ACTIVITY

Antioxidants in fruit crops can also be altered by exposure to high temperatures during the growing season. In 'Kent' strawberries grown in warmer nights (18–22 °C) and warmer days (25 °C) had a higher antioxidant activity than berries grown under cooler (12 °C) days. The high temperature conditions significantly increased the levels of flavonoids and, consequently, antioxidant capacity. Higher day and night temperatures had a direct influence in strawberry fruit color. Berries grown under those conditions were redder and darker reported by (McKeon et al., 2006; Wang and Zheng, 2001).

3.2.11   PHYSIOLOGICAL DISORDERS AND TOLERANCE TO HIGH TEMPERATURES

Frequent exposure of apple fruit to high temperatures, such as 40 °C, can result in sunburn, development of watercore and loss of texture (Fig. 3.4). Moreover, exposure to high temperatures on the tree, notably close to or at harvest, may induce tolerance to low-temperatures in postharvest storage (Buescher, 1979; Hickset et al., 1983).

FIGURE 3.4   Incidence of physiological disorders in apple, apricot, pomegranate and nectarine due to climate change.

3.2.12. EFFECTS OF HIGHER CO_2 AND GHG ON FRUIT YIELD AND QUALITY

Carbon dioxide is important because carbon atoms form the structural skeleton of the plant. A doubling of carbon dioxide levels may increase plant growth by 40–50% though continuous high levels saturate the plant's ability to use carbon dioxide and the benefits decrease with time. If other factors remain favorable, increased carbon dioxide concentrations will lead to greater rates of photosynthesis in plants. Current carbon dioxide concentrations limit plant photosynthesis. Growers of protected horticultural crops have already aware from so many years that artificially raising the concentration of carbon dioxide up to certain stage in greenhouses can substantially increase crop growth and yield.

Effect on timing of bud burst, cessation of growth, altered concentrations of carbohydrates and plant hormones in turn altered the dormancy status of trees thereby changing the timing of bud burst and the length of the active growing period. Flowering and fruiting of trees are likely to be hastened under conditions of elevated carbon dioxide. The evidence for an effect of carbon dioxide concentration on leaf senescence and leaf fall is rather contradictory and may be species dependent. Most predictions of the direct effects of carbon dioxide suggested that average yields will increase by about 40–50% with a doubling of carbon dioxide concentrations Leaves are able to detect and respond rapidly to carbon dioxide concentration. Stomata opening decreases in response to increased carbon dioxide concentration.

3.2.13 EFFECTS OF OZONE EXPOSURE FRUIT YIELD AND QUALITY

Higher concentrations of atmospheric ozone are found during summer due to increase in nitrogen species and emission of volatile organic compounds (Mauzerall and Wang, 2001). Concentrations are at maximum values in the late afternoon and at minimum values in the early morning hours, notably in industrialized cities and vicinities. The opposite phenomenon occurs at high latitude sites. Another potential source for increased levels of ozone in a certain region is via the movement by local winds or downdrafts from the stratosphere.

3.2.14 EFFECT ON INCIDENCE OF MOSSES AND ALGAE

The increase in temperature, especially if combined with wetter winters, causes increased incidence of mosses and algae, many of which have lower threshold temperatures for growth than those of most flowering plants. Hotter and drier summers may limit or counter this increase, or at least result in the mosses and algae adopting their dry resting state for a larger proportion of the summer.

3.2.15 EFFECT ON INCIDENCE OF INSECT, PEST AND DISEASE

Erratic changes in temperature and precipitation leads to more incidences of insect, pest and disease. In the last few years, the attack of apple scab, powdery mildew in apple, flee beetle in almost all the temperate fruit crops has been increased (Figs. 3.5 and 3.6).

FIGURE 3.5 Powdery mildew of apple at dry and warmer summers.

FIGURE 3.6 Flea beetle in warm and humid climate.

3.2.16   EFFECT OF TEMPERATURE ON FROST SUSCEPTIBILITY

Climate change will lead to earlier growth and therefore to greater susceptibility to, and damage from, late spring frosts (Fig. 3.7). Increases in winter temperatures, anticipated in all scenarios, will result in a very substantial increase in the number of days with temperatures above freezing, and above 5 °C, thus extending and advancing the growing season. The concern expressed is that such early onset of growth as a result of climate change may increase the risk of frost damage to plants. Although the incidence of spring frost damage to precocious growth is not expected to increase with climate change, there is some indication that autumn frosts may become more damaging. Reduced or delayed hardening of plants in the autumn combined with reduced cloud cover and an increased diurnal temperature range could lead to increased damage. Frost damage can also occur during the dormant period, so the ability of plants to withstand winter frosts may also be affected by climatic warming.

FIGURE 3.7   Frost injury in Walnut.

3.2.17   INTERACTION EFFECT OF TEMPERATURE AND CARBON

The combination of increased temperature and increased carbon dioxide predicted in all climate change scenarios suggests that for some species the growth stimulation may be greater than the 40–50%. The doubling of carbon dioxide concentration combined with a 3 °C increase in temperature could lead to 56% stimulation in growth.

3.3   POTENTIAL OF CROP IN CARBON SEQUESTRATION

Carbon sequestration plays an important role in the global carbon cycle. Green plants remove (sequester) carbon from the atmosphere through photosynthesis. They extract carbon dioxide from the air, separating the carbon and oxygen atoms, returning oxygen to the atmosphere, and using the carbon to make biomass in the form of roots, stems, and foliage. The tablebelow has shown different temperate fruit crop potential as sequesters (Table 3.2).

TABLE 3.2   Estimates of Standing Biomass per Area (t/ha or Mg/ha), and Calculated CO_2 Equivalents per ha for Perennial Crops

Crops	Total DW (t/ha)	Total CO_2 (t /ha)	Main source of data for calculations
Apple	36	66	Jiménez and Diaz, (2004), Palmer et al. (1991)
Peach and Nectarine	40	73	Chalmers and Van Den Ende (1975); Jiménez and Diaz, (2003)
Plums, Prunes	62	114	Kroodsma and Field, 2006
Almond	100	183	Kroodsma and Field, 2006
Walnuts	75	138	Kroodsma and Field, 2006

3.4   ADAPTATION STRATEGIES FOR MITIGATING EFFECT OF CLIMATE CHANGE

3.4.1   BREEDING STRATEGIES

- Pheno-typing of apple genetic wealth to enhancing temperature, moisture stress and genetic enhancement for tolerance to biotic and abiotic stress.
- Varieties and rootstocks will be evaluated under controlled temperate moisture stress, etc., gradient to identify suitable cultivars of apple % rootstocks Experiments on varietal evaluation will also be conducted under natural conditions at different altitudes (with natural variations in temperature and moisture falling under various agro-climatic zones of the state.
- Marker assisted selection and development of transgenic having resistance to biotic and abiotic resistance.
- Development of genotypes having resistance to heat and drought.
- Crop diversification.

3.4.2   AGRONOMIC MANAGEMENT STRATEGIES

- Assessment of the vulnerability and climate risks associated with temperate fruit production system in temperate region.
- Development of cropping systems under various agro-climatic conditions.
- Improvement in the irrigation and drainage systems.
- Development of appropriate tillage and intercultural operations.
- Integrated nutrient management.
- Integrated pest management.
- Integrated weed management.
- Development of water harvesting techniques.

3.4.3   BIOTECHNOLOGICAL INNOVATIVE STRATEGIES

- Molecular characterization for various traits in relations to biotic and abiotic stress.
- Transformation of plants from C_3 to C_4 plants.

- Gene pyramiding against biotic and abiotic stress.
- In vitro conservation of rare and useful species for future use.

3.4.4 MITIGATION STRATEGIES FOR HIGHER CO_2/GHG

- Assessment the carbon sequestration potential of perennial pome and stone fruit crops production system.
- To participate in the international dialog about greenhouse gas emissions management, global warming and sustainable energy development.
- Use of biofuel like diseal from *Jetropa* and *Pongamia* sp.
- The development of nuclear energy.
- The improvements in the efficiency of electricity generation, transmission and distribution.
- The use of fuels with lower carbon content, for example, natural gas, CNG, Gobber gas.
- Fuel switching, appliance efficiency and use of renewable energy;
- Tree planting and forest management;
- Waste processing

3.4.5 FUTURE RESEARCH STRATEGIES FOR OPTIMIZING PRODUCTION UNDER CHANGING CLIMATE SCENARIO

3.4.5.1. CROP IMPROVEMENT STRATEGIES

- Utilizing the current and future regional climatic scenarios of the temperate region a microlevel survey of agro-climatic zones of J&K, HP and UK should be conducted to identify sensitive regions with high vulnerability with respect to pome and stone fruit crop.
- Introduction of low chilling cultivars of pome, stone and nut fruits.
- Diversification with other high value fruit crops like peach, apricot, walnut, kiwi and olive.
- Development of new genotypes having resistance to high temperature and CO_2 concentrations.
- Marker assisted selection and development of transgenic having resistance to biotic and abiotic resistance.

- Development of genotypes having resistance to heat and drought.
- Biotechnological approaches for multiple stress tolerance will be standardized.

3.4.5.2   DEVELOPMENT OF AGRO-TECHNIQUES

- Assessment of the impact of elevated temperature and CO_2 on growth, development, yield and quality of commercial fruit crop using open top chamber (OTC) facilities and free air CO_2 enrichment system (FACE).
- Sensitive stages of crops to weather aberrations will be identified.
- The phenology of pome, stone and nut crops under changing climate will be monitored.
- In-situ soil moisture conservation practices including indigenous technical know-how will be validated to mitigate the impact of drought.
- Development of suitable agronomic adaptation measures for reducing the adverse climate related production losses.
- Study the impact of climate change and development of technologies on water productivity.
- Identify and develop good practices to enhance the adaptation of crop to increased temperature, moisture and nutritional stress.
- Identification and mapping of climate resilient as well as climatically vulnerable microniches in temperate regions.
- Extreme events, such as late spring frost or windstorm, may cause crop failure. Future climate may also increase occurrence of extreme impacts on crops, for example, weather conditions resulting in substantial reduction in yield and quality (for example severe drought or prolonged soil wetness).
- To develop a set of high-resolution daily based climate change scenarios, suitable for analysis of agricultural extreme events.
- To identify climatic thresholds having severe impacts on yield, quality and environment for representative crops and to assess the risks that these thresholds will be exceeded under climate change.

3.4.5.3   PLANT PROTECTION STRATEGIES

- Assessment of the pest and disease dynamics, study of disease triangle and development of prediction models.
- Strengthen surveillance of pest and diseases.
- Development of ecofriendly pest-ecologies and management strategies and early warning systems.

3.4.5.4   POST HARVEST MANAGEMENT STRATEGIES

- Development of cost effective storage techniques.
- Development of varieties having longer shelf life.
- Studies on mitigation of post harvest spoilage.

3.4.5.5   HRD AND CREATING AWARENESS

- Organize seminars/symposia/trainings and conduct field demonstrations, on effective climate resilient technologies.

3.5   CONCLUSION

Climate change impacts are to be looked not in isolation but in conjunction with all the aspect of agriculture and allied sectors. The effects of climate change on horticulture sector are still uncertain but a significant overhaul in preharvest and postharvest practices may be required as an adaptation measure to maintain productivity and fruit quality under changing climatic conditions. In the light of possible global warming, researchers should more emphasis on development of heat- and drought-resistance crops. Research is needed to define the current limits to these resistances and the feasibility of manipulation through modern genetic techniques. Both crop architecture and physiology may be genetically altered to adapt to warmer environmental conditions. At the regional level, those charged with planning for resource allocation, including land, water, and agriculture development should take climate change into account. The continuation of current and new initiatives to research potential minimize the effects of climate change at farm, regional, national and international level and will help to provide a more detailed picture of how world horticulture and agriculture could change. Only then may we see the implementation of policies and other adaptations in agricultural systems that would minimize the negative effects of climate change and exploit the beneficial effects.

KEYWORDS

- **Adaptation Strategies**
- **Antioxidant Activity**
- **Chilling Requirements**
- **Climate Change**
- **Physiological Disorders**
- **Quality Parameters**
- **Temperate Fruits**

REFERENCES

Buescher, R. W. (1979). Influence of high temperature on physiological and compositional characteristics tomato fruits. *Lebensmittel-Wissenenshaft*, 26, 237–268.

Chalmers, D. J. & B. Ende, V. D. (1975). Productivity of peach trees: Factors affecting dry-weight distribution during tree growth. *Annals of Botany*, 39, 423–432.

Jimenez, C. M. & Diaz, J. B. R. (2004). Statistical model estimates potential yields in 'Golden Delicious' and 'Royal Gala' apples before bloom. *Journal of the American Society for Horticultural Science*, 129, 20–25.

Jimenez, C. M., & Diaz, J. B. R. (2003). A statistical model to estimate potential yields in peach before bloom. *Journal of the American Society for Horticultural Science*. 128(3), 297–301.

Kjøhl, M., Nielsen, A., & Christian, S. N. (2011). Potential effects of climate change on crop pollination. Centre for Ecological and Evolutionary Synthesis (CEES), Department of Biology, University of Oslo, Norway.

Kroodsma, D. A., & Field, C. B. (2000). Carbon sequestration in California agriculture. *Journal of applied ecology*, 16(5):1975–85.

Mauzerall, D. L., & Wang, X. (2001). Protecting agricultural crops from the effects of tropospheric ozone exposure Reconciling science and standard setting in the United States, Europe, and Asia. *Annual Review of Energy and the Environment, Technology*, 12, 162–164.

McKeon, A. W., Warland, J., & McDonald, M. R. (2006). Long-term climate and weather patterns in relation to crop yield: A mini review. *Canadian Journal of Botany*, 84, 1031–1036.

Palmer, J. W., Cai, Y. L. & Edjamo, Y. (1991). Effect of part-tree flower thinning on fruiting, vegetative growth and leaf photosynthesis in 'Cox's Orange Pippin' apple. *Journal of Horticultural Science*. 66, 319–325.

Ranaa, R. S., Bhagata, R. M., Kaliaa, V., & Lal, H. (2004). Impact of climate change on shift of apple belt in Himachal Pradesh. ISPRS Archives XXXVIII–8/W3 Workshop Proceedings Impact of Climate Change on Agriculture.

Talbot, M. T., & Chau, K. V. (2002). Pre cooling strawberries agricultural and biological engineering department, florida cooperative extension service. Gainesville: Institute of Food and Agricultural Sciences, University of Florida [11 p, Bulletin 942].

Wang, S. Y., & Zheng, W. (2001). Effect of plant growth temperature on antioxidant capacity in strawberry. *Journal of Agricultural Food Chemistry*, 49, 4977–4982.

Woolf, A. B., Bowen, J. H., & Ferguson, I. B. (1999). Pre-harvest exposure to the sun influences postharvest responses of 'Hass' avocado fruit. *Postharvest Biology and Technology*, 15, 143–153.

CHAPTER 4

IMPACT OF CHANGING CLIMATE ON PRODUCTIVITY OF APPLE IN HIMALAYAS: URGENT NEED FOR MITIGATION OF HAIL DAMAGE

K. K. JINDAL

Dr. Y. S. Parmar University of Horticulture and Forestry, Nauni Solan - 173 230, India; E-mail: ecofriendlyhorticulture@gmail.com

CONTENTS

4.1   INTRODUCTION

The low productivity of apple has become a serious concern for the farmers, research workers and development agencies at national and state level for the last two decades. Several factors can be attributed to the declining trend in productivity like expansion of apple cultivation to marginal areas, monoculture of Delicious varieties, declining standards of orchard management and the fluctuating abnormal climatic conditions.

The productivity of temperate fruits especially apple in Himachal Pradesh is declining at a faster rate. Average yield of apple in India has been estimated at about 6 tons per hectare, which is far below the level of 30 tons per hectare in most advanced countries. The productivity has also not kept pace with the expansion in areas under temperate fruits due to various biotic and abiotic problems faced by the farmers in the Himalayas. This has caused a serious concern not only to the hill farmer community but also to researchers, development agencies and policy planners. With the global warming, the decline in productivity is being mainly attributed to changing climatic scenario.

4.2   LOW PRODUCTIVITY OF APPLE IN HP: WHY SUCH A SITUATION?

Causes of low productivity analyzed through various technical reports followed by a brain storming session with scientists and subject mater specialists were accessed in farmers' fields and in the experimental research stations of the different apple growing areas of Himalayan states in general and HP in particular. The below mentioned climatic and varietal biodiversity causes were identified on the basis of various inputs as most important factors for the low productivity in apple.

4.2.1   VARIETAL BIODIVERSITY

In Himachal Pradesh, Delicious group of varieties constitute about 83 percent of the total production of apples, the predominant varieties being Starking Delicious, Red Delicious and Richard. These varieties are self-unfruitful and require cross-pollination for fruitfulness. Moreover, these varieties have strong tendency of alternate bearing after a few years of commercial fruit production, which is also one of the reasons, which account for low production during the off-years. These varieties are highly susceptible to low temperature and frost and hail injury during flowering which also accounts for low production.

4.2.2 CLIMATIC CONSTRAINTS

Irregular bearing behavior of Starking Delicious is largely influenced by climatic conditions. The rains and hails during flowering adversely affects the fruit-set whereas moderate temperature of 20 °C with relatively low rains during flowering results in good fruit-set. Low temperature during prebloom or bloom stage can cause sub lethal injury to buds or flowers resulting in poor fruit-set and crop load. Early flower anthesis and full bloom were observed in good crop years. Long flowering period is also indication of good crop prospects.

4.3 IMPACT OF CHANGING CLIMATE CONDITIONS ON CHILLING UNITS, PHYSIOLOGICAL ATTRIBUTES AND PRODUCTIVITY OF APPLE IN HIMACHAL HIMALYAS: BACKWARD LINKAGE

At the University of Horticulture and Forestry Solan comparative investigations have been undertaken to study the influence of winter temperatures below 7 °C on effective chill units (ECU), growing degree hours Celsius (GDH°C) requirements and physiological changes associated with the bud dormancy of Starking Delicious apple under two locations viz., location A (ideal apple growing conditions with an altitude of 2286 in amsl) and location B (marginal apple growing conditions with an altitude of 1375 m amsl). Using the Utah Model, the effective chill unit requirements for location A and B were 1208 and 1130, respectively, whereas, the GDH°C requirements from rest completion to full bloom for the respective locations were 8893 and 9376. The quantitative analysis of physiological components indicated varying pattern during the course of dormancy. The effects of various chilling units on bud break and biochemical attributes in young potted apple plants were studied under controlled conditions. It has been observed that with the increase in chilling exposure, the days required for bud break were reduced. Further, the impact of climatic conditions during winter, spring and summer are being assessed for studying the relationship with fruiting parameters of Delicious apple.

4.3.1   HAIL STORM-A MAJOR CLIMATIC CONSTRAINTS FROM FLOWERING TO FRUIT GROWTH/HARVESTING

The productivity and quality apple is influenced by: (i) winter conditions (December–February); (ii) spring conditions during flowering (April); and (iii) postbloom summer conditions (May–June). The important climatic components affecting the productivity and quality parameters are temperature, rainfall, hails and frost.

The bud break to petal fall is the most sensitive stage when hail can reduce the prospective good crop year to almost 'off' year. It does not only inflict direct injury to buds, flowers and leaves but can causes sublethal injury to the developing fruits and spurs as well. Hails during the fruit development have more serious effects, however, the frequent occurrence of hailstorms in Himachal Pradesh has created havoc in fruit setting in apple. It is feared heavy crop loss for the last 3–4 years may be due to hails as computed in present studies. During 2009–2010 area affected by major calamity like hailstorm was 32,244 ha, 2010–2011 it was 106,467 ha, and in 2011–2012 it was 166,207 ha while in the current year the affected area is 24,350 hac with net loss of apple crop worth 1540 crores up to April, 2012.

4.4   MID SEASON CORRECTIONS TO MITIGATE LOSS DUE TO HAIL STORM

4.4.1   PLASTIC NETTING

Hail netting though fool proof method has not been readily adopted by fruit growers, mainly because of high cost, and partly because of concerns relating to tree growth and fruit quality under the plastic nets. It has been shown that shading from some netting materials decrease photosynthesis especially black may interfere with the development of red color in apple. White nets reduce radiation by 4–8% and black net by 33–37%. Hence, fruit coloring was less extensive under black net and little affected under white nets. The studies are in progress.

4.4.2 TRAINING SYSTEM

Central leader trees suffer the most total hail damage, while the modified central leader suffering the least as observed in RHRS, Shimla.

Bagging of fruits also reduces hail damage. However, it is very laborious. The spray of urea immediately after the hail storm helps in the repairing process of hail injury.

The injury caused by hail damage makes the fruits susceptible to attack of disease causing organisms. These diseases can be controlled effectively by using mild fungicides so as to reduce the loss of fruit quality on experiencing hail storm.

In order to heal the damage of hail on leaves and fruits and improve the size of quality fruits two plant hormones namely CPPU (a cytokinin derivative) and 30 to 60 ppm promalin (mixture of BA and GA 4+7) at different growth stages have been tried at Regional Research Station, Shimla. The impacts will be known after the harvesting of fruits in current season and flower bud initiation in the next season.

4.4.3 USE OF ANTI HAIL GUN

Himachal Pradesh State Department of Horticulture as a pilot project with the Dutch technology, installed three anti hail guns. Impacts on hail control are under study and standardization, these will be discussed detail.

4.5 OUTLOOK-FORWARD LINKAGES

Climate change per se will have impact on economically important species like apple but livelihoods of farmers are being threatened treatment due to incomplete chilling, longer GDH, erratic irruptive rainfall, and snow in winter, more frequent hail storms and enhanced abiotic and biotic stresses. However, measure to adapt to these climate changes is critical for sustainable production. Increased temperature and weather vagaries will have more effect on reproductive biology. The strategies had been identified and addressed to mitigate the adverse effects of weather and development of climate resilient plants species, like low chilling crops, culture practices and efficient use of water. Concerted and integrated efforts can convert challenges into opportunity.

ACKNOWLADGEMENT

Logistic support to conduct the present research studies by SILB, Solan (HPU), Dr. Y.S. Parmar University of Horticulture and Forestry Nauni-Solan and Regional research station (IARI, Shimla) is highly appreciated. The financial support by UGC in the form of Emeritus fellowship to Professor K. K. Jindal is thankfully acknowledged.

KEYWORDS

- **Apple Productivity**
- **Biodiversity**
- **Chilling Units**
- **Climatic Constraints**
- **Hail Damage**
- **Physiological Attributes**

CHAPTER 5

EMPIRICAL APPRAISAL OF SOME WEATHER PARAMETERS' DYNAMICS FOR THEIR POSSIBLE IMPLICATIONS ON MANGO PRODUCTION IN SOME IMPORTANT MANGO GROWING REGIONS WITH SPECIAL REFERENCE TO LUCKNOW REGION OF UTTAR PRADESH

H. RAVISHANKAR[1,3], TARUN ADAK[1], V. K. SINGH[1], B. K. PANDEY[1], A. K. SINGH[1], and B. R. SALVI[2]

[1]Central Institute for Subtropical Horticulture (ICAR), Rehmankhera, P.O. Kakori, Lucknow-226 101, Uttar Pradesh, India;

[3]E-mail: drhravishankar@gmail.com

[2]RFRS, Vengurle, Maharashtra, India.

CONTENTS

ABSTRACT

The Intergovernmental Panel on Climate Change (IPCC) has projected a temperature rise of 1–3 °C which may impact agriculture adversely, may reduce crop yields up to 10 percent by 2020 for Asia (IPCC, 2007). Some Indian studies have indicated that 5–7% decline in wheat yields for every degree Celsius increase in temperature given that current level of irrigation does not erode (Abdul Vahab Abdul Haris, 2013; Agarwal, 2009). Impacts and magnitude of climate change, a global phenomenon though not precisely understood as they become discernible at their tipping points, gradually increased risks originating from erratic weather, altering dynamics of pests and diseases have the potential to adversely impact crop yields and crop distribution in the country. Crop production dynamics in mango in different parts of the country in recent years is gripped with uncertainties due to occurrence of weather extremes, inadequacies of water and nutrient management and impacts of altered pests and diseases prevalence. The Konkan region of Maharashtra where the production of country's premier variety, '*Alphonso*' is under monoculture getting impacted by weather vagaries for the last four years under the influence of western disturbances needs elaborate studies for understanding in order to develop adaptation and mitigation strategies. Despite some regional studies undertaken to profile the effects of some weather parameters on response of mango crop in some agro-ecologies, a clear understanding of its vulnerability to different phenophases events *vis a vis* production constraints is yet to emerge. This is especially important as mango production in recent years is getting adversely impacted by such weather variables as both high as well as low temperatures, unseasonal rainfall, floods, sunshine hours, relative humidity, wind speed, etc. These variables potentially affected the different phenophases viz., vegetative growth dynamics, fruit bud differentiation, flowering, pests and diseases dynamics, fruit set and development, maturity, harvests and eventually the markets. Limited information however, is available on the understanding the impacts of weather components either individually or conjunctively influencing individual phenophases and their contribution to over all crop outputs in order to put in place effective horticultural interventions. The spatio-temporal relationships and relative effects of critical weather variables on phenophases and crop productivity when appraised on collective and regional basis will facilitate development of models that could form part of the Deci-

sion Support System (DSS) for empowering growers' communities. Long term adaptation and mitigation strategies however need to entail studies on carbon sequestration, development of climate resilient varieties, improved crop husbandry practices for risk management, productive use of water and nutrient resources and integrated management of pests and diseases. The present paper based on inputs received from Maharashtra, Karnataka, Gujarat, Odisha, Bihar, Uttar Pradesh unfolds effects of some key weather parameters of three fruiting seasons on critical phenophases on mango production in some important mango producing regions with special reference to Lucknow region of Uttar Pradesh.

5.1   INTRODUCTION

Mango (*Mangifera indica* L.), the '*King of fruits*' in India with rich variety diversity is grown in different agro-ecological regions of the country. Andhra Pradesh, Karnataka and Tamil Nadu in the south, Goa, Maharashtra and Gujarat in the west, Odisha, West Bengal and Bihar in the east and Uttar Pradesh in the north are the principal mango producing regions of the country. Among all the mango-producing states, Uttar Pradesh contributes 23.85% of the total mango production out of 11.6% area under mango in the country. The recent CII-McKinsey report (2013) has identified mango, banana, potato, soybean, and poultry as five items to drive the next wave of growth in Indian agriculture.

Uttar Pradesh falling under the '*Indo-Gangetic plains*' is categorized into nine agro-climatic zones. Eighty percent of the state's population is agrarian of which about 91 percent is represented by small farmers solely dependent on agriculture for their livelihoods. In the recent years, this sector is getting impacted by monsoon uncertainty, increasing duration of drought, floods in some areas and cold, and violent storms. Decrease in water table is progressing at an alarming rate and volume of water flow in major rivers is shrinking besides high levels of pollutants being discharged into the aquatic systems in their path. Soil sodicity is also a serious problem of the Indo-Gangetic plains in UP affecting the productivity and livelihoods of the people. It is estimated that approximately 1.3 million ha is affected by this problem accounting to more than 40 percent of the cultivated area. Further, secondary salinization is also assuming serious proportions. Majority of these soils are having pH > 10, exchangeable sodium percentage (ESP) >15 and varying electrical conductivity (EC). These soils could be ameliorated by providing a readily available source of calcium (Ca^{2+}), to replace excess Na^+ on the cation exchange complex. The research for the past 30 years in sodicity has revealed some important reclamation methodologies for the top 0–15 cm of soil with gypsum and subsequent cultivation of paddy and wheat for providing food security to the farmers of sodic soils. Mango production in such problem soils though is ruled out, perhaps with introduction of salt tolerant rootstock (s) viz., *M. indica* L. cv. M-13-1 and *M. zeylanica* Hooker f., having features of higher rates of CO_2 assimilation, root respiration, relatively low Na^+ and Clcon-tents along with higher K^+, Ca^{++} as well as Mg^{++} in the leaves, besides exploring new rootstocks tolerant to abiotic stresses and standardizing them for commercial cultivars could open up new vistas of horizontal expansion of area under mango in the region (Fig. 5.1).

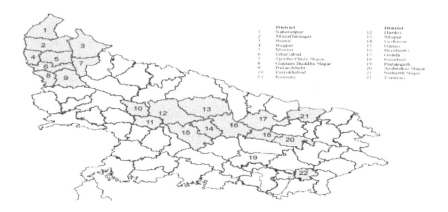

FIGURE 5.1   Mango map of Uttar Pradesh.

Based on agro-climate suitability, the following regions/districts in Uttar Pradesh have been identified for promoting mango cultivation:

Crop	Variety dynamics	Major Clusters / Districts
Mango	Dashehari, Langra, Chausa, Amrapali, Ramkela, Lucknow Safeda, Bombay Green	Saharanpur, Meerut, Bulandshahar, Bijnour, Moradabad, Muzaffarnagar, Sitapur, Hardoi, Lucknow, Unnao, Barabanki, Faizabad, Varanasi

Uttar Pradesh ranks third in fruit production among different states of the country with an area of 3.248 lakhs ha and production of 53.68 lakhs MT. There has been 7% and 15.2% increase in area under fruits in the state when compared with 1991–1992 and 2005–2006, respectively. Similarly, there has been 119% increase in fruit production when compared with 1991–1992 and 71.1% when compared with 2005–2006. Uttar Pradesh accounts for 23.9% of mango produced by the country and represent the major livelihood crop in both eastern and the western parts of the state. Eastern U.P. where small and big rivers exist, is a chronically flood prone area, it has now been witnessing flash floods and instances of water logging in mango orchards have become frequent which can affect soil properties and nutrient dynamics besides pests and diseases dynamics. Prolonged drought, floods and hailstorms and decline in annual rainfall has become a regular feature over the last 3–4 years in the state. In some regions of U.P. drought has become recurrent. This is manifested by, low and untimely rainfall, increased temperatures leading to enhanced rate of vaporization, depletion of soil moisture arisingdue to reduced carrying capacity of water in the soil. Duration of cold has also increased as it happened in 2012–2013 season, though its impact is yet to become discernible is also a case of uncertain weather dynamics. The mango belt of U.P. has been undergoing all these weather uncertainties the long-term effects of which needs delineation (Ravishankar et al., 2011). The short-term effects viz., disturbances to the phenophases, pests and diseases dynamics, delayed flowering (2013 fruiting season). Henceforth, analysis of extremism of weather components at farm scale is of great importance and the need for weather based agroadvisory forecasting system to sensitize the orchardists for adoption of appropriate strategies to avoid biotic and abiotic stresses in fruit production is increasingly felt (Adak et al., 2012).

5.2 WEATHER VARIABLESAND GROWTH CORRELATIONS

5.2.1 TEMPERATURE

5.2.1.1 VEGETATIVE GROWTH

The optimum temperature for tree growth has been reported to be 24–30 °C (Bruce Schaffer et al., 1994); though it can tolerate air temperatures

up to 48 °C, cold temperatures influence crop production. Sub-zero temperatures for even a single day or few hours severely damage or even kill
the trees the trees as could be seen during the 2013 fruiting season in Lucknow (portion of twig death) and Dehradun region of Uttarakhand where
nearly 10,000 trees have been reportedly dead. Trees manifest strong vegetative bias with increasing temperatures provided; water and nutrients are
not limiting (Whiley et al., 1989). They observed that the median (mean
of the maximum and minimum daily temperatures) daily temperature for
zero shoot growth of ten mango cultivars to be 15 °C which, was subsequently confirmed to be critical for critical daily mean temperature for
vegetative growth. Soil temperatures reportedly have a strong control on
synchronization of growth phases in the tree. The same authors found that
periodic shoot growth, changing between phase of activity and dormancy,
occurred when soil temperatures were held at 27 °C or 32 °C for 120 days
but at 21 °C an extended dormant period developed. Temperature also
influenced the number and size of leaves. An average of 7.1 leaves that
developed on shoots at 20/15 °C increased to 13.6 at 30/25 °C. Dry matter
partitioning to roots was the greatest in trees growing at 15/10 °C, that is,
temperatures that suppressed shoot growth. With increase in temperatures,
partitioning to roots declined but correspondingly increased for stem and
leaves. In 'Irwin,' there was a greater allocation of dry matter to roots at
higher temperatures of 25/20 °C and 30/25 °C, a reflection of its reduced
flushing and semidwarf architecture (Table 5.1).

TABLE 5.1 Competence of Important Phenophases in Respect of Key Vulnerable
Physiological Attributes in 'Dashehari' to Weather Dynamics

Type of shoot/stage	Rate of photosynthesis (PN) (μ mol m−2 s−1)	Leaf water potential (Ψw) (MPa)
Vegetative growth	9.25	−7.36
Reproductive growth	–	–
a) FBD	7.61	−7.95
b) Flowering	6.25	−8.46
c) Fruit set	7.27	−6.50

5.2.1.2   REPRODUCTIVE GROWTH

Studies under controlled conditions have shown, of the different weather variables, temperature exerts an important influence on different components of reproductive phenology starting right from flowering (Bruce Schaffer et al., 1994). They opined that mango gets affected when temperature falls below 0 °C hence low temperature factor more than others defines the range of its cultivation. A period of stress generally precedes flowering, which occurs following stress relief, and temperature, particularly in the subtropics provides the strong environmental stimulus for flower induction, the threshold being cultivar-specific. On the contrary, dry period preceding flowering appears necessary for satisfactory flowering in the tropics highlighting the role of drought stress to be the key environmental cue. Studies of Pongsomboon (1991) showed that no flowering occurred in trees held at 30/20 °C at either soil temperatures, while at 15/10 °C air or 12.5 °C soil temperature, all the trees had flowered after 16 weeks but only 40 percent trees flowered when soil temperatures were held at 25 °C. These results suggested that 'root signals' probably played significant role in induction of flowering in mango. Similar indications from the work of Ravishankar (1987) on Alphonso (alternate bearer) and Totapuri (regular bearer) strengthens this view. Export of cytokinins from roots impacting reproductive morphogenetic responses of shoots under the influence of temperature cannot be ruled out as high temperatures provided increased vegetative bias in mango (Whiley et al., 1989). The fact that foliar gibberellins applications suppressed flowering in mango despite trees being grown in inductive conditions (Kachru et al., 1971) emphasizes that interpolations of promoters and inhibitors under the influence of high soil temperatures perhaps play crucial role in coordinating shoot-root communication (Ravishankar, 1987) ultimately determining the flowering responses in mango whichhowever needs further investigation as major focus of research so far in mango focused on signals and responses generated in shoots (Chacko, 1991; Ravishankar, 1987).

Issarakraisila et al. (1992), reported that mean temperature of 12–15 °C sustained panicle growth in mangoes of the subtropics. Singh et al. (1965) found temperature dependency of male and hermaphrodite flowers dynamics of the panicles. Their study indicated that late emerging panicles of 'Dashehari' had seven times the percentage of perfect flowers as compared to early emerging panicles on the same trees with mean maximum/

minimum temperatures of 20.7 °C/3.7 °C and 27.3 °C/13.1 °C for early and late emerging panicles respectively signifying higher temperatures during panicle development are conducive. They also found much less percentage of hermaphrodite flowers in some south Indian cultivars when grown in northern India which was attributed to prevalence of lower maximum/minimum temperatures during panicle development as compared to south Indian conditions during the corresponding periods of panicle development.

Studies have shown, T_{min} (<15 °C) during the development of inflorescences resulted in deformation of flowers which were smaller as compared to those that developed at about 7 °C higher (Issarakraisila et al., 1992) and these abnormal flowers at such low temperatures had small ovary, dark and reduced style length, dark ovule and black another instead of the red one; pistil abnormalities ensued (Fig. 5.2). Issarakraisila and Considine (1994) noted, although mature pollen was tolerant to low temperatures even at 10 °C, this temperature being critical during meiosis, belowwhich the pollen viability falls to about 50%. They further observed, prevalence of high temperatures of 36 °C and above during the prevacuolate meiosis stage of microsporogenesis of pollen drastically reduced the pollen viability.

FIGURE 5.2 Inflorescence deformities due to low temperature effects (Tmin <15 °C) during 2013 fruiting season at CISH, Lucknow.

Besides the above, unseasonal rains especially at the time of fruit bud differentiation (FBD) could transform the already differentiated shoot meristems in favor of reproductive ones to vegetative ones, attributable

to their unfavorable intervention in the requirements of drought stress in the tropics while, low temperature in the subtropics; occurrence of frost (−0.2 °C to −1.2 °C on 9th and 10th January, 2013) as it happened at CISH, Lucknow howevervirtually killed a portion of the shoots and triggered the emergence of new vegetative flushes from the lateral buds subtending the dead portion of shoots, thus contributing to transgression of tree physiology from an expected reproductive one predominantly. Perhaps this was also compounded by the occurrence of hailstorm on 18th and 19th January 2013. Such a situation altered the dynamics of trips incidence in several cases, resurgence on new vegetative flushes despite spraying of thiomethoxam (0.02%) and in some instances, damages to the fruits at pea stage of development. Temperature dynamics together with soil moisture conditions (due to unseasonal rainfall that occurred on 18 and 19th January, 2013) that prevailed during 2013 witnessed however, below threshold levels incidence of mealy bugs, hoppers and powdery mildew. Summary responses of different phenophases to weather variables across standard meteorological weeks and their implications in inciting production constraints of '*Dashehari*' during 2010–12 are indicated in Table 5.2.

5.2.1.3   PHYSIOLOGICAL ATTRIBUTES

Under subtropical conditions, flowering in mango is influenced by cool temperatures, although the temperatures reported for cold induction in mango are much higher than those reported for vernalization. Vegetative growth in mango is through episodic flushing, which was more frequent as per temperature increases recorded during 2013 fruiting season in the Lucknow region.

Mango being a C_3 plant, the photosynthetic performance among cultivars is important for productivity under variable climatic conditions. Higher photosynthetic efficiency during reproductive phenophases especially flowering and fruit growth was identified as the important traits contributing to higher productivity. Under Lucknow conditions photosynthetic fluxes under different temperature regimes in 25 cultivars representing different agro-climatic regions of the country was evaluated (Singh and Rajan, 2009) where in the commercial cultivars of Lucknow region were found more efficient as compared to others. Precise understanding of effects of light interception patterns in mango canopies on productivity is seriously constrained by lack of sufficient quantitative data.

A key enzyme associated with abiotic stress responses of crops, *Rubisco* limited rate of photosynthesis is less sensitive to temperature as the decrease in carboxylation at low temperatures is partially compensated by suppression of photorespiration. *RUBP*-regeneration limited rate of photosynthesis, which depends upon much less on oxygenation however, is fairly sensitive to temperature. Observations at CISH, Lucknow on different varieties of regular (*Amrapali, Totapuri*) and biennial (*Langra, Chausa, Dashehari*) during different phenophases which experienced range of temperatures in 2012–2013,exhibited different levels of sensitivity to low temperatures for their photochemical reaction (*Personal observation* and Singh et al., 2012).

Higher amounts of PN with concomitantly high leaf water potential during periods of rank vegetative growth (March and September–November) coincided with periods characterized by the prevalence of high temperatures and soil moisture and hence they are found vulnerable FBD, a metabolically highly dynamic stage also maintained high PN levels obviously indicating high sink activities with correspondingly high leaf water potential, however a period of stress precludes this phenomenon and under subtropical conditions, this is provided by cold. This stage could turn out to be vulnerable, if intervened by heavy rainfall so that the reproductive primordia could revert to the vegetative one. Such an unfavorable event did not occur during the study period (Figs. 5.3–5.9). Standard meteorological weeks starting from September, 42–46). Decrease in PN in leaves during flowering is attributable to the decrease in stomata conductance. Reduction in this parameter could also be due to differences in the leaf water state of shoots. Thus it is possible that flowering results due to local water stress. Increase in PN during fruit set could be attributed to higher sink activity demanding increased mobilization of photosynthates to the developing fruit lets. Based on the pattern of flushing, the maximum flushes were recorded during March in both the years. However, in 2011 the significantly higher flush was also recorded during June-August (16.91–38.26%) in addition to March flushing, which is attributed to relatively lower intensity of reproductive phase of the previous fruiting season.

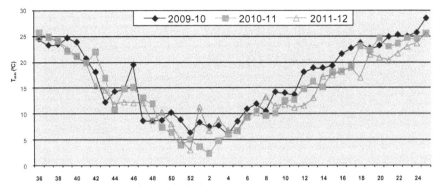

FIGURE 5.3   Weekly mean maximum temperature fluxes of the study period.

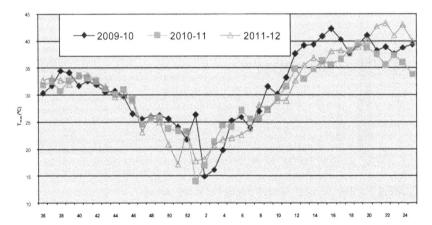

FIGURE 5.4   Weekly mean minimum temperature fluxes of the study period.

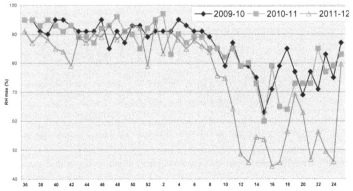

FIGURE 5.5   Weekly mean maximum relative humidity fluxes of the study period.

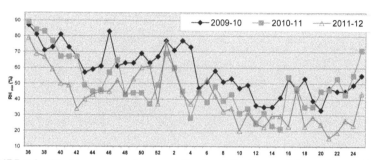

FIGURE 5.6   Weekly mean minimum relative humidity fluxes of the study period.

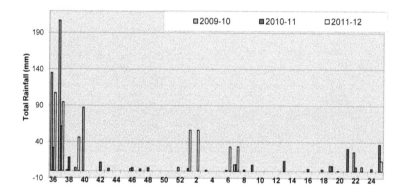

FIGURE 5.7   Weekly mean rainfall during the study period.

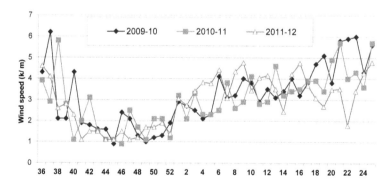

FIGURE 5.8   Weekly mean wind speed (above) and sunshine hours (below) of the study period.

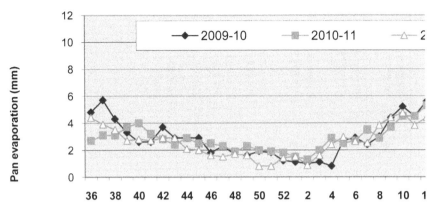

FIGURE 5.9 Weekly mean values of pan evaporation during the study period.

[**Initiation** (up to 20 days–30 days) – February–March, vulnerable to temperature and moisture dynamics; **Lag phase** 30–40 days March–April; vulnerable to high temperatures; low RH; moisture dynamics and under hormonal interpolations; fruit borer; preharvest fruit drop; **Grand growth phase** 40–70 days April–May vulnerable to hormonal interpolations; nutrients and water dynamics; fruit fly; preharvest fruit drop; shoulder browning if exposed to rainfall; consolidation of fruit quality principles **Climacteric** 70–80 days Mid June Mid July; high temperature effects on fruit size and quality; harvest maturity; jelly seed formation, exposure to rainfall; Senescence 80 days onwards high temperature effects; concentrated maturity; market gluts; exposure to rain fall; reduced profitability].

Fruits approaching maturity are greatly affected by high temperatures exceeding 35 °C as it happened in the Konkan region (day temperatures peaking to 40 °C and beyond in May) during 2010 fruiting season where crores worth '*Alphonso*' mangoes was lost while in 2011 fruiting season in Malihabad area of Lucknow witnessed sharp focused maturity and market gluts during the third week of June when the day temperatures peaked to 43 °C, a sudden increase of 2 °C over the previous week in a short span of a week's time.

TABLE 5.2 Summary Response of Different Phenophases in 'Dashehari' to Weather Variables During 2010–2012

Standard meteorological week (provisional)	Calendar month	*Vulnerable phenophases / production constraints							Meteorological / culture event
		Vegetative flush	FBD	Flowering	Fruit set	Development	Maturity	Harvesting	
42,43,44,45,46,47	October–November	* (Leaf webber; anthracnose)	*	-	-	-	-	-	High temperature; un-seasonal rainfall
		* (Leaf webber; anthracnose)	*	-	-	-	-	-	Un-seasonal rainfall; low temperature
48,49,50,51,52,1,2,3,4	Last week of November, December–January	*	*	*	-	-	-	-	Low temperatures and its duration
		*	-	* (Mealy bugs)	-	-	-	-	Low temperatures and frost
5,6,7,8,9,10,11,12	First week of February, March	* (Thrips; anthracnose)	-	* (Full bloom; inflorescence midge; hoppers; blossom blight; powdery mildew; extent; intensity; panicle configuration-percent hermaphrodite flowers; pollinators' activity)	* (Nubbin fruits)	*	-	-	Low / High temperatures, unseasonal rainfall/ excessive irrigations; high relative humidity

Year/Months	January	February	March	April	May	June	July	August	September	October	November	December
13,14,15,16,17,18,19,20,21	–	Last week of March, April–May	* (Thrips; anthracnose)	–	–	–	–	* (Pre-harvest fruit drop; fruit growth; fruit fly)	* (Pre-harvest fruit drop; fruit growth; fruit fly; fruit borer; fruit midge)	* (Pre-harvest fruit drop; size)	–	– Absence of rainfall / impaired water management; high temperatures
22,23,24,25..... up to 35		June–August	*	{Shoot gall psylla (Tarai and adjoining areas); leaf webber}			–	–	* (maturity; quantity and quality of harvest-jelly seed formation; anthracnose; effects on postharvest life)	–		High temperatures, high rainfall, high humidity, wind speed
36,37,38,39		September	*	{Shoot gall psylla (Tarai and adjoining area); leaf webber and anthracnose}			–	–		–		High temperatures, high humidity; excess application of nitrogen; deficiencies of micronutrients
New vegetative flush (%)												
2012	2.24	6.43	35.42	2.73	1.01	1.24	0.00	0.00	6.65	4.43	7.50	0.00
2011	2.28	5.23	30.23	2.58	2.34	38.26	16.91	23.72	7.31	3.34	8.27	1.23

5.2.1.4 SPATIO-TEMPORAL RELATIONSHIPS BETWEEN TEMPERATURE DYNAMICS AND PHENOLOGY AT SOME IMPORTANT MANGO PRODUCING REGIONS

Meteorological observations at CISH, Rehmankhera indicated that fruit development stages were vulnerable under high pan evaporation regimes (>6 mm/day to 10 mm/day) particularly during the last week of March and onwards. High pan evaporation values indicated higher rate of evapotranspiration resulting into drier soils with soil moisture stress developing silently within the soil and leading to abiotic stress complexes barring water and nutrient flow through xylem. Thus protective irrigations are required as an intervention strategy so as to maintain optimum soil moisture and leaf water potential in order to sustain fruit set, growth and development and minimizing preharvest fruit drop unless otherwise this period is intervened by unseasonal rains. Reports have indicated that mango is drought tolerant, due its ability to maintain turgor through osmotic adjustment attributable to the existence of lacticifer system. Mango, on the other hand, exhibit some degree of flood tolerance though production of hypertrophied lenticels that facilitate increased O_2 absorption besides offering sites for excretion of toxic by-products of anaerobic metabolism of roots (Bruce Schaffer et al., 1994). Perhaps flooding tolerance of mango trees in north Bihar could be explained to the above-mentioned phenomenon. Precise understanding of these features however, is constrained by the lack of quantitative data on water relations.

The vulnerability of mango to weather dynamics also emerged at some other important mango growing parts of the country. A recurrent flowering phenomenon, which has become common in the Konkan and Bangalore regions for the past four yearsappears to be a consequence of weather impacts which needs microanalysis with at least the Konkan case coming under the influence of western disturbances occurring over the Arabian sea. Meteorological data analysis at Bangalore indicated that T_{min} ranged between 13 and 16 °C, on majority of the days during January 2013 being 14–15 °C, a critical period for flowering. In the first fortnight of January 2013, T_{min} ranged between 10–16 °C. Hence during January 2013, attributable to this variation, flowering occurred in more than one flush, termed '*recurrent flowering.*' This kind of phenomenon was also reported in '*Kesar*' from Gujarat (*Personal communication*, 2013), which has led to occurrence of different stages of fruit growth and development stages on the same tree, consequently multiple harvests of varying quality may ensue affecting markets. During February 2013, the T_{max} was 28.7 °C and the T_{min} was 17.4 °C with the RH of 73%. The data recorded for two fortnights revealed that there was only very little difference in T_{max} and T_{min}. However, large variations in the T_{min} emerged during February 2013. On some days, the T_{min} was 21 °C and after a gap of 6 to 7 days, the T_{min} fell to 14 °C. Such variations in T_{min} could greatly affect the pollinator dynamics and their activity as was discernible during 2013 fruiting season at Lucknow and adjoining regions. This kind of T_{min} dynamics also probably triggered a second flush of flowering during 2013 at Bangalore region, infusing disturbed source-sink relationships as fruits of different growth and developmental stages became mutually competitive that could probably impactboth quantity as well as quality of harvests. February 2013, however, witnessed T_{min} fluctuating between 13 and 15 °C without much change.

Under the Konkan conditions of Maharashtra, new vegetative flushes emerged in the second week of November (2011–2012). The first flowering flush to the tune of about 5% occurred in the third week of December even though there were extended rains up to 29th November, 2011 in continuation to normal rainfall in the monsoon. The second flowering flush was noticed during January second week, which was to the tune of about 30%. The total flowering up to March, 2012 in 'Alphonso' was about 48% and in other varieties it was about 25%. The late showers received up to November end affected the fruit set of the first flowering flush as the flowering got delayed despite hermaphrodite flower percentage in the first flowering flush being normal up to 9.8%; the panicles emerging during second flowering flush were predominantly of mixed panicles having extra length (size-41 cm length) that contributed to low fruit set as was seen at Vengurle, Mulde, Deogad and Lanja in about 10.4% plants in both cases there was failure to set any fruits.

Since 2009, unpredictable weather conditions have been occurring in the premier mango growing belts of the country every year, like incidence of unseasonal rains (up to Dec. 23; March–May); cloudy weather, severe cold (8 to 12 °C), and extremes of T_{min} continuously for 60 to 70 days, (instead of 4 to 5 spells of 12 to 15 days cold period) and T_{max} prevailing during fruit growth, development and maturity. These aspects have come to sharp focus from 2009 onwards especially the Konkan region. The normal rainfall period of this region is June to September with 1 to 2 dry spells of 8 to 10 days period, with mean annual rainfall of 3000 mm received over an average of 105 days. However, deviations to this normal showed, the extended rainfall (15 to 42 days) with increased intensity (122 mm to 1500 mm). This situation leads to unavailability of sufficient physiological stress conditions, very much crucial for satisfactory induction of flowering in '*Alphonso*' under tropical conditions, thereby impacting flowering adversely.

Normally rainfall ceases by September end followed by onset of winter from November that continues up to February in the Konkan region. However, due to delayed and extended rains, this normal weather cycle is found disturbed and the much needed physiological stress preceding fruit bud differentiation are lacking and as a consequence, uncontrolled vegetative flushes have been occurring. Agro-climate, further impacted by extended cold periods (>122 days of <17 °C in 2011–2012), '*Alphonso*' manifested three reproductive flushes in the fruiting season despite favorable weather conditions prevailing subsequently. Similarly, continuance of severe cold for extra long periods (62 to 74 days) also prevailed during 2009 and 2010 fruiting seasons that resulted in altered phenophases dynamics in '*Alphonso*.' As a result the inflorescences emerging during this period were abnormal like extra long flowering panicles (42 to 45 cm); instead of normal (29 to 32 cm), mixed panicles and changed sex ratio. During 2010–2011, severe cold period (<17 °C) continued for 72 days (instead of 4 to 5 spells of 15 days period) which led to development of extra long inflorescences with altered sex ratio (4 to 5% hermaphrodite flowers) characterized by preponderance of male flowers (96%) adversely impacting fruit set; the reduced number and activity of pollinators further compounding the problem. Majority of the inflorescences of this period failed to set fruits due to inadequate pollination and hence subsequently dried up.

Intermittent flowering phenomenon in the peninsular region in the recent years since 2008 is posing serious problems. During January 2013, T_{min} of the region, ranged between 13 and 16 °C, but on majority of the days it was between 14 and 15 °C. In the first fortnight of January 2013, T_{min} recorded was between 11 to 18 °C and 10 to 16 °C. Hence, during 2013 fruiting season, the flowering flushes occurred more than twice attributable to this variation. During February 2013, the T_{max} was 28.7 °C and the T_{min} was 17.4 °C and the RH was 73%. The data recorded for two fortnights revealed very little difference between T_{max} and T_{min}. However, large variations in the T_{min} were seen during February. On some days the T_{min} was 21 °C and in some days after a gap of 6 to 7 days, the same declined to

14 °C. This probably triggered a wave of second flush of flowering in 2013 fruiting season (*Personal communication*, 2013). However, during February 2012, the minimum temperature always had hovered between 13 and 15 °C without much change. Similar observations of '*recurrent flowering*' phenomenon and occurrence of fruits of different developmental stages on the same tree have also been reported from Gujarat and Odisha (*Personal communication*, 2013).

A similar environmental cue prevailed in Lucknow region wherein the cold periods (<17 °C) were recorded from 15th October onwards in 2011–2012 and 2012–2013 (Fig. 5.10), The cold periods were found extended for at least a fortnight more in 2012–2013 (around 169 days) as compared to 2011–2012 (around 155 days). These extended cold periods impacted flowering phenology adversely; vegetative flushes emerging concomitantly with flowering, and altered pests and diseases dynamics in '*Dashehari*' (thrips; anthracnose) besides low number and activity of pollinators. Occurrence of frost (–0.2 °C to –1.2 °C on 9th and 10th January, 2013) was the other notable feature that killed the shoot apices including shoot portion of about 5–6 inches (Fig 5.10).

FIGURE 5.10   Frost damages to shoot apices, inflorescences and mango young plants in January 2013.

Early flowering in some mango cultivars during December was found subjected to frost injuries (–0.2 °C to –1.2 °C on 9th and 10th January,

2013) resulting in drying up of inflorescences giving charred appearance. Extended cold periods of 2012–2013 fruiting season as indicated (Fig. 5.11), have besides adversely impacting the flowering phenology across varieties including delayed flowering (late February-early March, 2013), may possibly also render fruit growth and development stages vulnerable to eventually increased hot and dry environmental cues, with adverse impacts on later stages of fruit development especially, maturity (during June to July). All these factors cumulatively may reduce the quantity and quality of fruit harvests in 2012–2013 fruiting season.

5.2.1.5   NUBBIN FRUIT FORMATION

'Dashehari' cultivar, being highly sensitive to temperature fluxes in respect of key reproductive phenophases viz., flowering and fruit set, also gets highly impacted by the abnormal weather conditions especially, the spatio-temporal relationships between T_{max} and T_{min}. The clustering (nubbin) of fruit lets phenomenon may be due to prevalence of high fluctuations in temperatures during February and March especially, the T_{min} of cold winters which coincided with the period of flowering and fruit development during 2013 fruiting season probably impacted by hormonal interpolations; this type of weather extremism despite good intensity of flowering and set, often ends up in clustering of fruit lets at the distal end of the inflorescence where, higher percentage of perfect flowers normally occurred in the panicle configuration. Such fruits invariably with aborted embryos (see the inset picture) prematurely drop off and are not carried to maturity, a direct loss to total crop outputs (Fig. 5.11).

Hormonal relationships also explained that, the normal fruits contained higher concentrations of gibberellins and cytokinins and lower concentrations of auxins and abscissic acid as compared to the nubbin fruits. These findings perhaps explained the reasons of occurrence nubbin fruits that have a slower growth rate than the seeded one and small in size, majority of these fruits dropped and failed to reach full size (Shaban and Ibrahim, 2009). Variety differences, effects of cold winter temperatures in this regard though have been implicated; exact reasons for this phenomenon however are not clearly understood (Campbell and Campbell, 1991).

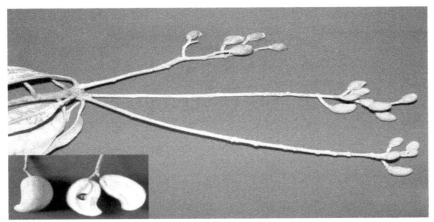

FIGURE 5.11 Clustering of fruits in mango (nubbin fruits formation); Inset: LS of fruits showing embryo abortion.

'*Jelly seed*' formation, a physiological disorder widely occurring in '*Dashehari*' affects the market prospects, appears to be related to reduced internal calcium allocation to fruits under the influence of high temperature holds ground in light of the observations of Batten et al. (1988) as its incidence increased in late harvested fruits exposed to high temperatures (T_{max} >40 °C). This view is also substantiated by the results of Singh et al. (2006) that highlighted the factors of late harvesting after a spell of rains, and association of fruit transpiration with internal break down in '*Dashehari*' Singh, (2007) found spraying of dehydrated Calcium chloride (2.0%) in combination with potassium sulfate (1.0%) along with soil application of borax (150–250 g per tree) in October–November, a month before harvesting of fruits controlled this disorder.

5.3 CONCLUSIONS

The flowering in mango whether in regular or irregular bearing varieties is the result of hormonal interpolations within the plant system under the strong influence of environmental factors especially the T_{min} though other components could also impact adversely the mango phenophases, about which sufficient quantified data are not available This environmental variable has been studied widely in mango by many researchers. Other environmental cues viz., light, water, sunshine hours, wind, salinity though exert considerable influence on productivity as has been reported in few cases have not been systematically investigated across agro-ecologies and varieties, hence delineation of their specified roles remains severely constrained. Experiences have shown that the above factors many times, either individually or synergistically influence productivity through their impacts on phenophases responses within a given fruiting season, their effects getting perpetuated in the following years. Studies to intervene and manipulate phenophases/tree physiology aberrations both in the country as well as in other mango producing countries of the world, by the use of *'Paclobutrazol'* which is antigibberellins, through its physiological action promotes flower induction at the expense of vegetative growth. Its uncontrolled use however, year after year without ensuring adequate foundation growth of active shoot mass per tree by appropriate crop husbandry practices, besides disturbing the phenological phases which, otherwise normally under intimate balance also has residue problems with attendant environmental concerns. This, prominently flags for focused research on the role of root dynamics in regulation of flowering in mango as *'root signals'* arising out down /up regulating of certain genes appear critical. The dynamism of temperature fluxes, unpredictablethough, as has been occurring in some parts of the country especially, in the regions of Karnataka, Maharashtra, Gujarat, Odisha, West Bengal in the recent years, appears to be adversely impacting productivity through induction of *'recurrent flowering'* over several short spells within the same fruiting season. This phenological aberration could have cascading effects on tree physiology in a perennial tree crop like mango, the details of whichhowever needs to be investigated critically. Meanwhile, it is of interest to note here that some of the exotic colored varieties viz., Tommy Atkins and Kent, etc., showing resilience to weather dynamics by such features as high percentage of hermaphrodite flowers, fruit set per panicle, fruit retention to final harvest, yield potential even under adverse weather dynamics as is being observed under the Konkan conditions, should prompt the researchers to delve into the association of genes responsible for color trait with the climate resilience trait(s). This empirical appraisal once again prominently flags the research gap in the valuation of the diverse genetic resources under different mango producing ecologies for their responses to varied changes of agro-climatic conditions so that appropriate gene products could be profiled in order to exploit biotechnological tools to address the challenges of developing climate resilient varieties. The economy and livelihoods of the growers since in many traditional regions are intricately linked to the mango productivity, undertaking macro and micro climate analysis will ultimately lead to the development of site-specific decision support systems for empowering local mango grower communities for adopting improved crop husbandry practices to mitigate the adverse effects of changing weather extremisms.

KEYWORDS

- **Climate Change**
- **Growth Correlations**
- **Mango**
- **Nubbin Fruit**
- **Spatio-Temporal Relationships**
- **Weather dynamics**

REFERENCES

Adak, T., Kumar, K., Shukla, R. P., & Ravishankar, H. (2012). Weather based agro-advisory inputs for decision support mechanism for improving mango production in Lucknow region of Uttar Pradesh. *In Proceedings of the National Seminar on "New Frontiers and Future Challenges in Horticultural Crops"* at PAU, Ludhiana India, held during 15–17th March, 2012, 56.

Agarwal, P. K. (2009). Global climate change and Indian agricultural case studies from ICAR network project, ICAR, New Delhi, 2009.

Abdul Vahab, Abdul Haris, Sandeep Biswas, Vandana Chhabra, Rajamanickam Elanchezhian and Bhagwati Prasad Bhatt, (2013). Impact of climate change on wheat and winter maize over a sub-humid climatic environment. *Current Science*, 104(2), 206–214.

Batten, D. J., Firth, D., & Miller, A. (1988). Sensation mango research. *Subtrop Fruit Grower*, 4, 13.

Bruce Schaffer, Anthony, W., & Crane, J. H. (1994). Mango *InHandbook of Environmental Physiology of Fruit Crops* (Eds.) Bruce Schaffer & Anderson, P. C. CRC Press Inc. Boca Raton, Florida, 33431. 165–196.

Campbell, R. J., & Campbell, C. W. (1991). The Parvin mango. *Proc. Fla. State Hort. Soc.* 104, 47–48.

Chacko, E. K. (1991). Mango flowering-still an enigma *Acta Hort* 291, 12.

IPCC (2007). Summary for policy makers. In: *Climate change2007. The Physical Science Basis,* Contribution of Working Group (WG) to the Fourth Assessment Report of the Intergovernmental Panel on Climate Change (eds. Solomon, S. et al.) Cambridge University Press, Cambridge, UK.

Issarakraisila, M., Considine, J. A., & Turner, D. W. (1992). Seasonal effects on floral biology and fruit set of mangoes in a warm temperate region of Western Australia. *Acta Hort.,* 321, 626.

Issarakraisila, M., & Considine, J. A. (1994). Effects of temperature on microsporogenesis and pollen viability in mango cv. Kensington. *Ann. Bot.,* 73, 231.

Kachru, R. B., Singh, R. N., & Chacko, E. K. (1971). Inhibition of flowering in mango (*Mangifera indica* L.) by gibberellic acid. *Hort Science*, 2, 140.

Pongsomboon, W. (1991). Effects of temperature and water stress on tree growth, flowering, fruit growth and retention of mango (*Mangifera indica* L.), Ph.D. Thesis, Kasetstart University, Bangkok, Thailand.

Ravishankar, H. (1987). Studies on the physiological and biochemical aspects into the causes and control of alternate bearing in mango (*Mangifera indica* L.) Ph.D. Thesis, University of Agricultural Sciences, Dharwad, 310.

Ravishankar, H., Adak, T., Kumar, K., & Shukla, R. P. (2011). Some aspects of weather dynamics influencing production and sustainability of mango (*Mangifera indica* L.) in Malihabad belt of Uttar Pradesh. In: *International Conference on Issues for Climate Change, Land Use Diversification and Biotechnological Tools for Livelihood Security,* October 8-10, 2011, SVPUA&T, Meerut, U.P., India, 208–209.

Shaban, A. E. A. & Ibrahim, A. S. A. (2009). Comparative study on normal and nubbin fruits of some mango cultivars. *Australian Journal of Basic and Applied Sciences*, 3(3): 2166.

Singh, R. N., Majumdar, P. K., & Sharma, D. K. (1965). Studies on the bearing behavior of some south Indian varieties of mango (*Mangifera indica* L.) under North Indian conditions. *Trop Agri.,* 42, 171.

Singh, V. K., Singh, D. K. & Pathak, S. M. (2006). Relationship of leaf and fruit transpiration rates to the incidence of softening of tissue in mango (*Mangifera indica* L.) cultivars. *American. J. Pl. Physiol.,* 1(1), 28–33.

Singh, V. K. (2007). Jelly seed in mango and its management. *ICAR News,* A Science and Technology Newsletter, ICAR, New Delhi, April–June, 2007, 13–14.

Singh, V. K. & Rajan, S. (2009). Changes in photosynthetic rate, specific leaf weight and sugar contents in mango (*Mangifera indica* L.). *The Open Horticulture Journal*, 2, 34–37.

Singh, V. K., Ravishankar, H. & Sridhar Gutam, (2012). Physiological attributes in mango (*Mangifera indica* L.) impacted by different CO_2 and temperature fluxes under Lucknow conditions. *In*: *National Dialogue on Climate Resilient Horticulture*, 28–29 January, 2012, Indian Institute of Horticultural Research, Hessaraghatta Lake Post, Bangalore, 147–153.

Whiley, A. W., Rasmussen, T. S., Saranah, J. B., & Wolstenholme, B. N. (1989). Effect of temperature on growth, dry matter production and starch accumulation in ten mango (*Mangifera indica* L.) cultivars. *J. Hort. Sci.,* 64, 753.

CHAPTER 6

PROSPECTS OF CASHEW CULTIVATION UNDER CHANGING CLIMATIC CONDITIONS

P. L. SAROJ[1] and T. R. RUPA[2]

[1]Director, [2]Principal Scientist (Soil Science) Directorate of Cashew Research Puttur – 574 202, Karnataka, India;
E-mail: dircajures@gmail.com, dircajures@yahoo.com

CONTENTS

ABSTRACT

Cashew is usually grown as a rainfed crop in ecologically sensitive areas such as coastal belts, hilly areas and areas with high rainfall and humidity, andhence its performance mainly depends on climate. Studies on suitability of cashew cultivation in India using GIS showed that cashew grows at an elevation ranging from 0 to 1000 m above MSL. However, the productivity is the highest up to the altitude of 750 m above MSL. The average annual rainfall distribution in cashew areas ranges from low rainfall (300–600 mm in Gujarat) to high rainfall (2700–3000 mm in west coast and NEH region) but the productivity is highest in regions with a mean annual rainfall distribution of 600–1500 mm. The productivity of cashew is higher in regions where the minimum temperature ranges from 10 to 22 C and is lower in regions where the minimum temperature drops below 10 °C. Unseasonal rains and heavy dew during flowering and fruiting periods are the major factors which adversely affect the nut yield. Heavy rains at the time of harvesting affects yield and quality of nuts. Cloudy conditions, high RH and heavy dewfall are favorable for outbreak of insect pests and diseases. To circumvent losses due to climate variability/change, adaptation and mitigation strategies are essential in affected areas. Some of the adaptation strategies include plant architecture, use of efficient technologies like drip irrigation, soil and moisture conservations measures, fertilizer management through fertigation, green manuring/intercropping, increase in input efficiency, pre and post-harvest management of economic produce cannot only minimize the losses but also increase the positive impacts of climate change. The flowering, fruiting, insect pest incidence in cashew crop, yield and quality of cashew nut and kernels are more vulnerable attributes for climate change. The sea water level rise due to the melting of glaciers as a result of increase in temperature may also pose problem for cashew cultivation since large proportion of cashew plantations exist in Eastern and Western Coastal regions of India. The perennial cashew crop has potential for carbon sequestration for mitigation of climate change.

6.1 INTRODUCTION

Cashew (*Anacardium occidentale* L) is a very important horticultural crop of India, which was introduced by Portuguese travelers in sixteenth century mainly for the purpose of soil and water conservation. Cashew, after its introduction to India five centuries back has naturalized well to the Indian climatic conditions and is presently grown in an area of about 9.91 lakh hectares with a production of 6.92 lakh tons of raw nuts per annum. The current productivity of cashew in the country is of 749 kg/ha. Low productivity of cashew in India is a matter of concern. The details of area, production and productivity of cashew is given in Table 6.1 (DCCD, 2012). Nigeria, India, Ivory Coast, Vietnam and Indonesia are leading cashew producing countries in the world (2010). In India, it is grown mainly in Maharashtra, Goa, Karnataka and Kerala along the west coast and Tamil Nadu, Andhra Pradesh, Odisha and West Bengal along the east coast. It is also grown to a limited extent in nontraditional areas such as Bastar region of Chattisgarh and Kolar (Plains) region of Karnataka, Gujarat, Jharkhand and in North Eastern Hilly region. India exports 0.926 lakh tons of cashew kernels per annum to over 65 countries and is a prominent trader for over a century. The major countries that import Indian cashew are United States of America, Netherlands, United Kingdom, United Arab Emirates, Japan, France, Saudi Arabia, Spain, Russia, Germany, Canada and Greece. Owing to its high nutritional value and increasing affordability by the consumers, demand for cashew continues to increase globally as well as domestically.

TABLE 6.1   Area, Production and Productivity of Cashew in India During 2011–2012

State	Area (ha)	Production (tons)	Productivity (kg/ha)
Kerala	83,000	73,000	948
Karnataka	1,21,000	60,000	517
Goa	58,000	25,000	455
Maharashtra	1,83,000	2,23,000	1282
Tamil Nadu	1,36,000	68,000	519
Andhra Pradesh	1,84,000	1,10,000	601
Odisha	1,58,000	97,000	683
West Bengal	20,000	5,000	500
Others	48,000	31,000	861
Total	9,91,000	6,92,000	749

Climate change is widely considered to be one of the greatest challenges to modern human civilization that has profound socioeconomic and environmental impacts. Over exploitation of fossil fuels ever since the beginning of industrial era led to increased concentration of green house gases (GHGs) *viz.,* carbon dioxide, methane, nitrous oxide, sulfur dioxide, etc. in the atmosphere. Increase in the concentration of these GHGs is responsible for global climate change. Climate change may be due to natural processes or external forcings, anddue to persistent anthropogenic changes in the composition of the atmosphere or inland use. The CO_2, concentration has increased from a preindustrial value of about 280 ppm to 395 ppm in 2012. Similarly, the global atmospheric concentration of methane and nitrous oxides and other important GHGs, has also increased considerably. This has resulted in warming of the climate system by 0.74 °C between 1906 and 2005 (IPCC, 2007). Climate change is often manifested in extreme events of precipitation, sea-level rise and temperature increase, leading to droughts, floods, forest fires and desertification. Cashew is usually grown as a rainfed crop in ecologically sensitive areas such as coastal belts, hilly areas and areas with high rainfall and humidity. Climate change therefore has a profound impact on cashew.

6.2   CLIMATIC SUITABILITY OF CASHEW CULTIVATION

Most of the regions where it is an economically important crop are between 15°South and 15°North. Cashew thrives at temperatures up to 40 °C. Damage to young trees or flowers occurs below the minimum temperature of 7 °C and above the maximum of 45 °C. Only prolonged cool temperatures will damage mature trees; cashew can survive temperatures of about 0 °C for a short time (Ohler, 1979). Low altitude areas with a mean rainfall of 1500 to 2000 mm are excellent for cashew. Environments with maximum temperature ranging from 28 °C to 32 °C, minimum winter temperature around 19 °C and 70–80% relative humidity are good for getting better results. Frost is detrimental to the crop. Mandal (1992) attempted to rate cashew growing environments as very good, good, fair and poor based on variation in altitude, rainfall, proximity to sea, maximum and minimum temperature, humidity and occurrence of frost. The ratings and the range of these parameters are indicated in Table 6.2.

TABLE 6.2 Environmental Rating for Growing Cashew

Parameter	Very Good		Good	Fair	Poor
	Class I	Class II	Class III	Class IV	Class V
Altitude (m)	20	20–120	120–450	450–750	
Rainfall (mm/year)	1500–2000		1300–1500	1100–1300	900–1100
Proximity to sea (km)	<80	80–160	160–240	240–320	
Maximum temperature (°C)	28–32	32–33	33–34	34–35	
Minimum temperature (°C)	19	18–19	17–18		15–17
Humidity (%)	70–80	65–70	60–65		50–60
Occurrence of frost	None	None	Very rare	Once in 5 years	

The growth and production of cashew is highly dependent on latitude, altitude, temperature, rainfall, relative humidity (RH), sunshine, wind and soil moisture content (Prasada Rao and Gopakumar, 1994). It has been reported that the maximum temperature, humidity and rainfall are the major climatic factorwhich determine the productivity of cashew. The RH during preflowering stage is the vital factor in explaining the yield variation in cashew plantations (Haldankar et al., 2003). Cashew is very sensitive to waterlogging and hence heavy clay soils with poor drainage conditions may not be suitable for its cultivation.

Studies conducted at Directorate of Cashew Research, Puttur to determine the suitability of cashew cultivation in India using Arc GIS showed that cashew is distributed along loamy red and lateritic soil, mixed red and black soil, coastal and deltaic alluvium derived soil. The elevation of the cashew growing areas ranged from 0 to 1000 m above mean sea level (MSL) and the productivity of cashew was higher in regions up to 750 m above MSL. It was observed that mean annual rainfall distribution in cashew area ranged from low rainfall (300–600 mm in Gujarat) to high rainfall (2700 to 3000 mm in West coast and NEH region). The productivity was highest in region with a mean annual rainfall distribution of 600 to 1500 mm. The overlay maps showed that cashew is cultivated along regions where the mean annual temperature ranged from 20.0 to even more than 27.5 °C and the productivity is higher in regions where the mean annual temperature ranged from 22.5 to 27.5°C. The productivity of cashew was higher in regions where the minimum temperature ranged from 10 to 22°C and was lower in regions where the minimum temperature drops below 10 °C.

The site suitability map for cashew revealed that Maharashtra, Goa, Kerala, West Bengal and Odisha are highly suitable, while Andhra Pradesh, Tamil Nadu, Karnataka, NEH regions and Gujarat are moderately suitable. Jharkhand, North Eastern Hilly (NEH) Region, Andaman & Nicobar Islands and Bastar region of Chhattisgarh are also moderately suitable for cashew cultivation (Fig. 6.1). In India, the area under cashew crop has increased from 5,65,420 ha in 1993–94 to 9,23,000 ha in 2010. Similarly, the production has also increased from 3,48,350 tons to 6,53,000 tons over the same period. In Maharashtra, the area and production of cashew crop has increased from 51,220 ha and 46,860 tons in 1993–94 to 1,75,000 ha and 1,98,000 tons in 2010, respectively.

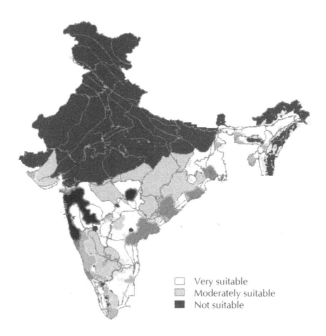

FIGURE 6.1   Crop suitability map for cashew.

6.3   IMPACT OF CLIMATE CHANGE ON CASHEW

Climatic variability like increase in temperature and moisture stress conditions during flowering, fruit setting and nut development cause heavy yield losses and adversely affect the nut quality even under better management conditions. Any change in weather has direct effect on reproductive phase of cashew. Cashew requires relatively dry atmosphere and mild winter (15 to 20 °C minimum temperature) coupled with moderate dew during night for profuse flowering. High temperature (>34.4 °C) and low RH (<20%) during afternoon cause drying of flowers and thereby reduction in yield. Unseasonal prolonged rainfall coupled with high wind velocity during flushing and flowering resulted in significant nut yield losses as large proportion of lateral branches remained as nonflowering laterals (Yadukumar et al., 2010). Unseasonal rainfall and heavy dew during flowering and fruiting also intensify the incidence of pests and diseases. In order to visualize the climate change scenario at DCR farm, temperature and humidity trends in last six years is furnished in Tables 6.3a and 6.3b.

TABLE 6.3A   Temperature and Relative Humidity Trends in Last Six Years at DCR

Month	2006–2007				2007–2008				2008–2009			
	Temp (°C)		RH (%)		Temp (°C)		RH (%)		Temp (°C)		RH (%)	
	Max	Min	FN	AN	Max	Min	FN	AN	Max	Min	FN	AN
September	25.4	17.8	75	53	30.5	-	94	75	32.1	-	94	72
October	33.0	21.8	93	59	32.0	-	93	67	34.7	-	93	57
November	33.3	17.3	90	43	33.8	-	91	60	35.4	-	90	56
December	34.5	17.1	95	51	33.9	-	86	45	34.8	-	91	56
January	34.9	18.4	93	45	34.3	-	94	58	35.4	-	92	52
February	36.2	22.3	92	45	34.8	-	91	52	37.3	-	92	46
March	36.8	29.4	88	46	34.5	-	93	52	38.0	-	90	44
April	36.8	-	90	49	35.1	-	91	54	37.8	23.8	89	49
May	34.5	-	90	59	35.0	-	92	58	36.6	23.6	93	55
June	29.3	-	96	79	29.6	-	96	87	32.8	23.1	96	79
July	28.7	-	96	86	31.1	-	94	78	29.9	22.5	96	80
August	29.8	-	94	79	31.0	-	94	73	31.3	22.9	97	80

TABLE 3B   Temperature and Relative Humidity Trends in Last Six Years at DCR

Month	2009–2010				2010–2011				2011–2012			
	Temp (°C)		RH (%)		Temp (°C)		RH (%)		Temp (°C)		RH (%)	
	Max	Min	FN	AN	Max	Min	FN	AN	Max	Min	FN	AN
September	31.3	22.9	96	86	31.8	22.7	96	79	31.9	22.3	96	80
October	34.1	22.3	94	62	32.4	22.6	96	74	35.2	22.6	95	60
November	33.9	21.9	94	71	33.7	22.2	95	66	33.2	19.5	93	53
December	34.5	20.5	92	65	34.2	19.0	94	57	33.7	17.4	92	46
January	34.7	18.7	92	60	36.1	16.0	92	47	33.7	17.4	93	41
February	36.6	19.9	91	51	36.8	17.5	90	43	36.1	19.3	92	40
March	37.7	23.0	90	46	38.0	21.4	91	39	36.2	21.3	90	41
April	37.1	23.7	90	48	37.8	23.0	92	45	35.5	22.5	91	49
May	36.6	23.4	90	52	36.6	23.2	93	53	35.3	22.7	90	50
June	32.0	22.8	95	84	31.3	22.8	98	89	31.3	23.0	93	79
July	29.8	22.6	98	94	30.6	22.6	98	91	30.8	23.2	94	79
August	30.5	22.9	96	90	31.2	22.9	97	85	28.0	22.6	93	81

Heavy rains at the time of harvesting affects yield and quality of nuts. Unseasonal rainfall of 201 mm received at DCR farm during 15–25 March, 2008 resulted in drastic reduction in nut yield (Tables 6.4 and 6.5). The peak harvesting of nuts was completed by 15 March, 2008. Rains at the time of harvesting resulted in blackening of nuts or nuts start germinating in the tree itself. Large quantity of nuts, which could not be picked up in time, germinated in the field. The collected nuts could not be dried in time due tononavailability of good sunshine hours resulting in poor quality of nuts. The excess moisture content in raw cashewnuts damages the kernel inside and changes its color from white to cream. It has been estimated that the nut yield losses due to unseasonal rains ranged from 50 to 65% in March, 2008.

Nut yield collected month-wise from cashew plantations at DCR for the past five years (2006 to 2010) (Table 6.6) showed that the quantity of nuts collected is highest during March followed by February or April

in all years except in 2010. During March, 2008 unseasonal rains caused damage to the flowers, tender nuts and matured nuts in the trees and also matured nuts fallen on the ground and finally resulted in deterioration of nut quality. A drastic reduction in nut yield was observed in March, 2009. It could be attributed to increase in mean monthly maximum temperature of about 1.2–1.6 °C in Mach, 2009 as compared to previous years (2005–2006 and 2006–2007). As a result, the actual nut yield was much lower than the expected nut yield.

TABLE 6.4 Unprecedented Rainfall Received in 2008

Date	Rainfall (mm)
15/3/2008	0.4
16/3/2008	14.8
17/3/2008	17.0
18/3/2008	11.0
19/3/2008	0
20/3/2008	12.0
21/3/2008	19.0
22/3/2008	25.4
23/3/2008	70.0
24/3/2008	17.0
25/3/2008	14.0
26/3/2008	0.4
Total	201

TABLE 6.5 Rainfall Pattern (mm) from September to August from 2006–2007 to 2011–2012

Month	2006–07	2007–08	2008–09	2009–10	2010–11	2011–12
September	1003	399	263	429.6	467.6	367.6
October	409	265	212	112.7	257.5	364.3
November	87	117	49	225.6	360.5	25.9

TABLE 6.5   *(Continued)*

Month	2006–07	2007–08	2008–09	2009–10	2010–11	2011–12
December	0	0	17	56.6	53.6	0
January	0	0	0	39.0	0.0	0
February	0	0	0	0.0	0.0	0
March	0	201	5.8	11.0	0.0	0
April	18	27.9	38.3	80.8	92.4	2.0
May	118	46.6	159.6	145.2	211.8	1.2
June	954	854.2	360.3	816.4	1085.6	30.0
July	1114	715	1828	1150	1187.8	17.0
August	690	465	430	647	662.4	33.0

TABLE 6.6   Production Trend of Raw Cashewnuts for the Last Five Years (2006 to 2010) at DCR Farm

Year	Month	Yield (kg)
2006	January	1700
	February	5925
	March	8252
	April	3961
	May	1162
	Total	21,000
2007	January	94
	February	2792
	March	10,242
	April	9547
	May	525
	Total	18,339
2008	January	488

TABLE 6.6   *(Continued)*

Year	Month	Yield (kg)
	February	7627
	March	10,049
	April	1832
	May	1000
	Total	20,995
2009	January	2384
	February	3977
	March	5530
	April	2967
	May	1567
	Total	16,425
2010	February	4167
	March	6043
	April	8281
	May	1535
	Total	20,026

Heavy rains coupled with high velocity wind during flowering phase resulted in flower drop and occurrence of fungal diseases. Pollinating insect activity also reduced to the minimum resulting in poor setting of nuts and low yield. At DCR farm, 10 days after the cessation of heavy rains (25 March, 2008), peak nut harvesting was partially coincided with the rainy period in mid- late varieties, which resulted in yield loss of about 30%. While in late varieties such as Ullal-1, Madakkathara-2 the yield loss was minimal. In farmers' fields, about 50% of the crop was lost in case of early and mid-late flowering varieties but the harmful effect was not noticed in late flowering varieties such as Ullal-1.

The total production of raw cashewnuts in Kannur and Kasargod districts of Kerala was lower during 1996–1997 (32,000 tons) and 1997–1998

(54,000 tons) as compared to 1995–1996 (70,000 tons) (Prasada Rao, 2002). In 1996–1997, moist weather due to moderate dew fall with mean minimum temperature of 20 °C during reproductive phase led to moderate flowering and enhanced the incidence of Tea Mosquito Bug (TMB) resulting in low nut yield. In 1997–1998, low dewfall with slightly higher mean minimum temperature of 21–22 °C during reproductive phase reduced flowering with consequent yield losses. Cashew genotypes distinctly vary in their heat units (day °C) requirement. Early variety 'Anakkayam-1' requires only 1953 day °C for reproductive phase, while late variety 'Madakkathara-2' requires 2483 day °C. Delayed flowering and fruiting in late varieties is likely to experience moisture stress, which can be managed by supplemental irrigation.

The mean maximum temperatures of 34.4, 36.1 and 36.2 °C were recorded during February 1992–1993, 2003–2004 and 2004–2005 and the corresponding RH in the afternoon was 33.0, 18.5 and 19.7%, respectively. Lower nut yield obtained during 2003–04 and 2004–05 as compared to 1992–93 was mainly due to drying of flowers. A rise in night temperature (>20 °C) coupled with fewer dew nights during flowering phase is detrimental for cashew flower production. Heavy cloudiness during flowering phase is detrimental for the opening of hermaphrodite flowers. Continuous rains without critical dry spells and late winter rains delay the bud break in cashew. A dry spell of 7 days is usually necessary 30 days prior to the bud break. Late and extended winter rains reduce the number of bright sunshine hours invariably which results in delaying of bud break and better availability of soil moisture during flowering (December and January). Incessant rains until November in 1998 delayed the bud break. The delay in bud break was prominent in early varieties. In late varieties, bud break was normal, because the required dry spell of 7 days was met 22–26 days before bud breaking (Table 6.7) (Prasada Rao et al., 2001).

TABLE 6.7 Time of Bud Break of Test Varieties at RARS, Pilicode from 1995–1996 to 1998–1999

Variety	Time of bud break (25%) of test varieties				
	1995–1996	1996–1997	1997–1998	1998–1999	Mean
Anakkayam-1	26 September	30 October	23 October	3 November	21October
Madakkathara-1	22 September	2 November	5 October	3 November	16 October

TABLE 6.7   *(Continued)*

| Variety | Time of bud break (25%) of test varieties | | | | |
	1995–1996	1996–1997	1997–1998	1998–1999	Mean
Kanaka	8 October	14 November	30 October	15 November	1 November
Madakkatha-ra-2	28 October	19 November	16 Dec	17 November	20 November
Mean	6 October	9 November	2 November	10 November	30 October

Peak flowering was early (8 December) during normal years (1995–1996) and late (29 December) in aberrant climate situations (1997–1998) which received late winter rains during November and December, 1997 (Table 6.8) Prasada Rao, 2002). There was a significant delay in peak flowering in Madakkathara 2 and Kanaka (13/1/98) during 1997–1998; however, early varieties flowered on normal dates. The delay in mean flowering date during 1997–1998 may be attributed to the late season types as they flowered very late.

TABLE 6.8   Time of Flowering of Test Varieties at RARS, Pilicode, from 1995–1996 to 1998–1999

| Variety | Time of flowering (50%) | | | | |
	1995–1996	1 9 9 6 – 1997	1997–1998	1998–1999	Mean
Anakkayam-1	28/11/95	04/12/96	02/12/97	07/12/98	3 December
Madakkath-ara-1	25/11/95	06/12/96	05/12/97	07/12/98	5 December
Kanaka	17/12/95	15/12/96	13/01/98	26/12/98	29 December
Madakkath-ara-2	22/12/95	25/12/96	2/02/98	21/12/98	1 January
Mean	8 December	13 December	29 December	17 December	17 December

The Tsunami in coastal Tamil Nadu on 26 December, 2004 devastated coastal parts of Kanyakumari and Cuddalore districts. Tsunami severely affected on cashew crop. Most of the standing crop was inundated with salty sea water ingressing in the mainland. Most of the cashew varieties are sensitive to salinity. Rise in sea water level due to climate change conditions may adversely affect the cashew plantation. Electrical conductivity of irrigation water 1.48 dS m^{-1} is a threshold tolerance for precocious cashew during the initial growth (Carneiro et al., 2002). Pot culture studies on the effect of salinity (1.2 to 20 d Sm^{-1}) on cashew seedlings by Valia and Patil (1997) showed a decrease in plant growth with increased salinity particularly >14 d Sm^{-1}. As salinity increased, all the nutrients (N, P, K, Ca, Mg, Zn, Fe, Mn and Cu) except S were depleted in leaves and roots. Also, chlorophyll content and transpiration rate decreased. 'Thane' cyclone with a very high wind speed created havoc in Cuddalore district of Tamil Nadu and Pondicherry on 30 December 2011. The entire cashew area of Panruti, Cuddalore and Kurinjipadi taluks of Cuddalore district were totally devastated. The extent of damage to cashew trees in Vridhachalam taluk was 60–70%due to decrease in the wind speed away from the coast.

Cashew is attacked by around 180 species of insect and noninsect pests in India resulting in substantial yield loss. Cashew stem and root borers (CSRB) and tea mosquito bug (TMB) are the major pests of cashew leading to economic losses. Tea mosquito bug (*Helopeltis antonii* Signoret) pest and inflorescence blight disease cause considerable damage to cashew. Incidence and severity of both are dependent on climate and weather factors. Prolonged and unseasonal rains coupled with heavy dew aggravate the incidence of pests and diseases resulting in loss of yield and quality of cashewnuts. Though cashew flowered profusely during 1995–1996 and 1998–1999 but there was a marked reduction in yield in 1998–1999 in Kannur and Kasaragod districts of Kerala. This could be ascribed to unprecedented incidence of pests aggravated by heavy dew in February, 1998. Rain received during flowering and harvesting in 2008 increased the incidence of pests and diseases resulting in very poor yield and quality (Prasada Rao, 2002).

Of the several foliage insect pests reported to infest cashew, the tea mosquito bug is of major importance. The production loss from the TMB alone is estimated to be about 30 percent. Its adults and nymphs feed on plant sap from the tender shoots, panicles and immature cashewnuts. The pest causes yield losses ranging from 30–90% depending on the extent of incidence. The pest population starts building up from September to October and gradually reaches a peak during February. The pest population is also synchronized with the phenological stages of the crop, which is also dependent on prevalent weather parameters. Various weather factors influence the pest build-up, of which, temperature and relative humidity have been reported to play a significant role.

Studies on the relationship between weather parameters and the population dynamics of TMB on cashew indicated that minimum temperature is negatively correlated with the pest incidence (Godse et al., 2005). The build-up of pest population commenced with the emergence of new flush during October after the cessation of monsoon shower and the pest remained in the field until January. The major outbreak was reported in November, coinciding with the emergence of panicles, reaching the peak in December. The minimum temperature plays a major role in the incidence of pest population. The favorable minimum temperature for TMB incidence ranged between 13 and 18 °C. Low temperature (12 °C) is antagonistic for pest build-up (Table 6.9) (Prasada Rao, 2002).

TABLE 6.9   Effect of Minimum Temperature on Pest Population of TMB

Night surface air temperature (^{0}C)	Intensity of TMB
14 to 18	Moderate to severe
12 to 14	Low to moderate
10 to 12	Low
<10	Nil

Fungal pathogens, *viz., Colletotrichum gleosporoides* and *Gleosporium mangiferae* have been reported to cause drying up of young shoots, inflorescence and immature nuts in cashew. The incidence of this disease is being reported from different new locations in which it was not prevalent earlier. High RH in forenoon during December–February both in 1997 and 1998 and the minimum temperature of 18–20 °C were found to be favorable for sporulation by fungi. A significant increase in dewfall was one of the most important factors, which favored the growth, sporulation and spread of fungi. Cloudiness leading to low bright sunshine hours (2 h/day) followed by dewfall triggered the growth, sporulation and spread of fungal pathogens causing inflorescence blight during 1998–1999 in Kannur and Kasaragod districts (Prasada Rao, 2002).

Seasonal incidence of cashew pest complex with weather parameters in Andhra Pradesh revealed that the incidence of shoot and blossom webber was severe from July to October and February to March with a peak during first fortnight of September exhibiting a negative correlation with maximum temperature, whilenonsignificant relationship with minimum temperature, relative humidity and rainfall. Leaf miner was severe from September to December with a peak during first fortnight of November. Relative humidity and rainfall had positive correlation with leaf miner incidence while maximum and minimum temperature had no significance. Weevil population was severe from first fortnight of August to second fortnight of October with a peak during first fortnight of September having positive correlation only with rainfall (Rama Krishna Rao and Haribabu, 2003).

6.4 ADAPTATION AND MITIGATION STRATEGIES TO MINIMIZE THE ADVERSE EFFECTS OF CLIMATE CHANGE

To circumvent losses due to climate change, adaptation and mitigation strategies are essential in affected areas. Some of the adaptation strategies include plant architecture, use of efficient technologies like drip irrigation, soil and moisture conservations measures, fertilizer management through fertigation, green manuring/intercropping, increase in input efficiency, pre and postharvest management of economic produce can not only minimize the losses but also increase the positive impacts of climate change. Mitigation is referred to the process in which the emissions of green house gases are reduced or they are sequestered. Improved agronomic practices for enhanced nutrient (nitrogen, inorganics and organics) use efficiency, water use efficiency for reduction of CHGs and ecofriendly disease and pest management strategies also form mitigation options.

Since climate change is projected to increase the frequency of extreme events like droughts, heavy rainfall events, etc., it is important to have proper soil and water conservation practices in place to minimize the adverse effects of climate change. Cashew can tolerate mild to moderate levels of moisture stress without affecting the growth of seedlings. Strong and severe water stress resulted in 20 and 22% reduction in number of scions, respectively (Shingre et al., 2003). Cashew experiences severe moisture stress from January to May, adversely affects its flowering and fruit set. In order to harvest the rainwater and to make it available to the cashew plant during critical period, *in situ*soil andwater conservation techniques are very important. Modified crescent bund and coconut husk burial are the two best soil and water conservation techniques for cashew orchard grown along medium to steep slopes. The available water during premonsoon, monsoon and post monsoon season can be efficiently harvested and used for higher yields (25–30% more than control) using proper soil and water conservation techniques.

Soil moisture stress in cashew orchard can be reduced by adopting various cultural practices such as mulching, green manuring or growing cover crops whichconserve soil health in term of soil moisture and nutrient conservation for crop production. Methods to minimize rise in soil temperature and maintaining optimum soil moisture may help in reducing the decomposition rate of the soil organic matter and thereby reducing the CO_2 emissions. Higher soil moisture content was observed in cashew orchard with glyricidia as green manuring crop (17.0 to 18.6% dry basis) compared to sunhemp (17.8 to 18.3% dry basis), sesbania (15.5 to 18.2% dry basis), cover crop (14.7 to 17.4% dry basis) and control (15.5 to 17.0% dry basis). Similarly, a significant increase in nut yield was observed with glyricidia as green manuring crop (2123 kg/ha/year) compared to control (1290 kg/ha/year) (Yadukumar et al., 2008). Mulching helps in many ways likeraising organic carbon content of the soil, improving water-holding capacity, reducing runoff and erosion, and making more water available for plants. The basin area of cashew plants can be mulched either with green leaves, dry leaves or weeds soon after planting. Black polythene mulch was helpful to conserve soil moisture (Nawale et al., 1985). Using coconut coir pith as soil mulch in cashew plantations resulted in 14.15% more water retention and suppression of weeds to an extent of 73.52% (Kumar et al., 1989).

Although cashew is considered to be a drought resistant plant, it does benefit from supplementary irrigation. Supplemental irrigation of 200 L of water/plant once in 15 days during January–March from ater collected in ponds through rainwater harvesting helps in flowering and nut development by improving the microclimate with increased humidity. It also leads to increased nut and kernel weight by reducing the flower and nut drying to some extent. Drip irrigation during fruit development may be helpful to rainfed cashew crop under climate change drought situations.

Forewarning the outbreak of the dreaded inflorescence blight disease in west coast region of India is of utmost importance for its effective management. Proper training of cashew trees from early stages and timely pruning of side and criss-cross branches may ensure that microclimate is not congenial for build-up of sucking pests and inflorescence blight disease.

Cashew crop offers immense scope for climate change mitigation through carbon sequestration and also by modification of microclimate. It has dense green leaves with good photosynthetic capacity and can also be grown in high-density planting system. Cashew is suitable crop for carbon sequestration. It can be grown in vast degraded/wasteland existing in cashew-growing regions. Based on research undertaken at DCR, it was found that, cashew genotype (VTH-174) trees of 7 years old sequestered about 2.2 fold higher carbon under high density planting system (625 trees/ha) as compared to normal density planting system (156 trees/ha). Carbon storage by cashew has been estimated as 32.25 and 59.22 t CO_2/ha at 5th and 7th years of growth, respectively under high density planting. The extent of carbon sequestered will depend on the amounts of C in standing biomass, age of the crop, tree density, variety, etc.

6.5   FUTURE RESEARCHABLE ISSUES

- Development of cashew varieties resistant to abiotic stresses like temperature, drought, salinity, floods, etc.
- Development of climate resilient agro-techniques in order to suit unfavorable abiotic stresses.
- Assessment of the impacts of elevated CO_2 and temperature on growth, development, yield and quality of cashew.
- Early warning weather forecasts on rains, storms, etc.
- Weather based pest-forecasting system.
- Quantification of carbon sequestration potential of cashew.
- Development of crop simulation models for cashew.
- Assessment of new pest and diseases in the scenario of climate change.
- Development of cost effective, ecofriendly approaches for management of emerging pests.
- To cope up with anticipated climate change impacts, mitigation and adaptation strategies for cashew need to be developed.

KEYWORDS

- **Adaptation**
- **Adverse Effects**
- **Cashew**
- **Climate Change**
- **Climatic Suitability**
- **Mitigation**

REFERENCES

Carneiro, P. T., Fernandes, P. D. & Gheyi, H. R. (2002). Germination and initial growth of pre-cocious dwarf cashew genotypes under saline conditions. *Revista Brasileira De Engenharia Agricola E Ambiental,* 6(2), 199–206.

DCCD (2012). Directorate of Cashew nut0 and Cocoa Development, Ministry of Agriculture, Kera bhavan, Kochi, Kerala.

Godse, S. K., Bhosle, S. R. & Patil, B. P. (2005). Population fluctuation studies of tea mosquito bug on cashew and its relation with weather parameters. *J. Agrometeorol.,* 7(1), 107–109.

Haldankar, P. M., Deshpande, S. B., Chavan, V. G & Rao, E. V. V. B. (2003). Weather associated yield variability in cashew nut. *J. Agrometeorol.,* 5(2), 73–76.

IPCC (Intergovernmental Panel for Climate Change) (2007). Summary for policymakers. In: Climate change 2007 physical science basis (Eds Solomon, S., Qin, D., Manning, M., Chen, Z., Marquis, M., Averyt, K. B., Tignor, M., Miller, H. L.). Contribution of Working Group 1, to the Fourth Assessment Report of the Intergovernmental Panel on Climate Change. Cambridge University Press, Cambridge, 996 pp.

Kumar, D. P., Subbarayappa, A., Hiremath, I. G., Khan, M., & Sadashivaiah, M. (1989). Use of coconut coir-pith a bio-waste as soil mulch in cashew plantations. *Cashew* (3), 23–24.

Mandal, R. C. (1992). Cashew Production and Processing Technology. Agro Botanical Publishers (India), Bikaner, 195.

Nawale, R. N., Sawke, D. P. & Salvi, M. J. (1985). Effect of black polyethylene mulch and supplemental irrigation on fruit retention in cashew nut. *Cashew Causerie* 7(3), 8–9.

Ohler, J. G. (1979). Cashew. Department of Agricultural Research, koninklijk Instituut voor de Tropen, Communication 71, Amsterdam. 260.

Prasada Rao, G. S. L. H. V. & Gopakumar, C. S. (1994). Climate and cashew. *The Cashew* 8(4), 3–9.

Prasada Rao, G. S. L. H. V. (2002). Climate and cashew. AICRP on Agro meteorology, Department of Agricultural Meteorology, College of Horticulture, Kerala Agricultural University, 100 pp.

Prasada Rao, G. S. L. H. V., Giridharan, M. P. & Jayaprakash Naik, B. (2001). Influence of weather factors on bud break, flowering and nut quality of cashew. *Ind. J. Agric. Sci.,* 71(6), 199–402.

Rama Krishna Rao, A., & Haribabu, K. (2003). Seasonal incidence of cashew pest complex in Tirumala hills of Andhra Pradesh.*Madras Agric. J.,* 90(4–6) 282–285.

Rejani, R., & Yadukumar, N. (2010). Soil and water conservation techniques in cashew grown along steep hill slopes. *Sci. Hortic.,* 126, 371–378. Doi: 10. 1016/j. scientia. 2010. 07. 032.

Shingre, D. V., Gawankar, M. S., & Jamadagni, B. M. (2003). Effect of irrigation and nitrogen on scion yield in cashew. *The Cashew* 17(1), 19–22.

Valia, R. Z., & Patil, V. K. (1997). Growth, physiological and nutritional status as influenced by soil salinity in cashew. *J. Plantation crops* 25(1), 62–67.

Yadukumar, N., Rejani, R., & Nandan, S. L. (2008). Studies on green manufacturing in high density cashew orchards. *J. Plantation Crops*, 36(3), 265–269.

Yadukumar, N., Raviprasad, T. N. & Bhat, M. G. (2010). Effect of climate change on yield and insect pests incidence on cashew. In Challenges of Climate Change Indian Horticulture (Eds Singh, H. P., Singh, J. P. & Lal, S. S.). Westville Publishing House, New Delhi. 224.

CHAPTER 7

PROTECTED CULTIVATION TECHNOLOGIES FOR VEGETABLE CULTIVATION UNDER CHANGING CLIMATIC CONDITIONS

BALRAJ SINGH[1] and R. K. SOLANKI[2]

NRC on Seed Spices, Tabiji, Ajmer-305206, Rajasthan, India.
Email: drbsingh2000@yahoo.com

[2]Department of Horticulture (Vegetables and Floriculture), Bihar Agricultural College, Bihar Agricultural University, Sabour (Bhagalpur) – 813 210, Bihar, India.

CONTENTS

7.1   INTRODUCTION

Agriculture is highly dependent on environment, and it's very difficult to get favorable climatic conditions for crop growth and development as per crop need. Agriculture is basically climate/season based, a hot and humid climatic conditions characterized in rainy and post rainy season is most favorable for both crop and crop enemies. To raise a healthy disease free crop spring-summer seasons was counted as most suitable. But, fast climatic changes happening across the globe has changed climatic characteristics of a season, which has resulted in untimely rains and other fluctuations in the spring-summer season, raising the challenge to develop climate resilient technologies. Not even that, with time extreme hot and cold temperature stresses have been noticed in geographically varied locations where it was not supposed to be earlier based on various geographical factors deciding the climatic conditions of that area. Therefore, there is need to develop suitable varieties and technologies to sustain these challenges whichmay come up in form of various biotic and abiotic factors (Singh, 2013). Vegetable cultivation is an awesome business in India, but under open field conditions by following traditional cultivation practices it is difficult to manage various abiotic and biotic stresses. These stresses not only reduce productivity levels but they are also responsible for poor quality specifically during rainy and post rainy season. Mostly to manage biotic stresses farmers spray large amount of different chemicals, this not only enhances the cost of cultivation but it also increases residual toxicity in the freshly produced vegetables, which is ultimately hazardous to human health.

How to address these issues, can we manipulate the climatic conditions or can we provide protection to the crops against climatic fluctuations and various other related stresses. Yes, protected cultivation technology has the answer to this but, it's a tricky technology highly depending upon intelligent implementation of protected structures for vegetable cultivation by having a knowhow on "*What, When, Where and Why*" *to* implement. Every protected structure has its own limitations and advantages (Singh and Kalia, 2005), but the basic benefit is its extra protective shelter restricting or minimizing the exposure of the crops to various adverse factors, which are high in open conditions. Even though the application of chemicals for controlling biotic stresses is also low under protected structures which gives a high quality safe vegetables for human consumption. By using protected structures, it is also possible to raise a offseason and long dura-tion vegetables crop of high quality (Sabir and Singh, 2013). Vegetable farming in agri-entrepreneurial models targeting various niche markets of the big cities is inviting regular attention of the vegetable growers for di-versification from traditional ways of vegetable cultivation to the modern methods. Under the new era of FDI (Foreign Direct Investment) in retail, these kinds of models posses high potential for enhancing the income of farmers opting for quality and offseason vegetable cultivation through protected cultivation (Singh et al, 2012). Production of vegetable crops under protected conditions provides high water and nutrient use efficiency under varied agro climatic conditions of the country. This technology has very good potential especially in peri-urban areas adjoining to the major cities which is a fast growing markets of the country, since it can be prof-itably used for growing high value vegetable crops like, tomato, cheery tomato, colored peppers, parthenocarpic cucumber, healthy and virus free seedlings in agri-entrepreneurial models (Singh et al., 2010). But protect-ed cultivation technology requires careful planning, attention and details about timing of production and moreover, harvest time to coincide with high market prices, choice of varieties adopted to the off season environ-ments, andable to produce economical yields of high quality produce, etc.

All kind of protected technologies may not be suitable and economical to all group of Indian farmers, because of their very high initial basic cost of fabrication, running and maintenance cost, but some protected technol-ogies are not only low cost and simple but they are energy efficient and are highly profitable under different Indian conditions and more specifically for peri-urban areas, which can be adopted by Indian farmers for produc-tion of diverse high value vegetable crops and nursery raising in profit-able agri-business models (Singh et al., 2012). Some of the technologies likelow-pressure drip irrigation and low cost nursery raising technology are highly suitable for livelihood security even in tribal areas of the coun-try. These two technologies have already been demonstrated successfully and have proved their suitability, efficiency and profitability in tribal area of Jharkhand, Madhya Pradesh, Gujarat and other states of the country. Similarly these technologies can also be replicated in other parts of the country among the small and marginal farmers. Protected vegetable cul-tivation technology is highly relevant under the era of changing climatic conditions and can be well adopted in big way under model vegetable farm concept or in cluster approach in peri-urban areas of big cities for not only supplying high quality vegetables in the niche markets of these cities but also stabilizing the huge fluctuations in market prices of fresh vegetables in these cities almost every year in India.

7.2   PLUG TRAY NURSERY RAISING TECHNOLOGY FOR VEGETABLES

Quality seed and planting material always plays a vital role for improving the productivity of vegetable crops. The lower productivity of vegetable could be attributed due to inadequate availability of virus and disease free healthy seedlings and unavailability of grafted planting material having resistant root stocks against soil borne diseases in vegetable crops.

To ensure high productivity and high quality in vegetables, raising of high quality virus free seedlings through the use of good planting material at right time and at a appropriate place is one of the cheapest but most important way. Most of the Indian farmers are raising their vegetable seedlings under open field conditions, which are always inferior in quality, as the seedlings are infected with virus when raised in open during rainy and post rainy season. On one side soil borne fungus and nematodes create severe problem for raising the seedlings in soil media in open fields during hot summers and rainy season but on the other hand the very high cost of hybrid seeds in vegetables has also warranted the farmers to improve or change their traditional nursery raising method to increase the productivity and quality of vegetables. Plug tray nursery raising in vegetable crops has already working as a well established industry in several developed countries. Under this system seedlings are raised in plastic multicelled plug-trays in artificial soil-less media in especially designed greenhouses or other protected structures (Singh et al., 2005). A large number of virus free healthy seedlings of different vegetables can be raised in a small area of green house in plastic plug-trays either for main season or for their off season cultivation. With the use of this technology it is now almost possible to raise healthy vigorous seedlings of almost all vegetables. Unemployed educated youths of our country can start nursery production as a agri entrepreneur in major vegetable growing pockets of the country very successfully. By this way the vegetable growers can get virus free or off-season healthy nursery as per their requirement and it will also generate additional employment in agriculture sector. Therefore, this is the first and most important step for enhancing productivity, quality and profitability in vegetables through protected cultivation. One thousand square meters nursery green house is capable to produce nearly 15–20 lakhs of vegetable seedlings in 6–8 batches resulting in an net income of Rs. 2.5–3.0 lakh and moreover, it can also provide employment to nearly 4–5 youths for conducting different operations in nursery raising throughout the year.

7.3   OFF-SEASON CULTIVATION OF VEGETABLE CROPS UNDER PLASTIC LOW TUNNELS

Plastic low tunnels are flexible transparent coverings that are installed over single or multiple rows of vegetables to enhance the plant growth by warming the air around the plants in the open field during winter season when the temperature is below 8 °C. Plastic low tunnels are often used to promote the growth of plants during the period of winter season. Low tunnels are supported above the plants by using hoops of GI wire and a clear or transparent plastic of 20–30 micron is covered/stretched over the hoops and the sides are secured by placing in soil. The plastic is having vented or silted during the growing season as the temperature increase within the tunnels. The farmers can grow different varieties of summer squash (round fruited, long fruited), which is a emerging crop along with cultivation of netted muskmelon varieties in place of traditional varieties (Singh and Kumar, 2009). Bitter gourd and round melon are two other crops with increasing demand and which usually fetches very high price during off-season and can be grown successfully by using the plastic low tunnel technology. This technology is highly suitable and profitable for the farmers living in northern plains of India (Singh et al., 2004).

7.4   OFF SEASON VEGETABLE CULTIVATION UNDER WALK IN TUNNELS

Walk in tunnels are temporary structures erected by using G.I. pipes and transparent plastic. They are generally used for complete off season cultivation of vegetables like bottle gourd, summer squash, cucumber, etc. during peak winter season (Dec-mid February) the basic objective and utility of walk in tunnels is to fetch high price of the complete off season produce to earn more profit per unit area. The ideal size of a walk in tunnel can be of 4.0 m width and 30 m length (120 m²) and cost of fabrication may range from Rs.160–180 per m². The ideal size has been standardized for optimum cross ventilation, to have a single piece coverage of above sized structure with plastic commonly manufactured by firms of dimension 7 × 30 m or 7 × 36 m and a length of nearly 25–30 m is very suitable for honey bees to fly from one end to other for pollination (Singh et al., 2007).

Walk-in-tunnel technology is a simple and profitable technology for off-season cultivation of cucurbits during the winter season in northern plains of our country. Crops like summer squash can be grown as a complete off-season crop, whereas other cucurbits like muskmelon, round melon, bottle gourd, cucumber, bitter gourd, watermelon are mainly grown during off-season. Basically these are temporary and really low cost structures sincethe fixed investment made on plastic can last for 5–6 years and the investment made on GI pipes can last for more than 20 years if proper care is taken.

7.4.1 INSECT PROOF NET HOUSES FOR SAFE VEGETABLE CULTIVATION

The basic objective of Insect proof net house vegetable cultivation is to minimize the use of pesticides in fresh vegetable cultivation for producing safe vegetables and for production of quality seeds either of the seasons. Since, vegetables are highly prone to large number of viruses and other insect pests, which occur high during hot humid conditions of rainy and post rainy season. To counter the problem it was suggested to raise the crop in summers for getting disease and virus free seed crops. But due to the dynamic changes in the climatic conditions untimely rains and temperature fluctuations are being experienced in spring and summer seasons. So, an approach for taking safe commercial and seed production vegetable crop has failed taking advantage of seasonal conditions in the present scenario. The only solution is to reduce the affect of climatic conditions by using protected structures. Pests and other vectors population can be checked effectively by using insect proof net houses by creating a physical barrier between crop and open environment (Singh et al., 2005). Usually the farmers are growing their vegetable crops like tomato, chili, sweet pepper, okra which are highly affected by viruses and other pests like borers and fruit-flies under open fields during rainy and post rainy season The farmers are using several insecticides for several sprays to control these vectors, even though they could not effectively control these vectors. Huge amount of insecticidal spray increases the residual toxicity which can be minimized by using insect proof net houses of 40 or 50 meshes covered walk in tunnels. This technology can be used for raising crops like tomato, chili, sweet pepper, etc., but for growing these crops under insect proof net houses, it is prerequisite to raise virus free healthy seedlings of the crop either in the greenhouse or by covering the nursery beds with insect proof net (Singh et al., 2010; Singh, 2011). These structures can be fabricated with a cost of Rs.350–400/m^2 having 40–50% shading net covering during critical summer months (April–June) and with transparent plastic covering during critical winter months (Dec–Feb) under arid and semiarid climatic conditions. High value vegetables like tomato, cherry tomato (crop duration 7–8 months), two crops of parthenocarpic cucumber (summer and post rainy season) and capsicum (crop duration 7–8 months) can be produced (Singh and Tomar, 2007).

7.5 OFF SEASON VEGETABLE CULTIVATION UNDER SHADE NET HOUSES

Shade nets are perforated plastic materials used to cut down the solar radiation and prevent scorching or wilting of leaves caused by marked temperature increases within the leaf tissue from strong sunlight. These nets are available in different shading intensities ranging from 25% to 75%. Leafy vegetables and ornamental greens are recommended to be grown under shade nets whose growth rates are significantly enhanced compared to un-shaded plants when sunlight is strong. The basic objective of shade net is to reduce radiation and temperature up to some extent during critical summer months (May–Sept.). Black color shade nets are most efficient in reduction of temperature compared to other colors like green, white or silver, etc. as the black color in the maximum absorbent of heat. Mostly leafy vegetables like beet leaf and green coriander are preferred to be grown under shade nets, but it is also suitable for growing early cauliflower and radish cultivation during June to September months.

7.6   ZERO ENERGY NATURALLY VENTILATED GREENHOUSES FOR HIGH VALUE VEGETABLE CULTIVATION

Naturally ventilated greenhouses are the protected structures where no heating or cooling devices are provided for climate control. They are simple and medium cost greenhouses which are erected with a cost of Rs.700–800 per sq. mt. and these greenhouses can be used successfully and efficiently for growing year round parthenocarpic slicing cucumber, off season muskmelon, tomato and sweet pepper crops for 8–9 months duration. These structures are having a manually operated natural ventilation system, as and when required (Singh and Kumar, 2006). Looking to the year round increasing demand of high quality vegetables like parthenocarpic slicing cucumber, standard tomatoes, cherry tomatoes and colored peppers, muskmelon in urban areas of the country specially in up markets of the metro viz. stared hotels, shops of embassies or high commissions of various countries situated in Delhi and other big cities has raised the opportunities for peri-urban farmers to fetch very high price of the season and off-season high quality produce. Greenhouse vegetable production is a highly intensive enterprise requiring substantial labor and strong commitment, which restrict the adoption of this technology. But now the time has come when unemployed educated youths are required be motivated and trained in various parts of the country to use naturally ventilated greenhouse technology for cultivation of high value vegetables for high profits.

The basic prerequisite of implementing the technology is to depending upon selection of appropriate design based on the climatic conditions, market available and the type of vegetable crop. Under arid and semi arid conditions maximum ventilation up to 40–50% is required to make the structure efficient and successful to raise vegetable crops. Under extreme hot periods (May–July) rooftops of the greenhouses should be covered with shade nets (preferably with black color) allowing a space between the shade net and roof surface for air movement. Such greenhouses can be equipped with low-pressure drip irrigation system to make them energy efficient ecofriendly model (Singh and Hasan, 2011).

7.7 CONCLUSIONS

Keeping in view the huge fluctuations in market prices of fresh vegetables and the increasing demand due to urbanization and changing life styles with increasing purchasing power of people has created vegetable cultivation as big enterprises near big cities. There is an urgent need for diversification from the traditional agriculture by production of high value vegetable crops under protected conditions for increasing their productivity and quality not only for getting high returns but also in a Horti-Business/Veggie-business model. Nursery raising under protected cultivation is highly suitable for such Horti-business model or can be adopted as a agri entrepreneur business in major vegetable growing areas of the country by unemployed educated youths. Low cost protected technology like plastic low tunnels or walk in tunnels, shade net houses can be used for off-season vegetable cultivation for getting high returns from off season produce. Similarly, insect proof net houses can be used on a large-scale safe vegetable cultivation by way of minimizing the use of pesticides in vegetable cultivation and virus free quality seed production in large number of vegetables in horti-business models. Some of the technologies likelow-pressure drip irrigation and low cost nursery raising technology can also be replicated in other parts of the country even among the small and marginal farmers. Looking to the climatic conditions found suitable for promoting protected cultivation in the mid eastern parts of the country covering states of Bihar, Odisha, Chhattisgarh and Jharkhand, vegetable cultivation can be highly rewarding for the unemployed educated youths and farmers.

KEYWORDS

- Adverse effect
- Climate change
- High Value Vegetable
- Improper pollination
- Off-season Cultivation
- Temperature

REFERENCES

Singh, B., Tomar, B. S. & Kumar, M. (2004). Plastic low tunnel a profitable and sustainable technology for peri-urban areas. *Intensive Agriculture*, 42(7–8), 18–20.

Singh, B., & Kalia, P. (2005). Protected cultivation of vegetable crops: Problem, potential and prospects in India. *Indian Journal of Fertilizers,* 1(4), 93–97.

Singh, B., Kumar, M., & Yadav, H. L. (2005). Plug-tray nursery raising technology for vegetables. *Indian Horticulture,* 49(4), 10–12.

Singh, B., Kumar, M., & Singh, V. (2005). Nylon Mesh Screens Reduce Incidence of Leaf Curl Virus and Improves Yield in Sweet Pepper. Journal of Vegetable Science (USA). 12(1), 65–70.

Singh, B., & Kumar, M. (2006). Techno-economic feasibility of Israeli and indigenously designed naturally ventilated greenhouses for year round cucumber cultivation. *Acta Horticulturae*. 710, 535–538.

Singh, B. & Tomar, B. S. (2007). Keet avrodhi nylon net dwara sabjion kee vishanu rog rahit swasth paudh utpadan. Phal Phool 29(5), cover page 1&3.

Singh, B., & Kumar, M. (2009). Evaluation of Summer Squash varieties under plastic low tunnels during their off season cultivation. *Indian J. Hort.* 66(1), 135–136.

Singh, B., Singh, A. K., & Tomar, B. S. (2010). In Peri Urban Protected cultivation technology to bring prosperity. *Indian Horticulture.* 55(4), 31–32.

Singh, B., Tomar, B. S. & Thakur, S. (2010). Quality seed Production of Parental lines of Pumpkin (C. moschata) under Insect Proof net house. *Acta Horticulturae* (ISHS) 871, 275–278.

Singh, B. (2011). Insect proof net house for cultivation for high value vegetables. *ICAR News.* 17(3), 6.

Singh, B., & Hasan, M. (2011). Low pressure drip irrigation system: A success story *(In Hindi).* Agriculture Extension Review. 21(2), 11–12.

Singh, B., Singh, A., & Kumar, M. (2012). *Pari nagriya shetroo me: Aiise hogi sabjiyo kee sanrakshit kheti.* Phal Phul. 33(3), 3–6.

Singh, B., Tomar, B. S., & Kumar, M. (2004). Plastic low tunnel: A profitable and sustainable technology for peri-urban areas. *Intensive Agriculture*, 42(7–8), 18–20.

Singh, B., & Kalia, P. (2005). Protected cultivation of vegetable crops: Problem, potential and prospects in India. *Indian Journal of Fertilizers,* 1(4), 93–97.

Singh, B., Kumar, M., & Yadav, H. L. (2005). Plug-tray nursery raising technology for vegetables. *Indian Horticulture,* 49(4), 10–12.

Singh, B., Kumar, M., & Singh, V. (2005). Nylon Mesh Screens Reduce Incidence of Leaf Curl Virus and Improves Yield in Sweet Pepper. Journal of Vegetable Science (USA). 12(1), 65–70.

Singh, B., & Kumar, M. (2006). Techno-economic feasibility of Israeli and indigenously designed naturally ventilated greenhouses for year round cucumber cultivation. *Acta Horticulturae*. 710, 535–538.

Singh, B., & Tomar, B. S. (2007). Keet avrodhi nylon net dwara sabjion kee vishanu rog rahit swasth paudh utpadan. Phal Phool 29(5), cover page 1&3.

Singh, B., & Kumar, M. (2009). Evaluation of Summer Squash varieties under plastic low tunnels during their off season cultivation. *Indian J. Hort.* 66(1):135–136.

Singh, B., Singh, A. K., & Tomar, B. S. (2010). In Peri-Urban: Protected cultivation technology to bring prosperity. *Indian Horticulture.* 55(4), 31–32.

Singh, B., Tomar, B. S., & Thakur, S. (2010). Quality seed Production of Parental lines of Pumpkin (C. moschata) under Insect Proof net house. *Acta Horticulturae* (ISHS) 871, 275–278.

Singh, B. (2011). Insect proof net house for cultivation for high value vegetables. *ICAR News.* 17(3), 6.

Singh, B., & Hasan, M. (2011). Low pressure drip irrigation system: A success story *(In Hindi)*. Agriculture Extension Review. 21(2), 11–12.

Singh, B., Singh, A., & Kumar, M. (2012). *Pari nagriya shetroo me: Aiise hogi sabjiyo kee sanrakshit kheti.* Phal Phul. 33(3), 3–6.

Singh, B. (2012). Seed Production of summer squash in North Indian plains. ICAR News 18(2), 14.

Sabir, N., & Singh, B. (2013). Protected cultivation of vegetables in global arena A review. Indian Journal of Agricultural Sciences. 83(2), 123–135.

Singh, B., Tomar, B. S. & Kumar, M. (2004). Plastic low tunnel: A profitable and sustainable technology for peri-urban areas. *Intensive Agriculture*, 42(7–8), 18–20.

Singh, B., & Kalia, P. (2005). Protected cultivation of vegetable crops: Problem, potential and prospects in India. *Indian Journal of Fertilizers*, 1(4), 93–97.

Singh, B., Kumar, M., & Yadav, H. L. (2005). Plug-tray nursery raising technology for vegetables. *Indian Horticulture*, 49(4), 10–12.

Singh, B., Kumar, M., & Singh, V. (2005). Nylon Mesh Screens Reduce Incidence of Leaf Curl Virus and Improves Yield in Sweet Pepper. Journal of Vegetable Science (USA). 12(1), 65–70.

Singh, B., & Kumar, M. (2006). Techno-economic feasibility of Israeli and indigenously designed naturally ventilated greenhouses for year round cucumber cultivation. *Acta Horticulturae*. 710, 535–538.

Singh, B., & Tomar, B. S. (2007). Keet avrodhi nylon net dwara sabjion kee vishanu rog rahit swasth paudh utpadan. Phal Phool 29(5), cover page 1 &3.

Singh, B., Kumar, M., & Sirohi, N. P. S. (2007). Protected cultivation of cucurbits under low cost protected structures. A sustainable technology for peri-urban areas of Northern India. Acta Horticulturae. (ISHS) 731, 267–272.

Singh, B., & Kumar, M. (2009). Evaluation of Summer Squash varieties under plastic low tunnels during their off season cultivation. *Indian J. Hort.* 66(1), 135–136.

Singh, B., Singh, A. K., & Tomar, B. S. (2010). In Peri-Urban: Protected cultivation technology to bring prosperity. *Indian Horticulture*. 55(4), 31–32.

Singh, B., Tomar, B. S., & Thakur, S. (2010). Quality seed Production of Parental lines of Pumpkin (C moschata) under Insect Proof net house. *Acta Horticulturae* (ISHS) 871, 275–278.

Singh, B. (2011). Insect proof net house for cultivation for high value vegetables. *ICAR News.* 17(3), 6.

Singh, B., & Hasan, M. (2011). Low pressure drip irrigation system: A success story *(In Hindi)*. Agriculture Extension Review. 21(2), 11–12.

Singh, B., Singh, A., & Kumar, M. (2012). *Pari nagriya shetroo me: Aiise hogi sabjiyo kee sanrakshit kheti.* Phal Phool. 33(3), 3–6.

Singh, B. (2013). Protected Cultivation Technologies for Biotic and Abiotic Stress Management in Vegetables. Lead paper presented in National Symposium on Abiotic and Biotic Stress Management in Vegetable Crops held from April 4–6 at IIVR, Varanasi (India)

Sabir, N., & Singh, B. (2013). Protected cultivation of vegetables in global arena A review. Indian Journal of Agricultural Sciences 83(2), 123–135.

CHAPTER 8

CLIMATE CHANGE AND ITS IMPACT ON PRODUCTIVITY AND BIOACTIVE HEALTH COMPOUNDS OF VEGETABLE CROPS

PRITAM KALIA[1] and R. K. YADAV

Division of Vegetable Science Indian Agricultural Research Institute, New Delhi, India; [1]Email: pritam.kalia@gmail.com

CONTENTS

ABSTRACT

Vegetable crops are playing pivotal role in ensuring food, nutritional and livelihood security due to their high production per unit area and time besides high level of biologically active compounds that impart health benefits beyond basic nutrients. High intake of vegetables has been associated with a lesser incidence of degenerative diseases. With the increasing health consciousness among masses, demand for these crops is rising day by day. In order to enhance their suitability beyond fresh, efforts are going on to develop bioactive compound rich varieties in respective crops so as to harness their industrial potential. These compounds have also been reported to enhance storability and delays senescence. Abrupt rise and fall in temperature due to Climate change will definitely pose a threat to production and productivity of perishable vegetable crops besides disturbing their endogenous biochemical level and taste. The cool season vegetables are more sensitive to adverse weather than warm season. Abiotic stresses like extreme temperature (low/high), soil salinity and drought are detrimental for vegetable production. Thus, high temperatures and limited soil moisture are the major causes of low yields in vegetables. The different development phases like vegetative growth, flowering and fruit development are significantly influenced by climatic vagaries. The effects of the elevated temperature, unpredictable and irregular precipitation can disrupt the normal growth and development of plant that ultimately affect the crop productivity. It is projected that Crop yields will decline in Asia by 2.5–10% from 2020 onwards and by 5–30% after 2050. However, this decline is further expected to compound in South and Central Asia.

Concerning the possible impact of climatic change on vegetable production as well as national economy, a holistic approach is required to over stress tolerance rather than a single method. A systems approach, where all available options are considered in an integrated manner, will be effective and sustainable under a variable climate. For this to succeed, adequate and long-term funding is necessary, scientific results have to be delivered, best approaches has to be used so as to deliver global public goods for impact involving public and private sector together. In order to mitigate adverse effect of climate change it is necessary to identify germplasm of vegetable crops tolerant to drought, high temperature and other environmental stresses. Besides, this must also have ability to maintain yield in marginal soils to serve as sources of these traits for both public and private vegetable breeding programs. This germplasm will include both cultivated and wild accessions possessing genetic variation unavailable in extensively grown cultivars. Genetic populations are being developed to introgression and identify genes conferring tolerance to stresses and at the same time generate tools for gene isolation, characterization, and genetic engineering. Furthermore, agronomic practices that conserve water and protect vegetable crops from suboptimal environmental conditions must be continuously enhanced and made easily accessible to farmers in the developing world.

8.1 INTRODUCTION

Global climate system warmed up by 0.74 °C between 1906 and 2005 due to increased concentration of greenhouse gases (GHGs). The trends of rise in temperature, heat waves, droughts and floods, and sea level observed by the Indian scientists are in line with the Inter-Governmental Panel on Climate Change (IPCC), though the magnitude of changes varies. Agriculture in India, where nearly 60% of area is rain fed, has been a highly risky venture with vagaries of monsoon besides the interplay of other abiotic and biotic factors. Climate change is further compounding the daunting complex challenges already being faced by agriculture. Concerted efforts are, therefore, required for mitigation and adaptation to counter the adverse impacts of climate change and making it more resilient.

A significant change has been observed in the gaseous composition of earth's atmosphere during the last few decades which is attributable to the increased emissions from energy, industry and agriculture sectors widespread deforestation as well as fast changes in land use and land management practices mainly active gases, viz. carbon dioxide (CO_2), methane (CH_4) and nitrous oxide (N_2O), popularly known as the 'greenhouse gases' (GHGs). These GHGs trap the outgoing infrared radiations from the earth's surface and, thus, raise the temperature of the atmosphere. The global mean annual temperature at the end of the twentieth century, as a result of GHG accumulation in the atmosphere, has increased by 0.4–0.7°C above that recorded at the end of the nineteenth century. An increasing trend in temperature at 0.13 °C/decade has been observed during the past 50 years, however the rise has been much higher during the past one and half decades.

8.2   EMISSION OF GREENHOUSE GASES

The three major GHGs are carbon dioxide, methane and nitrous oxide, besides chlorofluorocarbons. A brief description about their sources and sinks is given below.

8.2.1   CARBON DIOXIDE

The main sources of carbon dioxide emission are decay of organic matter, forest fires, eruption of volcanoes, burning of fossil fuels, deforestation and land-use changes.

8.2.2   METHANE

Methane is about 25 times more effective as a heat-trapping gas than CO_2. The main sources of methane are: wetlands, organic decay, termites, natural gas and oil extraction, biomass burning, rice cultivation, cattle and refuse landfills.

8.2.3   NITROUS OXIDE

As a greenhouse gas, nitrous oxide is 298 times more effective than CO_2. Forests, grasslands, oceans, soils, nitrogenous fertilizers, and burning of biomass and fossil fuels are the major sources of nitrous oxide, while it is removed by oxidation in the Stratosphere. Soil contributes to the largest amount of nitrous oxide emission. The major sources are soil cultivation, fertilizer and manure application, and burning of organic material and fossil fuels. From an agricultural perspective, nitrous oxide emission from soil represents a loss of soil nitrogen, reducing the nitrogen-use efficiency.

8.3   IMPACTS OF CLIMATE CHANGE ON AGRICULTURE

Global climatic changes can affect agriculture through their direct and indirect effects on the crops, soils; livestock and pests. An increase in atmospheric carbon dioxide level will have a fertilization effect on crops with C3 photosynthetic pathway promoting, thereby, their growth and productivity. The increase in temperature, depending upon the current ambient temperature, can reduce crop duration, increase crop respiration rates, alter photosynthate partitioning to economic products, affect the survival and distribution of pest populations, hasten nutrient mineralization in soils, decrease fertilizer-use efficiencies, and increase evapo-transpiration rate. Indirectly, there may be considerable effects on land use due to snow melt, availability of irrigation water, frequency and intensity of inter and intraseasonal droughts and floods, soil organic matter transformations, soil erosion, changes in pest profiles, decline in arable areas due to submergence of coastal lands, and availability of energy. Major impacts of climatic change are as follow:

Reduction in crop yield, shortage of water, irregularities in onset of monsoon, drought, flood and cyclone, rise in sea level, decline in soil fertility, loss of biodiversity, increasing in diseases, insect pests and weeds may be the probable adverse effect of climate change.

8.3.1 EFFECT OF CLIMATE CHANGE ON VEGETABLE CROPS

Vegetables being rich sources of calories, health-building substances ensure food, nutritional and livelihood security. Optimum consumption of vegetables lowers incidence of degenerative diseases. Bioactive compounds besides their health benefits, affects storability of vegetables that contain them. They also play an important role in senescence of vegetables by combating relative oxygen in species (Hodges and Forney, 2003). Therefore, a product with a high concentration of antioxidants is well protected against oxidation and may thereby retain its quality longer. Freshness of vegetables, to some extent, is a marker for its food value as regards the content of bioactive compounds. Climate change may have adverse small and marginal farmers who are mainly dependent on vegetables (FAO, 2009). Moreover, the cool season vegetables are more sensitive to adverse weather than warm season. Abiotic stresses like extreme temperature (low/high), soil salinity and drought are detrimental for vegetable production. Thus, high temperatures and limited soil moisture are the major causes of low yields in vegetables. The different development phases like vegetative growth, flowering and fruit development are significantly influenced by climatic vagaries. The effects of the elevated temperature, unpredictable and irregular precipitation can disrupt the normal growth and development of plant that ultimately affect the crop productivity. Crop yields in Asia are expected to decline by 2.5–10% from 2020 onwards and by 5–30% after 2050, with declines worst in South and Central Asia (Cruz et al., 2007). Different research initiatives have been undertaken in the wake of climate change like selection of better adaptable genotypes, genetic manipulation to overcome extreme climatic stresses, approaches to improve water & nutrient use efficiency and biological nitrogen fixation as well as to exploit the beneficial effects of CO_2 enhancement on crop growth to mitigate the challenges. The present study reviews the impact of global climate change on vegetable crop growth and yield and strategies to overcome the harmful consequences. It also reviews the effect of climate on different quality aspects/bioactive compounds of vegetable crops that may occur under changed climate. In this regard, it is necessary to under-stand the effect of climate change on nutritive value of some important vegetables especially with respect to antioxidant property.

8.3.2   STATUS OF VEGETABLE PRODUCTION

Vegetables are the cheapest resource of nutrients, especially vitamins and mineral and helpful for overcoming malnutrition and provide farmers much higher income per unit area per unit time than staple crops and generate more employment. The worldwide production of vegetables has doubled over the past quarter century and the value of global trade in vegetables now exceeds that of cereals. According to NHB 2011, World's vegetable acreage is 54 million hectare and total production is 1013 million tons. India is the second largest producer of vegetable in world after China with total production of 147 million tons from 8.5 million hectare area. Produc-tion of vegetables has increased at an annual rate of 3.6% in 2009–2010 over 2008–2009. The productivity of vegetables has also increased up to 17.3t/ha in 2010–2011 from 10.5t/ha in 1991–1992. Though more than 60 vegetable crops are grown commercially, however potato, onion, tomato, chili, okra, brinjal, garden pea, cabbage, cauliflower occupy larger area in India. The leading vegetable producing states are West Bengal, Uttar Pradesh, Bihar, Andhra Pradesh and Orissa (NHB, 2011). Most vegetables prefer mild temperatures, thus productivity is low in the hot and humid areas of India. Vegetables are generally sensitive to environmental extremes, and thus high temperatures and limited soil moisture are the major causes of low yields and will be further magnified by climate change (Table 8.1).

TABLE 8.1 Vegetables Sources of Bioactive Compounds and Antioxidants

Vegetable	Phytochemicals/bioactive compounds	Benefits
Beas	Flavonoids (saponins)	Protect against cancer, lower cholesterol
Broccoli, cabbage, kale, etc.	Indoles, Isothiocyanates, glucosinolates	Protect against cancer, heart disease and stroke
Carrots	Beta-carotene (orange/red)	Antioxidant
	Anthocyanin (black/purple), lignin	
Tomato	Lycopene, Vitamin C, Flavonoids	Protect against cancer, fight infection
Onion & Garlic	Allyl sulfides, Quercetin, myricetin	Protect against certain cancers and heart disease, boost the immune system
Watermelon	Lycopene, citrulline	Protect against cancer, improve immunity
Bitter gourd	Momordicin and Charantin	Diabetes, blood purifier, Hypertension,
Radish	Isothiocyanates, sinigrin	Jaundice, Liver infection, Piles
Chili/pepper	Capsaicin	Antirheumatic
Artichoke	Silymarin	

8.3.3 ENVIRONMENTAL CONSTRAINTS LIMITING VEGETABLE PRODUCTIVITY

Environmental stress is the primary cause of the production of most of the vegetables worldwide; reducing average yields for most major vegetables as under optimum climatic condition, the productivity of vegetables is three-four folds. Climatic changes will influence the severity of environmental stress imposed on vegetable crops. Moreover, increasing temperatures, reduced irrigation water availability, flooding, and salinity will be major limiting factors in sustaining and increasing vegetable productivity. Plants may respond similarly to avoid one or more stresses through morphological or biochemical mechanisms (Capiati et al., 2006). Environmental interactions may make the stress response of plants more complex or influence the degree of impact of climate change. Measures to adapt to these climate change-induced stresses are critical for sustainable vegetable production. There is a need to do more research on how vegetable crops are affected by increased abiotic stresses as a direct potential threat from climate change. Some of the important environmental stresses, which affect vegetable production at great extent, have been reviewed below.

8.3.4 HIGH TEMPERATURE

Heat stress due to increase temperature is a major agricultural problem in many areas in the world. Exposure to elevated temperatures can cause morphological, anatomical, physiological, and ultimately, biochemical changes in plant tissues and as a consequence, can affect growth and development of different plant organs (Table 8.2). These events can cause drastic reductions in commercial yield. Constantly high temperature can affect the seed germination, plant growth, flower shedding, pollen viability, gametic fertilization, fruit setting, fruit size, fruit weight, etc. The optimum temperatures for tomato cultivation are between 25 and 30 °C during the photoperiod and 20 °C during the dark period. However only 2–4 °C increase in optimal temperature adversely affected gamete development and inhibited the ability of pollinated flowers into seeded fruits and thus reduced crops yields (Peet et al., 1997; Firon et al., 2006). High temperatures also interfere with floral bud development due to flower abortion. High temperatures can cause significant losses in tomato productivity due to reduced fruit set, and smaller and lower quality fruits (Stevens and Rudich, 1978). High temperatures in tomato causes bud drop, abnormal flower development, poor pollen production, dehiscence, and viability, ovule abortion and poor viability, reduced carbohydrate availability, and other reproductive abnormalities. In addition, significant inhibition of photosynthesis occurs at temperatures above optimum, resulting in considerable loss of potential productivity. Lettuce celery and cauliflower grown under higher temperature matures earlier than the same crops grown under lower temperature (Phelps, 1996) (Table 8.3).

TABLE 8.2 Symptom of Heat Injury on Different Vegetable Has Been Summarized in Following Table (Moretti et al., 2010)

Crop	Symptom
Snap bean	Brown and reddish spots on the pod; spots can coalesce to form a water-soaked area
Cabbage	Outer leaves showing a bleached, papery appearance; damaged leaves are more susceptible to decay
Lettuce	Damaged leaves assume papery aspect; affected areas are more susceptible to decay; tip burn is a disorder normally associated with high temperatures in the field; it can cause soft rot development during postharvest
Muskmelon	Characteristic sunburn symptoms: dry and sunken areas; green color and brown spots are also observed on rind
Potato	Black heart: occur during excessively hot weather in saturated soil; symptoms usually occur in the center of the tuber as dark-gray to black discoloration
Tomato	Sunburn: disruption of lycopene synthesis; appearance of yellow areas in the affected tissues
Bell pepper	Sunburn: Yellowing and in some cases, a slight wilting.

TABLE 8.3 Effect of High Temperature on Quality of Vegetables

Tomato	Threshold temperature 35°C above reduction in quality
Lettuce	Above 30°C temperature lettuce tip burn disorder, bolting (premature seed bud production, Less anthocyanin production).
Capsicum	Red color development inhibits above 27°C temperature
High temperature and low humidity	Poor pollination in lettuce, cucurbits, carrot
High temperature coupled with water stress	Reduced the quality in terms of vitamins antioxidants and minerals.
	Affect the photosynthesis reproduction growth and mineral uptake resulting in poor growth and lower nutritive value of produce.

8.3.5 HIGH TEMPERATURE INCREASE IN DISEASES, INSECT PESTS AND WEEDS

As temperature increases, the insect-pests will become more abundant through a number of interrelated processes, including range extensions and phenological changes, as well as increased rates of population development, growth, migration and overwintering. The climate change is likely to alter the balance between insect pests, their natural enemies and their hosts. The rise in temperature will favor insect development and winter survival. Rising atmospheric carbon dioxide concentrations may lead to a decline in food quality for plant-feeding insects, as a result of reduced foliar nitrogen levels. The epidemiology of plant diseases will be altered. The prediction of disease outbreaks will be more difficult in periods of rapidly changing climate and unstable weather. Environmental instability and increased incidence of extreme weather may reduce the effectiveness of pesticides on targeted pests or result in more injury to nontarget organisms.

- Insect pest will increase with increase temperature and pest activity period will be longer for example DBM in cauliflower and *Heliothis* in tomato.
- There will be negative growth of predator like *Trichoderma*
- Weed and pest will have opportunity to spread where they have currently unable to establish because of temperature constraints, for example, white fly.
- Poor effectiveness of biological pesticides, NPV for tomato (effective under cool moist condition).

8.3.6 LOW TOLERANCE IN VEGETABLES

The cultivated tomato genotype (*Solanum lycopersicum* earlier known as *Lycopersicon esculentum* L.) displays limited growth and development at temperatures under 12°C (Hu et al., 2006). At temperatures between 0 and 12°C, plants are damaged by chilling stress. The severity of damage is proportional to the length of time spent in this temperature range.

8.3.6.1 WATER STRESS

Water availability is expected to be highly sensitive to climate change and severe water stress conditions will affect crop productivity, particularly that of vegetables. In combination with elevated temperatures, decreased precipitation could cause reduction of irrigation water availability and increase in evapotranspiration, leading to severe crop water-stress conditions (IPCC, 2001). Vegetables, being succulent products by definition, generally consist of greater than 90% water (AVRDC, 1990). Thus, water greatly influences the yield and quality of vegetables; drought conditions drastically reduce vegetable productivity. Drought stress causes an increase of solute concentration in the environment (soil), leading to an osmotic flow of water out of plant cells. This leads to an increase of the solute concentration in plant cells, thereby lowering the water potential and disrupting membranes and cell processes such as photosynthesis. Water stress, particularly at the tuber forming stage, can also lead to a higher susceptibility of potatoes to postharvest development of black spot disorder (Hamouz et al., 2011).

Vegetable crop water requirements range from about 6 inches of water per season for radishes to 24 inches for tomatoes and watermelons. Precise irrigation requirements can be predicted based on crop water use and effective precipitation values. Lack of water influences crop growth in many ways. The effect depends on the severity, duration, and time of stress in relation to the stage of growth. Nearly all vegetable crops are sensitive to drought during two periods: during flowering and two to three weeks before harvest.

8.3.6.2 LEAFY VEGETABLES

Cabbage, lettuce, and spinach are generally planted at or near field capacity. Being shallow rooted, these crops benefit from frequent irrigation throughout the season. As leaf expansion relates closely to water availability, these crops, especially cabbage and lettuce, are particularly sensitive to drought stress during the period of head formation through harvest. Overwatering or irregular watering can result in burst heads. Broccoli and cauliflower, although not grown specifically for their leaves, respond to irrigation much as the leafy vegetables do. Broccoli and cauliflower are sensitive to drought stress at all stages of growth, responding to drought with reduced growth and premature heading.

8.3.6.3   ROOT, TUBER, AND BULB VEGETABLE CROPS

In sweet potatoes, potatoes, carrots, and onions, yield depends on the production and translocation of carbohydrates from the leaf to the root or bulb. The most sensitive stage of growth generally occurs as these storage organs enlarge. Carrots require an even and abundant supply of water throughout the season. Stress causes small, woody, and poorly flavored roots. Uneven irrigation can lead to misshapen or split roots in carrots, second growth in potatoes, and early bulbing in onions.

8.3.6.4   FRUIT AND SEED VEGETABLE CROPS

Cucumbers, melons, pumpkins and squashes, lima beans, snap beans, peas, peppers, sweet corn, and tomatoes are most sensitive to drought stress at flowering and as fruits and seeds develop. Fruit set on these crops can be seriously reduced if water becomes limited. An adequate supply of water during the period of fruit enlargement can reduce the incidence of fruit cracking and blossom-end rot in tomatoes. Irrigation is often reduced as fruit and seed crops mature.

Plant growth stage also influences the susceptibility of crops to drought stress. Irrigation is especially useful when establishing newly seeded or transplanted crops. Irrigation after transplanting can significantly increase the plant survival rate, especially when soil moisture is marginal and the evapotranspiration rate is high. Irrigation can also increase the uniformity of emergence and final stand of seeded crops. For seeded crops, reduce the rate of application and the total amount of water applied to avoid crusting. If crusting is present, use low application rates and small amounts of irrigation water to soften the crust while seedlings are emerging.

8.3.6.5   SALINITY

Salinity is also a serious problem that reduces growth and productivity of vegetable crops in many salt affected areas. Furthermore, it is estimated that about 20% of cultivated lands and 33% of irrigated agricultural lands worldwide are afflicted by high salinity (Foolad, 2004; Ghassemi et al., 1995; Szabolcs, 1992). Moreover, the salinized areas are increasing at a rate of 10% annually; low precipitation, high surface evaporation, weathering of native rocks, irrigation with saline water, and poor cultural practices are the major contributors to the increasing soil salinity. In spite of the physiological cause ion toxicity, water deficit, and/or nutritional imbalance, high salinity in the root area sternly inhibits normal plant growth and development, resulting in reduced crop productivity or total crop failure (Ghassemi et al., 1995). Young seedlings and plants at anthesis appear to be more sensitive to salinity stress than the mature stages (Lutts et al., 1995). Onions are sensitive to saline soils, while cucumbers, eggplants, peppers, beet palak and tomatoes are moderately sensitive. Tomatoes grown under high salinity will produce smaller fruit with higher soluble solids (Mizrahi, 1982). Smaller fruit will have higher surface area to volume ratios, hence greater susceptibility to postharvest water loss (i.e., desiccation stress) (Shibairo et al., 1997). While there is no direct information in the literature to confirm that smaller tomato fruit from saline growing conditions would be subject to greater desiccation stress postharvest, firmness declines for tomatoes grown under 3 and 6 dS m^{-1} salinity levels were increased by 50 to 130%, respectively, at two weeks holding at 20 °C compared with control fruit (Mizrahi et al., 1988).

8.3.6.6 FLOODING

Vegetables production starts at commercial level in many areas on onset of monsoon, but production occurs in both dry and wet seasons. However, production is often limited during the rainy season due to excessive moisture brought about by heavy rain. Most vegetables are highly sensitive to flooding and genetic variation with respect to this character is limited, particularly in tomato, early cauliflower and *karif* onion. In general, damage to vegetables by flooding is due to the reduction of oxygen in the root zone, which inhibits aerobic processes. Flooded tomato plants accumulate endogenous ethylene that causes damage to the plants (Drew, 1979). The rapid development of epinastic growth of leaves is a characteristic response of tomatoes to water-logged conditions and the role of ethylene accumulation has been implicated (Kawase, 1981). The severity of flooding symptoms increases with rising temperatures; rapid wilting and death of tomato plants is usually observed following a short period of flooding at high temperatures (Kuo et al., 1982).

8.3.7 EFFECT OF ATMOSPHERIC GASES ON VEGETABLE PRODUCTION IN CLIMATE CHANGE SCENARIO

8.3.7.1 ELEVATED CO2 AND QUALITY OF VEGETABLES

Carbon dioxide (CO2), also known as the most important greenhouse gas, and ozone (O3) concentrations in the atmosphere are changing during the last decade and are affecting many aspects of fruit and vegetable crops production around the globe (Felzer, Cronin, Reilly, Melillo, and Wang, 2007; Lloyd and Farquhar, 2008). Due to global warming atmospheric carbon dioxide (CO_2) concentration has increased and it may reach 550 µmol mol^{-1} by 2050. This increase in CO_2 could benefit the crop by reducing the losses in agricultural production caused by increased drought and temperature. Elevated CO_2 effects on physiology and quality of vegetables have been summarized by Moretti et al., 2010. Elevated CO_2 has improved the vitamin C, Sugar, acids and carotenoids in tomato (Shivashankar, 2013). Positive effects of CO_2 was observed on total antioxidant capacity, phenols and anthocyanin, however elevated CO_2 reduced protein and mineral content of produce since more carbon in fixed in relation to other nutrients. There is a possibility that the additional carbon fixed by plant due to high CO_2 may be inverted in protective antioxidant compounds such as ascorbate and phenolics. Low organic acids and more ascorbic acid and sugar were found in tomato due to elevated Co_2 (Islam et al., 1946). Carbon dioxide accumulation in atmosphere has directly effects on postharvest quality causing tuber malformation occurrence of common seeds and changes in reducing sugar contents in potato. High concentration of ozone can potentially cause reduction in photosynthetic process growth and biomass accumulation. Ozone enriched atmosphere increases vitamin C content. Tomato exposed to ozone (0.005 to 1.0 µmol/mol) had a transient increase in β-carotene, lutein and lycopene contents (Moretti et al., 2010).

8.3.8   MITIGATION STRATEGIES TO CLIMATE CHANGE

The strategies for mitigation nitrous oxide and carbon di oxide in vegetable cultivation would be to reduce nitrous oxide emission is site-specific, efficient nutrient management. The emission could also be reduced by nitrification inhibitors such as nitrapyrin and dicyandiamide (DCD). There are some plant-derived organics such as neem oil, neem cake and karanja seed extract which can also act as nitrification inhibitors. Mitigation of CO_2 emission from agriculture can be achieved by increasing carbon sequestration in soil through manipulation of soil moisture and temperature, setting aside surplus agricultural land, and restoration of soil carbon on degraded lands. Soil management practices such as reduced tillage, manuring, residue incorporation, improving soil biodiversity, micro aggregation, and mulching can play important roles in sequestering carbon in soil. Some technologies such as intermittent drying, site-specific N management, etc. can be easily adopted by the farmers without additional investment, whereas other technologies need economic incentives and policy support (Wassmann and Pathak, 2007).

8.3.9   ADAPTATION STRATEGIES TO CLIMATE CHANGE

To deal with the impact of climate change, the potential adaptation strategies are: developing cultivars tolerant to heat and salinity stress and resistant to flood and drought, modifying crop management practices, improving water management, adopting new farm techniques such as resource conserving technologies (RCTs), crop diversification, improving pest management, better weather forecasting and crop insurance and harnessing the indigenous technical knowledge of farmers. Some of these strategies are discussed in the following sections.

8.3.10.   GENERAL ADAPTATION STRATEGIES FOR CLIMATE CHANGE

8.3.10.1   DEVELOPING CLIMATE-READY CROPS

Development of new crop varieties with higher yield potential and resistance to multiple stresses (drought, flood, salinity) will be the key to maintain yield stability. Improvement in germplasm of important crops for heat-stress tolerance should be one of the targets of breeding program. Similarly, it is essential to develop tolerance to multiple abiotic stresses as they occur in nature. The abiotic stress tolerance mechanisms are quantitative traits in plants. Germplasm with greater oxidative stress tolerance may be exploited as oxidative stress tolerance is one example where plant's defense mechanism targets several abiotic stresses. Similar to the research efforts on conversion of C3 to C4 crop for improvement in radiation-use efficiency.

Improvement in water-use and nitrogen-use efficiencies is being attempted since long. These efforts assume more relevance in the climate change scenarios as water resources for agriculture are likely to dwindle in future. Nitrogen-use efficiency may be reduced under the climate change scenarios because of high temperatures and heavy precipitation events causing volatilization and leaching losses. Apart from this, for exploiting the beneficial effects of elevated CO_2 concentrations, crop demand for nitrogen is likely to increase. Thus, it is important to improve the root efficiency for mining the water and absorption of nutrients. Exploitation of genetic engineering for 'gene pyramiding' has become essential to pool all the desirable traits in one plant to get the 'ideal plant type' which may also be 'adverse climate-tolerant' genotype.

Farmers need to be provided with cultivars with a broad genetic base. Their adaptation process could be strengthened with availability of new varieties having tolerance to drought, heat and salinity and thus, minimize the risks of climatic aberrations. Similarly, development of varieties is required to offset the emerging problems of shortening of growing season and other vagaries of production environment. Farmers could better stabilize their production system with basket of technological options.

8.3.10.2   CROP DIVERSIFICATION

Diversification of crop, including replacement of plant types, cultivars and hybrids intended for higher drought or heat tolerance, are being advocated as having the potential to increase productivity in the face of temperature and moisture stresses. Diversity in the seed genetic structure and composition has been recognized as an effective defense against disease and pest outbreak and climatic hazards. Moreover, demand for high-value food commodities, such as fruits, vegetables, dairy, meat, eggs and fish is increasing because of growing income and urbanization. Diversification from rice-wheat towards high-value commodities will increase income and result in reduced water and fertilizer use. However, there is a need to quantify the impacts of crop diversification on income, employment, soil health, and water use and greenhouse gas emissions. A significant limitation of diversification is that it is costly in terms of the income opportunities that farmers forego, i.e., switching of crop can be expensive, making crop diversification typically less profitable than specialization. Moreover, traditions can often be difficult to overcome and will dictate local practices.

8.3.10.3 CHANGES IN LAND-USE MANAGEMENT PRACTICES

Adjusting the cropping sequence, including changing the timing of sowing, planting, spraying, and harvesting, to take advantage of the changing duration of growing seasons and associated heat and moisture levels is another option. Altering the time at which fields are sowed or planted can also help farmers regulate the length of the growing season to better suit the changed environment. Farmer adaptation can also involve changing the timing of irrigation or use of other inputs such as fertilizers.

8.3.10.4 ADJUSTING CROPPING SEASON

Cropping systems may have to be changed to include growing of suitable cultivars (to counteract compression of crop development), increasing crop intensities (i.e., the number of successive crop produced per unit area per year) or planting different types of crops. Farmers will have to adapt to changing hydrological regimes by changing crops.

8.3.10.5 EFFICIENT USE OF RESOURCES

The resource-conserving technologies (RCTs) encompass practices that enhance resource- or input-use efficiency and provide immediate, identifiable and demonstrable economic benefits such as reduction in production costs; savings in water, fuel and labor requirements; and timely establishment of crops, resulting in improved yields.

8.3.10.6 IMPROVED PEST MANAGEMENT

Changes in temperature and variability in rainfall would affect incidence of pests and disease and virulence of major crops. It is because climate change will potentially affect the pest/weed-host relationship by affecting the pest/weed population, the host population and the pest/weed-host interactions. Some of the potential adaptation strategies could be: (i) developing cultivars resistance to pests and diseases; (ii) adoption of integrated pest management with more emphasis on biological control and changes in cultural practices; (iii) Insect pest forecasting using recent tools such as simulation modeling; and (iv) developing alternative production techniques and crops, as well as locations, that are resistant to infestations and other risks. Management of pests and diseases with use of resistant varieties and breeds; alternative natural pesticides; bacterial and viral pesticides; pheromones for disrupting pest reproduction, etc. could be adopted for sustainability of agricultural production process. Bio- agents have a crucial role in pest management, hence practices to promote natural enemies like release of predators and parasites; improving the habitat for natural enemies; facilitating beetle banks and flowering strips; crop rotation and multiple cropping should be integrated in pest management practices. Reduction in use of pesticides will also help in reducing carbon emissions.

8.3.10.7 BETTER WEATHER FORECASTING AND CROP INSURANCE SCHEMES

Weather forecasting and early warning systems will be very useful in minimizing risks of climatic adversaries. Information and communication technologies (ICT) could greatly help the researchers and administrators in developing contingency plans. Effective crop insurance schemes should be evolved to help the farmers in reducing the risk of crop failure due to these events. Both formal and informal, as well as private and public, insurance programs need to be put in place to help reduce income losses as a result of climate-related impacts. However, information is needed to frame out policies that encourage effective insurance opportunities.

Growing network of mobile telephony could further speed up SMS-based banking services and help the farmers in having better integration with financial institutions. However, compared to microfinance, microinsurance innovations and availability is limited. There is a need to develop sustainable insurance system, while the rural poor are to be educated about availing such opportunities.

8.3.11 SPECIFIC STRATEGIES FOR ADAPTATION TO CLIMATE CHANGE IN VEGETABLES

Adaption of effective and efficient measures are required to mitigate the adverse effects of climate change on vegetable productivity, and particularly on vegetable production, quality and yield. Current, and new, technologies being developed through plant stress physiology research can potentially contribute to mitigate threats from climate change on vegetable production. As most of the farmers in developing countries are small-holders, have fewer options and must rely heavily on resources available in their farms or within their communities. Thus, technologies that are simple, affordable, and accessible must be used to increase the resilience of farms in less developed countries. Many institutes have been working to address the effect of environmental stress on vegetable production. Germplasm of the major vegetable crops, which are tolerant of high temperatures, flooding and drought has been identified and advanced breeding lines are being developed in many institutions. Efforts are also underway to identify nitrogen-use efficient germplasm. In addition, development of production systems geared towards improved water-use efficiency and expected to mitigate the effects of hot and dry conditions in vegetable production systems are top research and development priorities. Some of these strategies are discussed below.

8.3.11.1 HIGH TEMPERATURE

This problem can be minimized by the improvement of cultural practices and breeding approaches. So there are different types of morphological traits, which help for heat tolerance in conventional breeding approaches.

- Long root length which has good ability to uptake water and nutrients from the soil surface.
- Short life span which help to minimize the temperature effect on plant.
- Hairiness which provide partially shade to cell wall, cell membrane and repel sun rays.
- Small leave size that resist evaporation due to shortening of stomata.
- Leaf orientation enhances the photosynthetic activity and produce tolerance against heat stress.
- Leaf glossiness and waxiness which repel sunlight.

With the improvement through conventional breeding two cultivars which both desire characters i.e. heat tolerance and high yielded genes are selected and hybridized for selection of desirable plants from segregating generation. However molecular and biotechnological strategies are also used for the purpose of heat tolerance in tomato by using such techniques i.e. use of genetic engineering for transfer of heat tolerant gene in plant cell which have all other desirable characters. High temperature tolerance has been genetically engineered in plants mainly by overexpressing the heat shock protein genes or indirectly by altering levels of heat shock transcription factor proteins. Apart from heat shock proteins, thermo tolerance has also been altered by elevating levels of osmolytes, increasing levels of cell detoxification enzymes and through altering membrane fluidity. It is suggested that HSPs may be directly implicated in thermo tolerance as agents that minimize damage to cell proteins. Tissue culture technique may be used to make plant from transformed cell. Molecular / genetic marker used to identify the gene in plant cell or in F_2 generation that has ability to tolerance against high temperature. Now a day's biotechnology contributed significantly to better understanding of heat tolerant gene and heat tolerance mechanism in tomato plants.

8.3.11.2   LOW TOLERANCE

Many vegetable viz. broccoli, Brussels sprout, cabbage, kale, knol-khol, onion, mustard, parsley, spinach, turnip which can tolerate very low temperature can be grown in frost/chilling sensitive area successfully. Commercial cultivated tomatoes are planted in the field at later dates to avoid excessively low temperatures and minimize the risk of chilling damage.

Cold resistant cultivars could be planted earlier in the season leading to an earlier harvest. Unlike cultivated tomatoes, wild tomato species such as *S. habrochaites* S. Knapp & D.M. Spooner, *S. chilense* (Dunal) Reiche and *S. peruvianum* L., recover rapidly after exposure to suboptimal temperatures. These genotypes can be grown at low temperature at high elevation where temperature remains below 10°C. Wild species have been used for constructing genetic maps and identifying genes of agronomic importance. Through backcrosses and selection assisted by molecular markers, cold resistant genes from wild species can be bred into cultivated tomato varieties (Goodstal et al., 2005). Following strategies can be used.

- These wild genotypes can be introgressed in cultivated by somatic and sexual hybridization as well as chloroplast exchange. The cultivated tomato (*L. esculentum*) is used as female in crosses due to inability of *esculentum* to fertilize ovules of most of the wild species (unilateral incongruity).
- Evaluate the off springe for chilling tolerance.
- Identify the chromosome regions which may be associated with chilling tolerance in the wild species.

With development of chilling tolerant tomato the tomato growing season can be extended and produced round the year and area of adaptation may be broaden for glass house production, cold tolerant genotype may reduce the energy consumption in horticulture and contribute a reduction in CO_2 emissions and greenhouse effect on global warming, in general cold tolerant genotypes have earliness, water use efficiency, adaptability and high yield when grown under suboptimal temperature. Physiological traits of chilling tolerance in tomato are as follows.

- Thin stem, short dense glandular hairs and narrow leaflets (*L. peruvianum*)
- Densely hairy stem, leaves and fruits (*L. hirsutum*), high photosynthetic rate and seedling can survive at 0oC temperature.
- *L. chilense* can survive on rock as having deep root system and tolerate moisture stress.

8.3.11.3   SALINITY

- Selection of salt tolerant crops and varieties may be an effective strategy to overcome the problem of salinity. Cucumbers, eggplants, peppers, beet palak and tomatoes are moderately sensitive to salinity. One of the most effective ways to overcome salinity problems is the use of tolerant species and varieties (Yilmaz et al., 2004). The response of plants to increasing salt application may differ significantly among plant species as a function of their genetic tolerance.

8.3.11.4 WATER MANAGEMENT

Since vegetables contain very high amount of water and many vegetables are eaten raw, therefore use of quality water remains a major concern. The quality and efficiency of water management determine the yield and quality of vegetable products. Too much or too little water causes abnormal plant growth, predisposes plants to infection by pathogens, and causes nutritional disorders. If water is scarce and supplies are erratic or variable, then timely irrigation and conservation of soil moisture reserves are the most important agronomic interventions to maintain yields during drought stress. There are several methods of applying irrigation water and the choice depends on the crop, water supply, soil characteristics and topography. Surface irrigation methods are used in more than 80% of the world's irrigated lands yet its field level application efficiency is often 40–50% (Von et al., 2004). To generate income and alleviate poverty of the small-holder farmers, promotion of affordable, small-scale drip irrigation technologies are essential. Drip irrigation minimizes water losses due to run-off and deep percolation and water savings of 50–80% are achieved when compared to most traditional surface irrigation methods. Crop production per unit of water consumed by plant evapo-transpiration is typically increased by 10–50%. Thus, more plants can be irrigated per unit of water by drip irrigation, and with less labor. The water-use efficiency by chili pepper was significantly higher in drip irrigation compared to furrow irrigation, with higher efficiencies observed with high delivery rate drip irrigation regimes (AVRDC, 2005). For drought tolerant crop like watermelon, yield differences between furrow and drip irrigated crops were not significantly different; however, the incidence of Fusarium wilt was reduced when a lower drip irrigation rate was used. In general, the use of low-cost drip irrigation is cost effective, labor-saving, and allows more plants to be grown per unit of water, thereby both saving water and increasing farmers' incomes at the same time.

8.4 CULTURAL MANAGEMENT

Simple, affordable and accessible technologies like, mulching and the use of shelters and raised beds help to conserve soil moisture, prevent soil degradation, and protect vegetables from heavy rains, high temperatures, and flooding. The use of organic and inorganic mulches is common in high-value vegetable production systems. These protective coverings help reduce evaporation, moderate soil temperature, reduce soil runoff and erosion, protect fruits from direct contact with soil and minimize weed growth. In addition, the use of organic materials as mulch can help enhance soil fertility, structure and other soil properties. Rice straw is abundant in rice-growing areas and generally recommended for summer tomato production. Polythene and *Sarkanda* (*Saccharum* spp. and *Canna* spp.) can also be used as mulching materials. In the areas where temperatures are high, dark-colored plastic mulch is recommended in combination with rice straw (AVRDC, 1990). Dark plastic mulch prevents sunlight from reaching the soil surface and the rice straw insulates the plastic from direct sunlight thereby preventing the soil temperature rising too high during the day. During the hot rainy season, vegetables such as tomatoes suffer from yield losses caused by heavy rains. Simple, clear plastic rain shelters prevent water logging and rain impact damage on developing fruits, with consequent improvement in tomato yields (Midmore et al., 1992). Fruit cracking and the number of unmarketable fruits are also reduced. Another form of shelter using shade cloth can be used to reduce temperature stress. Planting vegetables in raised beds can ameliorate the effects of flooding during the rainy season (AVRDC, 1979, 1981).

8.4.1 GRAFTING OF VEGETABLES FOR STRESS MANAGEMENT

Grafting of susceptible plant (scion) on tolerant plant (rootstock) helps to grow plant successfully under stress condition especially under salt and drought stress condition. Grafting vegetables originated in East Asia during the twentieth century and it has been used primarily to control soil-borne diseases affecting the production of vegetables such as tomato, eggplant, and cucurbits. However, it can provide tolerance to soil-related environmental stresses such as drought, salinity, low soil temperature and flooding if appropriate tolerant rootstocks are used. Grafting of eggplants was started in the 1950s, followed by grafting of cucumbers and tomatoes in the 1960s and 1970s (Edelstein, 2004). Romero et al. (1997) reported that melons grafted onto hybrid squash rootstocks were more salt tolerant than the nongrafted melons. However, tolerance to salt by rootstocks varies greatly among species, such that rootstocks from *Cucurbita* spp. are more tolerant of salt than rootstocks from *Lagenaria siceraria* (Matsubara, 1989). Grafted plants were also more able to tolerate low soil temperatures. *Solanum lycopersicum* x *S. habrochaites* rootstocks provide tolerance of low soil temperatures (10 °C to 13 °C) for their grafted tomato scions, while eggplants grafted can be grafted on wild brinjal (*S. integrifolium*) as rootstocks to overcome low temperature (18 °C to 21 °C). Most of the vegetables are unable to tolerate excessive soil moisture. Tomatoes in particular are considered to be one of the vegetable crops most sensitive to excess water. Until now, genetic variability for tolerance of excess soil moisture is limited or inadequate to prevent losses. Many accessions of eggplant are highly tolerant of flooding (Midmore et al., 1997), thus, can be grafted to improve the flood tolerance of tomato using eggplant rootstocks which were identified with good grafting compatibility with tomato and high tolerance to excess soil moisture. In addition to protection against flooding, some eggplant genotypes are drought tolerant and eggplant rootstocks can therefore provide protection against limited soil moisture stress.

8.4.2 DEVELOPMENT OF STRESS TOLERANT VARIETIES

8.4.2.1. HEAT AND COLD TOLERANT GENOTYPES

Till today, several heat tolerance genotypes have been developed in vegetables particularly in tomato. AVRDC, Taiwan has made significant contributions to the development of heat-tolerant tomato and Chinese cabbage lines (*Brassica rapa* subsp. *pekinensis* and *chinenesis*) adapted to hot and humid climate. The key to achieving high yields with heat tolerant cultivars is the broadening of their genetic base through crosses between heat tolerant tropical lines and disease resistant temperate or winter varieties (Opena and Lo, 1981). The heat tolerant tomato lines were developed using heat tolerant breeding lines and landraces from the Philippines (e.g.,

VC11-3-1-8, VC 11-2-5, Divisoria-2) and the United States (e.g., Tamu Chico III, PI289309) (Opena et al., 1992). However, lower yields in the heat tolerant lines are still a concern. More heat tolerant varieties are required to meet the needs of a changing climate, and these must be able to match the yields of conventional, nonheat tolerant varieties under non-stress conditions. A wider range of genotypic variation must be explored to identify additional sources of heat tolerance, for example AVRDC's breeding line, CL5915, has demonstrated high levels of heat ranges from 15%–30% while there is complete absence of fruit set in heat-sensitive lines in mean field temperatures of 35 °C. Now, new breeding lines have been developed from CL5915 and other sources that exhibit increased heat tolerance. A CL5915 line is considered best combiners for percentage fruit set and total yield in hybrids developed for heat-tolerance (Metwally et al., 1996). Similarly for cold tolerance several genotypes have shown very good tolerance like, PI-120256, a primitive tomato from Turkey; LA-1777 (*Solanum habrochaites*) from AVRDC, Taiwan and *Lycopersicon hirsutum*. LA3921 and LA3925 both *Solanum habrochaites* from AVRDC, Taiwan has also shown chilling tolerance. Similarly EC-520061 (*Solanum habrochaites*) can set fruits both under high (40±2 °C) and low (10±2 °C) temperature. These lines can be used for development of cold tolerance in various backgrounds.

Presently Division of Vegetable Science, Indian Agricultural Research Institute has developed some varieties of vegetables to mitigate the harmful effect of heat. Tomato varieties Pusa Sadabahar and Pusa Sheetal and one hybrid Pusa Hybrid-1 have been developed. They are tolerant to high and low temperature. Radish variety, Pusa Chetaki has been developed having better root formation under high temperature regime i.e. April-August. Similarly carrot variety, Pusa Vrishti can form root at high temperature and high humidity i.e. March-August. Early cauliflower variety, Pusa Meghna has been developed which can form root at high temperature. These varieties can be used directly for mitigating effect of high temperature as well as for future breeding program. There is need to transfer tomato leaf curl virus (TYLCV), early and late blight resistance in heat tolerant lines through gene pyramiding using wild relatives, that is, *S. habrochiates* and *S. pimpinellifolium* for their wide adaptability.

8.4.2.2   DROUGHT TOLERANCE

Most of the vegetables are sensitive to drought, however brinjal, cowpea, amaranth, tomato can tolerate drought to certain extent. Genetic variability for drought tolerance has been found in wild tomato (*S. lycopersicum)* is limited and inadequate. The best source of resistance is from other species in the genus *Solanum*. Wild accessions of tomato i.e. *S. cheesmanii, S. chilense, S. lycopersicum, S. lycopersicum* var. *cerasiforme, S. pennellii, S. peruvianum* and *S. pimpinellifolium* have stress tolerance. *S. chilense* and *S. pennellii* produce small green fruit and have an indeterminate growth habit. *S chilense* is adapted to desert areas and often found in areas where no other vegetation grows (Rick, 1973, Maldonado et al., 2003). *S. chilense* has finely divided leaves and well-developed root system (Sanchez-Pena, 1999). *S. chilense* has a longer primary root and more extensive secondary root system than cultivated tomato (O'Connell et al. 2007). Drought tests show that *S. chilense* is five times more tolerant of wilting than cultivated tomato. *S. pennellii* has the ability to increase its water use efficiency under drought conditions unlike the cultivated *S. lycopersicum* (O'Connell et al., 2007). It has thick, round waxy leaves, is known to produce acyl-sugars in its trichomes, and its leaves are able to take up dew (Rick, 1973). Transfer and utilization of genes from these drought resistant species will enhance tolerance of tomato cultivars to dry conditions, although wide crosses with *S. pennellii* produce fertile progenies, *S. chilense* is cross-incompatible with *S. lycopersicum* and embryo rescue through tissue culture is required to produce progeny plants.

8.4.2.3   SALT TOLERANCE

Conventional breeding programs have shown very limited success in improvement of salt tolerance due to the genetic and physiologic complexity of this trait (Flowers, 2004). Success in breeding for salt tolerance requires effective screening methods, existence of genetic variability, and ability to transfer the genes to the species of interest. Screening for salt tolerance in the field is not a recommended practice because of the variable levels of salinity in field soils. Screening should be done in soil-less culture with nutrient solutions of known salt concentrations (Cuartero and Fernandez-Munoz, 1999). Few vegetables like, beet palak, tomato can tolerate salt to some extent. Most commercial tomato cultivars are moderately sensitive to increased salinity and only limited variation exists in cultivated species. Genetic variation for salt tolerance during seed germination in tomato has been identified within cultivated and wild species. Yildirim and Guvenc (2006) reported that pepper genotypes Demre, Ilica 250, 11-B-14, Bagci Carliston, Mini Aci Sivri, Yalova Carliston, and Yaglik 28 can be useful as sources of genes to develop pepper cultivars with improved germination under salt stress. In Tunisia, pepper cultivar 'Beldi' significantly out-yielded than other test cultivars at high salt treatments. *S. esculentum* accession (PI174263) showed that the ability of tomato seed to germinate rapidly under salt stress (Foolad & Jones 1991).tomato genotypes, LA1579, LA1606 both *S. pimpinellifolium*) and LA4133 (*S. lycopersicum* var *cerasiforme*) from AVRDC, Taiwan have shown salt tolerance. Wild tomato species *S. cheesmanii, S. peruvianum, S pennelii, S. pimpinellifolium*, and *S. habrochaites are* the potential source of salt tolerance (Cuartero et al., 2006; Flowers, 2004; Foolad, 2004). Attempts to transfer quantitative trait loci (QTLs) and elucidate the genetics of salt tolerance have been conducted using populations involving wild species. Elucidation of mechanism of salt tolerance at different growth periods and the introgression of salinity tolerance genes into vegetables would accelerate development of varieties that are able to withstand high or variable levels of salinity compatible with different production environments.

8.4.3   USE OF BIOTECHNOLOGY TOOLS IN STRESS MANAGEMENT

Use of molecular technologies has revolutionized the process of traditional plant breeding. Combining new knowledge from genomic research with traditional breeding methods enhances our ability to improve crop plants. The use of molecular markers as a selection tool provides the potential for increasing the efficiency of breeding programs by reducing environmental variability, facilitating earlier selection, and reducing subsequent population sizes for field testing. Molecular markers facilitate efficient introgression of superior alleles from wild species into the breeding programs and enable the pyramiding of genes controlling quantitative traits. Thus, enhancing and accelerating the development of stress tolerant and higher yielding cultivars for farmers in developing countries. Several QTLs have been identified to stress tolerance in tomato i.e. for water use efficiency in *S. pennellii and S. pimpinellifolium* as source of salt tolerance. Only a few major QTLs account for the majority of phenotypic variation indicating the potential for marker-assisted selection (MAS) for salt tolerance. Integration of QTL analysis with gene discovery and modeling of genetic networks will facilitate a comprehensive understanding of stress tolerance, permit the development of useful and effective markers for marker-assisted selection, and identify candidate genes for genetic engineering.

8.4.4   CLIMATE CHANGE BRINGING NEW AVENUES

Due to increase in temperature the area under tropical and subtropical vegetables will increase and it can be grown in those areas where it was not grown earlier. Thus will be opportunity to introduce new vegetables in these areas. However, the temperature vegetables which required chilling temperature like beet, cabbage temperature carrot, etc. may be shifted to higher altitude.

8.4.4.1   THRUST AREA

To mitigate the effect of climate change in vegetable crop following strategies are to be adopted:
  1. Screening of germplasm of major vegetable crops for stress tolerance (high/low temperature, drought, flood, etc.).
  2. Development of cold and hot set varieties of tomato with better consumer's acceptance.
  3. Development of low chilling requirement varieties in temperate vegetables like cabbage, carrot, beet root, turnip, knol-khol, brussel sprout, etc.
  4. Development of leafy vegetables like palak, spinach, coriander, etc., with wider adaptability and greater tolerance to high temperature.
  5. There is also need to increase 'N' use efficiency and water use efficiency of vegetable crops through better production system.

A sustainable program has been made during last decade in breeding of tomato, radish, cauliflower, carrot that is tolerant to high temperature and produce satisfactory yield. As greater resistance to insect and disease attack is bred into the heat tolerant varieties, higher and more stable yields no doubt will be obtained in the years ahead. Thus there is scope for breeding varieties of many vegetable crops better adapted to the high temperature and more resistant to attack by insect and diseases.

KEYWORDS

- **Bioative molecules**
- **Climate change effect**
- **Greenhouse Gases**
- **High Temperature**
- **Specific Strategies**
- **Vegetable Productivity**

REFERENCES

Abdalla, A. A., & Verderk, K. (1968). Growth, flowering and fruit set of tomato at high temperature. The Neth J Agric. Sci., 16, 71–76.

AVRDC. (1979). Annual Report. Asian Vegetable Research and Development Center. Shanhua, Taiwan. 173 p.

AVRDC. (1981). Annual Report. Asian Vegetable Research and Development Center. Shanhua, Taiwan. 84, p.

AVRDC. (1990). Vegetable Production Training Manual. Asian Vegetable Research and Training Center. Shanhua, Tainan, 447.

AVRDC. (2005) Annual Report. AVRDC the World Vegetable Center. Shanhua, Taiwan.

Capiati, D. A., País, S. M., & Téllez-Iñón, M. T. (2006). Wounding increases salt tolerance in tomato plants evidence on the participation of calmodulin like activities in cross-tolerance signaling. J Exp Bot 57, 2391–2400.

Cruz, R. V., Harasawa, H., Lal, M., Wu, S., Anokhin, Y., Punsalmaa, B., Honda, Y., Jafari, M., Li, C., & Huu, N. (2007). Asia Climate Change 2007 Impacts, Adaptation and Vulnerability. 469–506. In: Parry, M. L., Canziani, O. F., Palutikof, J. P., van der Linden, P. J. & Hanson, C. E. (eds.), Contribution of Working Group II to the Fourth Assessment Report of the Intergovernmental Panel on Climate Change, Cambridge Univ. Press, Cambridge.

Cuartero, J., & Fernandez-Munoz, R. (1999). Tomato and salinity. Sciencia Horticulturae 78:83–125.

Cuartero, J., Bolarin, M. C., Asins, M. J. & Moreno, V. (2006). Increasing salt tolerance in tomato. J Exp Bot 57:1045–1058.

Drew, M. C. (1979). Plant responses to anaerobic conditions in soil and solution culture. Curr Adv Plant Sci. 36:1–14.

Edelstein, M. (2004). Grafting vegetable crop plants Pros and Cons. Acta Horticulturae 65.

FAO (2001). Climate variability and change A challenge for sustainable agricultural production. Committee on Agriculture, Sixteenth Session Report, 26–30 March, 2001. Rome, Italy.

FAO (2009). Global Agriculture towards 2050. Issues brief. High level expert forum. Rome, 12–13 October 2009. www.fao.org/wsfs/forum2050/wsfs-background-documents/hlef-is-sues-briefs/en/Accessed March 2010.

FAOSTAT (2010). Crop production. faostat. fao. org/site/567/default. aspx#ancor Accessed August 2010.

Firon, N., Shaked, R., Peet, M. M., Phari, D. M., Zamsk, E., Rosenfeld, K., Althan, L., & Pressman, N. E. (2006). Pollen Grains of Heat Tolerant Tomato Cultivars Retain Higher Carbohydrate Concentration Under Heat Stress Conditions. Scientia Horticulturae: 109, 212–217.

Flowers, T. J. (2004). Improving crop salt tolerance. J Exp Bot 55, 307–319.

Foolad, M. R. (2004). Recent advances in genetics of salt tolerance in tomato. Plant Cell Tissue Organ Culture 76:101–119.

Foolad, M. R., & Jones, R. A. (1991). Genetic analysis of salt tolerance during germination in *Lycopersicon*. Theor. Appl. Genet. 81, 321–326.

Ghassemi, F., Jakeman, A. J. & Nix, H. A. (1995) Salinisation of land and water resources: human causes, extent management and case studies. Canberra, Australia, the Australian National University, Wallingford, Oxon, UK: CAB International, Wallingford, England. 526.

Goodstal, F., Kohler, G., Randall, L., Bloom, A., & D St., Clair. (2005). A major QTL introgressed from wild *Lycopersicon hirsutum* confers chilling tolerance to cultivated tomato (*Lycopersicon esculentum*). Theoretical and Applied Genetics 111, 898–905.

Hu, W. H., Zhou, Y. D., Du, Y. S., Xia, X. J. & Yu, J. Q. (2006). Differential response of photosynthesis in greenhouse and field ecotypes of tomato to long-term chilling under low light. Journal of Plant Physiology163, 1238–1246.

IPCC (2001). Climate change 2001 Impacts, adaptation and vulnerability. Intergovernmental Panel on Climate Change. New York, USA.

Kawase, M. (1981). Anatomical and morphological adaptation of plants to water logging. Hort Science 16, 30–34.

Kuo, D. G., Tsay, J. S., Chen, B. W. & Lin, P. Y. (1982). Screening for flooding tolerance in the genus *Lycopersicon*. Hort Science 17(1), 76–78.

Lutts, S., Kinet, J. M., & Bouharmont, J. (1995). Changes in plant response to NaCl during development of rice (Oryzasativa L.) varieties differing in salinity resistance. Journal of Experimental Botany, 46, 1843–1852.

Maldonado, C., Squeo, F. A., & Ibacache, E. (2003). Phenotypic response of *Lycopersicon chilense* to water deficit. Revista ChilenaHistoria Natural 76, 129–137.

Matsubara, S. (1989). Studies on salt tolerance of vegetables-3. Salt tolerance of rootstocks. Agric Bull, Okayama Univ 73, 17–25.

Metwally, E., El-Zawily, A., Hassan, N., & Zanata, O. (1996). Inheritance of fruit set and yields of tomato under high temperature conditions in Egypt. First Egypt-Hung. Hort Conference, Vol I. 112–122.

Midmore, D. J., Roan, Y. C. & Wu, D. L. (1997). Management practices to improve lowland subtropical summer tomato production: yields, economic returns and risk. Exptl Agric 33:125–137.

Midmore, D. J., Roan, Y. C., & Wu, M. H. (1992). Management of moisture and heat stress for tomato and hot pepper production in the tropics. In Kuo CG (Ed) Adaptation of food crops to temperature and water stress. AVRDC, Shanhua, Taiwan 453–460.

Moretti, C. L., Mattos, L. M., Calbo, A. G., & Sargent, S. A. (2010). Climate changes and potential impacts on postharvest quality of fruit and vegetable crops: A review. Food Research International, 43, 1825–1832.

NHB (2011). National Horticulture Database, 2010, Ministry of Agriculture, Govt. of India, Gurgaon, 262p.

O'Connell, M. A., Medina, A. L., & Sanchez Pena Pand Trevino, M. B. (2007). Molecular genetics of drought resistance response in tomato and related species. In Razdan, M. K., & Mattoo, A. K. (eds). Genetic Improvement of Solanaceous Crops, Vol. 2, Tomato, Science Publishers, Enfield USA 261–283.

Opena, R. T., & LO, S. H. (1981). Breeding for heat tolerance in heading Chinese cabbage. In: Talekar, N. S., Griggs, T. D. (eds). Proceedings of the 1st International Symposium on Chinese cabbage. AVRDC, Shanhua, Taiwan.

Opena, R. T., Chen, J. T., Kuo, C. G., & Chen, H. M. (1992). Genetic and physiological aspects of tropical adaptation in tomato. In: Kuo, C. G. (ed). Adaptation of food crops to temperature and water stress. AVRDC, Shanhua, Taiwan 321–334.

Peet, M. M., Willits, D. H., & Gardner, R. (1997). Response of ovule development and post pollen production processes in male-sterile tomatoes to chronic, sub-acute high temperature stress. J. Experimental Botany. 48(306) 101–111.

Rick, C. M. (1973). Potential genetic resources in tomato species: clues from observation in native habitats. In: Srb AM (ed.) Genes, enzymes and populations, Plenum Press, New York, 255–269.

Romero, L., Belakbir, A., Ragala, L., & Ruiz, M. J. (1997). Response of plant yield and leaf pigments to saline conditions effectiveness of different rootstocks in melon plant (*Cucumis melo* L). Soil Sci Plant Nutr 3, 855–862.

Sánchez Peña, P. (1999). Leaf water potentials in tomato (*L. esculentum* Mill.) *L chilense* Dun. and their interspecific F1. MSc Thesis, New Mexico State University, Las Cruces, NM, USA.

Shivashankara, K. S., Rao, N. K. S., & Geetha, G. A. (2013). Impact of climate change on fruit and vegetable quality. H. P Singh (eds.). In climate resilient Horticulture: adaptation and mitigation strategies. Springer India. 237–244.

Stevens, M. A., & Rudich, J. (1978). Genetic potential for overcoming physiological limitations on adaptability, yield, and quality in tomato. Hort Science 13, 673–678.

Szabolcs, I. (1992). Salinisation of soils and water and its relation to desertification. In Genetic Improvement of Solanaceous Crop (Maharaj K. Razdan, and Autar K. Mattoo, eds.), 521–590. Beltsville, MD, USA.

Von Westarp, S., Chieng, S., & Scheier (2004). A comparison between low-cost drip irrigation, conventional drip irrigation, and hand watering in Nepal. Agricultural Water Management. 64, 143–160.

Yildirim, E., & Guvenc, I. (2006). Salt tolerance of pepper cultivars during germination and seedling growth. Turk J Agric Forestry 30, 347–353.

Yilmaz, K., Akinci, I. E., & Akinci, S. (2004). Effect of salt stress on growth and Na, K contents of pepper (*Capsicum annuum* L.) in germination and seedling stages. Pakistan Journal of Biological Science, 7(4), 606–610.

CHAPTER 9

FUNCTIONAL PHYSIOLOGY IN DROUGHT TOLERANCE OF VEGETABLE CROPS—AN APPROACH TO MITIGATE CLIMATE CHANGE IMPACT

A. CHATTERJEE[1] and S. S. SOLANKEY[2]

[1]Molecular Biology Section, Centre of Advanced Study in Botany, Banaras Hindu University, Varanasi - 221 005 (U.P.), India.

[2]Department of Horticulture (Vegetables & Floriculture), Bihar Agricultural College, Bihar Agricultural University, Sabour (Bhagalpur) - 813 210, Bihar, India; E-mail: shashank.hort@gmail.com

CONTENTS

ABSTRACT

Moisture stress is one of the greatest environmental factors reducing yield in arid and semiarid crops. Drought is often accompanied by relatively high temperatures, which promote evapotranspiration, (ET) and affects photosynthetic kinetics, thus intensifying the consequences of drought and further reducing crop yield. About two third of the geographical area of India receives low rainfall (less than 1000 mm), which is also characterized by uneven and erratic distribution. Out of net sown area of 140 million hectares about 68% is reported to be vulnerable to drought stress and about 50% of such vulnerable area is classified as 'severe,' where frequency of drought is almost regular. Vegetables being succulent in nature, are sensitive to drought stress, particularly during flowering to seed development stage. Moreover, the legume vegetables, for instance cowpea, vegetable pea, Indian beans, etc., grown in arid and semiarid regions are generally affected by drought at the reproductive stage. Drought stress triggers drought tolerance mechanisms involving certain morphological, physiological and biochemical traits in vegetables, which are considered to be adaptive in nature. These traits are investigated thoroughly to serve as screening tools in developing drought resistance varieties with greater potential to maximize use of stored soil water and increased economic yield per unit water use. Starting from germination potential, shoot and root architecture and root anatomy, shoot and root fresh and dry weight ratio, total leaf area, stability to flowering processes are needed to be analyzed. Important physiological traits such as osmotic adjustment, cell membrane stability, (CMS), photosynthetic and transpiration rate, stomatal conductance, water use efficiency, (WUE) chlorophyll fluorescence parameters, photosynthetic pigment content, plant canopy temperature simultaneously with biochemical attributes including level of antioxidant enzymes, non enzymatic antioxidant and pyridine nucleotides involve in drought tolerance mechanism.

9.1 INTRODUCTION

Drought may be defined as a climatic hazard, which implies the absence or very low level of rainfall for a week, month or years, long enough to cause moisture depletion in soil with a decline of water potential in plant tissues.

Drought is often accompanied by relatively high temperatures, which promotes ET and affects photosynthetic kinetics, thus intensifying the effects of drought and further reducing crop yields (Mir et al., 2012). Drought is an inevitable feature of climate that occurs in virtually all climate regimes. Although global figures for the trends in economic losses associated with drought do not exist, an UNDP Bureau of Crisis Prevention and Recovery (2004) indicates that annual losses associated with natural disasters increased from US$75.5 billion in the 1960s to nearly US$660 billion in the 1990s. Agricultural regions affected by drought can experience yield loss up to 50% or more. Over 35% of the world's land surface is considered arid or semiarid, experiencing precipitation that is inadequate for most agricultural uses. Over exploitation of natural resources has endangered water resources, biodiversity and soil quality, more than 1.2 billion people in over 110 countries are already affected by the social and environmental effects of the land degradation in dry lands, which leads to declining biological and economic productivity worldwide (Pervez et al., 2009). The threat of global warming may further increase the frequency and severity of extreme climate events in the future. India is also one of the countries, which have agriculture system challenged with climate change. Vegetables are basically succulent in nature and generally consist of more than 90% of water. Thus, water stress, mostly at critical period of growth may drastically reduce productivity and quality of vegetables (Table 9.1). Moisture stress during vegetative or early reproductive growth phase usually reduces yield by reducing the number of fruits/ seeds in vegetables, while during flowering and fruit setting stage drought stress reduces fruit quality, number of fruits, size of fruits and finally yield loss. Drought also reduces seed number, viability, and vigor. Physiological processes mainly responsible for plant growth and development are affected by water deficit condition, and plant exhibit various defense mechanisms against drought stress at the molecular, cellular and whole plant levels.

TABLE 9.1 Critical Stages of Drought Stress and its Impact on Vegetable Crops

Vegetable crops	Critical period for watering	Impact of water stress
Tomato	Early flowering, fruit set, and enlargement	Flower shedding, lack of fertilization, reduced fruit size, fruit splitting, puffiness and development of calcium deficient disorder i.e. blossom end rot (BER), poor seed viability

TABLE 9.1 *(Continued)*

Vegetable crops	Critical period for watering	Impact of water stress
Brinjal	Flowering and fruit development	Reduces yield with poor color development in fruits, poor seed viability
Chilli and Capsicum	Flowering and fruit set	Shedding of flowers and young fruits, reduction in dry matter production and nutrient uptake, poor seed viability
Potato	Tuberization and tuber enlargement	Poor tuber growth and yield, splitting, internal brown spot
Okra	Flowering and pod development	Considerable yield loss, development of fibers, high infestation of mites, poor seed viability
Cauliflower, cabbage and broccoli	Head/ curd formation and enlargement	Tip burning and splitting of head in cabbage; browning and buttoning in cauliflower
Carrot, radish and turnip	Root enlargement	Distorted, rough and poor growth of roots, strong and pungent odor in carrot, accumulation of harmful nitrates in roots
Onion	Bulb formation and enlargement	Splitting and doubling of bulb, poor storage life
Cucumber	Flowering as well as throughout fruit development	Deformed and nonviable pollen grains, bitterness and deformity in fruits, poor seed viability
Melons	Flowering and evenly throughout fruit development	Poor fruit quality in muskmelon due to decrease in TSS, reducing sugar and ascorbic acid, increase nitrate content in watermelon fruit, poor seed viability
Summer squash	Bud development and flowering	Deformed and nonviable pollen grains, misshapen fruits

TABLE 9.1 *(Continued)*

Vegetable crops	Critical period for watering	Impact of water stress
Leafy vegetables	Throughout growth and development of plant	Toughness of leaves, poor foliage growth, accumulation of nitrates
Asparagus	Spear production and fern (foliage) development	Reduce spear quality through reduced spear size and increased fiber content, leading to tougher, lower grade spears.
Lettuce	Consistently throughout development	Toughness of leaves, poor plant growth, tip burning
Vegetable pea	Flowering and pod filling	Reduction in root nodulation and plant growth, poor pod filling, poor seed viability
Lima bean	Pollination and pod development	Leaf color takes on a slight grayish cast, blossom drop, poor seed viability
Snap bean	Flowering and pod enlargement	Blossoms drop with inadequate moisture levels and pods fail to fill, poor seed viability
Sweet corn	Silking, tasseling and ear development	Crop may tassel and shed pollen before silks on ears are ready for pollination, lack of pollination may result in missing rows of kernels, reduced yields, or even eliminate ear production, poor seed viability
Sweet potato	Root enlargement	Reduced root enlargement with poor yield, growth crack

Source: Bahadur et al. (2011); Kumar et al. (2012).

The identification of suitable plant characters for screening large numbers of genotypes, in a short time at critical stages of crop growth, with the aim of selecting drought tolerant cultivars, remains a major challenge to the plant breeders. Drought tolerance is the interactive result of diverse morphological, physiological and biochemical traits and thus, these components could be used as strong selection criteria to screen out appropriate plant ideotypes. Implications of developing an effective screening procedure for drought tolerance have been realized using different procedures (Table 9.2). Traditionally, plant physiologists have addressed the problem of environmental stress by selecting for suitability of performance over a series of environmental conditions using extensive testing (Blum, 1988).

TABLE 9.2   Screening Procedure for Drought Tolerance

S. No.	Instruments/techniques used	Purpose of screening	References
1	Infrared thermometry	Efficient water uptake	Blum et al., 1982
2	Banding herbicide metribuzin at a certain depth of soil, and use of iodine-131 and hydroponic culture under stress of 15 bar	Root growth	Robertson et al., 1985; Ugherughe, 1986
3	Adaptation of psychometric procedure	Evaluation of osmotic adjustment	Morgan, 1980; 1983
4	Diffusion porometry technique	Leaf water conductance	Gay, 1986
5	Mini-rhizotron technique	Root penetration, distribution and density in the field	Bohm, 1974
6	Infrared aerial photography	Dehydration postponement	Blum et al., 1978
7	Carbon isotopediscrimination	Increased water-use efficiency	Farquhar and Richards, 1984
8	Drought index measurement	Total yield and number of fruits	Clarke et al., 1984; Ndunguru et al., 1995
9	Visual scoring or measurement	Maturity, leaf molding, leaf length, angle, orientation, root morphology and other morphological characters	Mitra, 2001

Source: Kumar et al. (2012).

9.2 RESPONSES OF PHYSIOLOGICAL TRAITS UNDER DROUGHT STRESS

The screening of more number of genotypes for drought tolerance would be accelerated with a greater understanding of physiological traits related to water stress. Many physiological characters responsible for continued growth under water stress have been identified. For example, osmotic adjustment is considered to be an adaptation to drought stress by which an increase in the solute content of cells can lead to maintenance of turgor and turgor-related processes at low water potential (Kumar and Elston, 1992). Variation in water use WUE, leaf area, specific leaf area (SLA), leaf area ratio (LAR) and leaf gas exchange (i.e., carbon assimilation (A_N), transpiration, stomatal conductance (gs) and internal CO_2 concentration) in response to water deficit are important parameters to serve as an index for drought tolerance (Anyia and Herzog, 2004). Drought avoidance by maintaining high leaf water content is negatively associated with leaf area as well as SLA. High assimilation rate under water deficit is associated with high relative leaf water content (RWC). Decline in assimilation rate are due mainly to stomata closure, however, evidences of nonstomatal regulation were also found. Instantaneous WUE (iWUE, a molar ratio of assimilation to transpiration) and leaf internal CO_2 (Ci) are negatively correlated, while Ci is moderately related with SLA. Drought tolerant genotypes had the higher leaf water retention, CMS, RWC, and the lower Relative Water Loss in comparison to drought sensitive genotypes. The interacting effect of all above parameters depends on growth strategy of the species considered, and their ability to adjust during drought stress. Some of the physiological parameters, which are noticeably affected by water deficit conditions, are discussed below:

Physiological traits relevant for the responses to water deficits and/or modified by water deficits span a wide range of vital processes (Table 9.3).

TABLE 9.3 Response of Physiological Traits to Drought Conditions

Plant traits	Effects relevant for yield	Modulation under stress	References
Stomatal conductance/leaf Temperature	More/less rapid water consumption. Leaf temperature reflects the evaporation and hence is a function of stomatal conductance	Stomatal tolerance increases under stress	Jones (1999), Lawlor and Cornic (2002)

TABLE 9.3 *(Continued)*

Plant traits	Effects relevant for yield	Modulation under stress	References
Photosynthetic capacity	Modulation of concentration of Calvin cycle enzymes and elements of the light reactions	Reduction under stress	Lawlor and Cornic (2002)
Timing of phenological phases	Early/late flowering. Maturity and growth duration, synchrony of silk emergence and anthesis, reduced grain number	Wheat and barley advanced flowering, rice delayed, maize asynchrony	Slafer et al. (2005), Richards (2006)
Anthesis-Silking interval (ASI) in maize	ASI is negatively associated with yield in drought conditions	Drought stress at flowering causes a delay in silk emergence relative to anthesis	Bolanos and Edmeades (1993), Edmeades et al. (2000)
Starch availability during ovary/embryo development	A reduced starch availability leads to abortion, reduced grain number	Inhibition of photosynthetic activity reduces starch availability	Boyer and Westgate, (2004)
Partitioning and stem reserve utilization	Lower/higher remobilization of reserves from stems for grain-filling, effecting kernel weight	Compensation of reduced current leaf photosynthesis by increased remobilization	Blum (1988), Slafer et al. (2005)
Stay green	Delayed senescence	-	Rajcan and Tollenaar (1999)
Single plant leaf area	Plant size and related productivity	Reduced under stress (wilting, senescence, abscission)	Walter and Shurr (2005)
Rooting depth	Higher/lower tapping of soil water resources	Reduced total mass but increased root/shoot ratio, growth into wet soil layers, regrowth on stress release	Hoad et al. (2001), Sharp et al. (2004)

TABLE 9.3 *(Continued)*

Plant traits	Effects relevant for yield	Modulation under stress	References
Cuticular tolerance and surface roughness	Higher or lower water loss, modification of boundary layer and reflectance	-	Kerstiens (1996)
Photosynthetic pathway	C3/C4/CAM, higher WUE and greater heat tolerance of C4 and CAM	-	Cushman (2001)
Osmotic adjustment	Accumulation of solutes: ions, sugars, poly sugars, amino acids, glycine betaine	Slow response to water potential	Serraj and Sinclair (2002)
Membrane composition	Increased membrane stability and changes in aquaporin function	Regulation in response to water potential changes	Tyerman et al. (2002)
Antioxidative defense	Protection against active oxygen species	Acclimation of defense systems	Reddy et al. (2004)
Accumulation of stress-related proteins	Involved in the protection of cellular structure and protein activities	Accumulated under stress	Ramanjulu and Bartels (2002), Cattivelli et al. (2002)

Source: Cattivelli et al. (2008); Kumar et al. (2012).

9.3 PHYSIOLOGICAL PARAMETERS HELP IN SCREENING DROUGHT RESISTANT VARIETEIES

Screening of genotypes for drought tolerance would be accelerated with a greater understanding of physiological traits involved in drought tolerance. Many physiological characters responsible for continued growth under water stress have been identified. The interacting effect of all physiological parameters depends on growth strategy of the species considered, and their ability to adjust during drought stress. Some of the physiological parameters that are noticeably affected by water deficit conditions and can be used as screening parameters to develop drought resistant varieties are discussed in the following subsections.

9.3.1   STOMATAL CONDUCTANCE AND PHOTOSYNTHESIS

One of the basic mechanisms for reducing the impact of drought is early stomatal closure at the beginning of the period of water deficit. Stomatal closure not only reduces water loss, but also reduces the gas exchange between the plant and the ambient air. The reduced CO_2 intake then results in reduced photosynthesis (Chaves et al., 2002). As plant water potential falls due to water deficit, the sensitivity of stomatal conductance and photosynthesis rate reduced. Water deficit causes reduction in photosynthesis mainly due to decreased stomatal conductance. Stomatal closure has been reported in tomato at leaf water potential (Ψ_{leaf}) between −0.7 to 0.9 MPa (Duniway, 1971), in pepper −0.58 to −0.88 MPa (Srinivasa Rao and Bhatt, 1988), however eggplant can withstand a greater drought than the most other vegetables. Srinivasa Rao and Bhatt (1990) observed that drop in photosynthesis of eggplant with decreasing Ψ_{leaf} was less than tomato and capsicum. Bahadur et al. (2009) observed significant reduction in photosynthesis rate and stomatal conductance in spring-summer okra when water stress was imposed for 10 or 12 days. Stomatal conductance is the major limitation to photosynthesis under drought conditions in cowpea; however, a pronounced nonstomatal limitation can occur under severe drought stressed conditions that may also lead to impairment of photosynthetic activity (Singh and Reddy, 2011). If perpetual decline in photosynthesis is more than the transpiration, then nonstomatal factors contribute more to the reduction of photosynthesis than stomatal effects. This is because stomatal resistance accounts for a smaller portion of total resistance in CO_2 pathway. During this nonstomatal control of photosynthesis intercellular resistance for CO_2 from the intercellular space of the chloroplasts plays an important role. Thus, a decrease of the photosynthesis rate under water deficit condition can be attributed to both stomatal and nonstomatal limitations. Non-stomatal photosynthesis limitation has been attributed to the reduced carboxylation efficiency, ribulose-1, 5-bisphospate (RuBP) regeneration, amount of functional Rubisco, or to the inhibited functional activity of PSII. Flexas et al. (2002) have shown that drought induced changes in many photosynthetic parameter are more related to variations

in maximum daily stomatal conductance than to variations in the most commonly used water status parameters, like leaf water potential or relative water content. The drought-tolerant species control stomatal function to allow some carbon fixation at stress, thus improving WUE, or open stomata rapidly when water deficit is relieved. In fact stomatal conductance can be used as an integrative parameter to reflect the severity of water stress. It is well established that in conditions of moderate water deficit, the photosynthetic apparatus is not damaged but continue to function, however, under severe water deficit, the photosynthetic capacity is reduced which could be reflected in enhancement of the internal CO_2.

Mesophyll conductance (g_m) and biochemical limitation (b_l) (often termed as nonstomatal limitations) to photosynthesis mainly under severe water stress has also gained importance in the recent years and their relative importance to photosynthesis limitation has been subjected to long-standing debate. (Keenan et al., 2010) In drought stress, solute potential (g_s) has been shown to relate well and exhibit a specific pattern over almost all the important photosynthetic parameters similarly (Rouhi et al., 2007). Earlier, Srinivasa Rao et al. (1999) also reported that cultivar Arka Meghali has better ability to cope with water stress at various crop stages exhibiting better osmotic adjustment, photosynthesis, RWC and other physiological traits under mild and severe drought stress.

9.3.2 OSMOTIC ADJUSTMENT AND MAINTENANCE OF CELL TURGOR

Osmotic adjustment (OA) has been considered as an important physiological adaptation character associated with drought tolerance and it has drawn much attention during the past years. Osmotic adjustment (OA) is defined as the active accumulation of organic solutes in plant tissues in response to an increasing water deficit. It is considered as useful process for maintaining cell turgor when tissue water potential declines. OA has been shown to maintain stomatal conductance and photosynthesis at lower water potentials, delayed leaf senescence and death, reduced flower abortion, improved root growth and increased water extraction from the soil as water deficit develops (Turner et al., 2001). OA involves the net accumulation of solutes in a cell in response to fall in water potential of the cell's environment. As a consequence, the cell's osmotic potential is diminished which in turn attracts water into the cell by tending to maintain turgor pressure. Earlier research indicate that compatible solutes like sugars, glycerol, amino acids such as proline or glycinebetaine, polyols, sugar alcohols (like mannitol and other low molecular weight metabolites) would also contribute to this process. In addition, Hessini et al. (2009) argued that these compounds benefit stressed cells in two ways: (1) by acting as cytoplasmic osmolytes, thereby facilitating water uptake and retention, and (2) by protecting and stabilizing macromolecules and structures (i.e., proteins, membranes, chloroplasts, and liposomes) from damage induced by stress conditions. Osmotic adjustment allows the cell to decrease osmotic potential and, as a consequence, increases the gradient for water influx and maintenance of turgor. Osmotic adjustment has been assessed as a capacity factor (rate of change in solute potential (Ψ_s) with RWC), as described by Kumar et al. (1984). Physiological indices such as leaf water potential (Ψ_{leaf}), solute potential (Ψ_s), relative water content, turgor potential (Ψ_p), osmotic adjustment, leaf diffusive conductance (K_l), difference between canopy and air temperature (T_c–Ta) and water loss from excised leaves can be used as a screening tool. A study conducted by Kumar and Singh (1998) on *Brassica* genotypes revealed that higher osmotic adjustment extracted relatively more water from the deep soil layer (90–180 cm) than genotypes with lower osmotic adjustment (ranging from 50 mm to 69 mm). High-osmotic adjustment genotypes maintained full turgor down to a Ψ_{leaf} of −2.4 MPa, but turgor potential (Ψ_p) fell more rapidly with decreasing Ψ_{leaf} in genotypes showing low osmotic adjustment. The decrease in Ψ_{leaf} with RWC was smaller in low than high-osmotic adjustment genotypes of *Brassica* species. Osmotic adjustment was linearly, but negatively, related to water loss from leaves and positively related to K_l and T_c–T_a. Plants with higher osmotic adjustment transpired more water (greater K_l) and therefore, had cooler canopies (lower canopy temperature and greater T_c–T_a difference) than the plants with lower osmotic adjustment (Kumar and Singh, 1998). Ψ_s in low-osmotic adjustment plants fell linearly and more rapidly with decrease in Ψ_{leaf}, whereas it was not related to Ψ_s in high osmotic adjustment genotypes (Kumar et al., 1984). The relationship between Ψ_s and RWC revealed that high-osmotic adjustment genotypes maintained higher RWC as water deficits increased, with a greater decrease in Ψ_s. Even at low water potential the leaves maintained greater turgor and this may have contributed to the maintenance of higher K_l and photosynthetic activity. High-osmotic adjustment genotypes maintained higher K_l and transpirational cooling (higher T_c-T_a) but showed lower water loss than low-osmotic adjustment genotypes.

Osmotic adjustment could play a significant role in maintaining turgor potential and turgor-related processes, such as opening of the stomata, photosynthesis, shoot growth and extension of roots in deeper soil layers. Continued root growth leads to greater exploration of soil volume and an enhanced water supply to the plant. Genotypic variability for osmotic adjustment exists in vegetable crops. Srinivasa Rao and Bhatt (1992) noticed better OA in tomato cvs. Arka Saurabh, Pusa Early Dwarf and Sioux, thereby relatively higher yield in these cultivars under moisture deficit condition. Furthermore, Srinivasa Rao et al. (1999) reported that OA in four cultivars of tomato did not show any significant variation during first week of drought stress, but after three weeks of stress the maximum OA of 0.17 MPa during flowering stage and 0.47 MPa during fruiting stage was observed in Arka Meghali. During the vegetative stage, better recovery of osmotic potential was observed in RFS-1 followed by Arka Meghali and Pusa Ruby, however, during the fruiting stage, recovery was better in cvs. Pusa Ruby, Arka Meghali and RFS-1.

9.3.3 CHLOROPHYLL FLUORESCENCE

Drought stress is known to inhibit photosynthetic activity in tissues due to an imbalance between light capture and its utilization (Foyer and Noctor, 2000). The decrease in the maximum quantum yield of PSII photochemistry (F_V/F_M) implies a decrease in the capture and conversion rate of excitation energy by PSII reaction centers and so, a reduction in PSII photochemical efficiency indicating the disorganization of PS II reaction centers under water stress conditions. In general, the harvested energy in excess of that consumed by the Calvin Cycle must be dissipated to avoid oxidative stress and may lead to decreased PSII performance (Wilhelm and Selmar, 2011). (F_V/F_M) was not affected by drought in Calluna, but a small (1.5%) yet significant decrease was seen in *Deschampsia* across season. Photosystem II (PSII) is highly sensitive to light and down regulation of photosynthesis under drought stress causes an energy imbalance in the PSII reaction center leading to photoinhibition (Pastenes et al., 2005). Mechanisms have evolved in the plant to protect from photoinhibition, such as non photochemical quenching, transport to molecules other than

CO_2, particularly to oxygen, which leads to photorespiration and/or Mehler reaction (Flexas et al., 2002) nonradiative energy dissipation mechanisms (Souza et al., 2004) and chlorophyll concentration changes (Pastenes et al., 2005). However, these processes ultimately lead to the lower quantum yield of PSII (Govindjee et al., 1999). Measurements of F_v/F_M may provide rapid indication of change in current plant productivity in response to water change, andmay be a good tool in genetic improvement or programs enabling genotypes with particular characteristics to be selected at an early stage, but further work is required to examine its potential. Chlorophyll fluorescence measurements allow the discrimination among the tolerant and sensitive genotypes. Under water deficit condition, the tolerant genotypes maintain a higher photosynthetic activity than the sensitive. Studies conducted in tomato by Srinivasa Rao et al. (1999) and Bahadur et al. (2010) indicated that PSII activity (F_v/F_M) of drought tolerant genotypes was less decreased with imposing water stress than susceptible genotypes.

9.3.4 WATER USE EFFICIENCY (WUE)

WUE is traditionally defined either as the ratio of dry matter accumulation to water consumption over a season or as the ratio of photosynthesis (A) to transpiration (E) over a period of time (IWUE). It is among one of traits that has been studied a lot because it can give an idea of the variation among genotypes in ability to use water efficiently under limited water supply. The large assemblage of literature on crop WUE as derived from research on carbon isotope discrimination allows some conclusions on the relations between WUE on the one hand, and drought tolerance and yield potential (YP) on the other. Briefly, apparent genotypic variations in WUE are expressed mainly due to variations in water use. Higher WUE is generally achieved by specific plant traits and environmental responses that reduce YP. Under most dry land situations where crops rely on unpredictable seasonal rainfall, the maximization of soil moisture use is a crucial component of drought resistance (avoidance), which is generally expressed in lower WUE (Blum, 2005). It is now well documented that high YP and high yield under water-limited conditions are generally associated with reduced WUE mainly because of high water use. Features linked to low YP, such as smaller plants (Martin and Ruiz-Torres, 1999) or short growth duration (Lopezcastaneda and Richards, 1994), ascribe high WUE because of reduced water use. Genotypic variation in WUE was driven mainly by variations in water use rather than by variations in plant production or assimilation per unit of water use. If low water use is the breeder's target, it is highly probable that selection for the same can be achieved by directly selecting for these plant traits, without measuring WUE (Blum, 2005). The enhancement of biomass production under drought stress can be achieved primarily by maximizing soil water capture while diverting the largest part of the available soil moisture towards stomatal transpiration. This is defined as effective use of water, and it is the major engine for agronomic or genetic enhancement of crop production under a limited water condition. High WUE is a critical characteristic of drought-tolerant species, and is a water-saving strategy of plants in arid regions. However, there are many relative physiological traits affecting leaf WUE expressing wide variations in leaf WUE under normal and water stress conditions. Intrinsic water use efficiency (IWUE) estimated as a ratio of photosynthesis/transpiration has been recognized as a measure of carbon gain per unit of water loss and found to be inversely proportional to the ratio of intercellular and ambient CO_2 concentrations (C_i/Ca) (Martin et al., 1992). Large variability in WUE has been reported among several species as well as cultivars within a species including cowpea (Condon et al., 2002). Because higher rates of leaf photosynthesis are often associated with faster crop growth rates, a combination of higher photosynthesis and improved WUE mayplay a vital role for yield enhancement of crops under drought stress conditions (Parry et al., 2005).

9.3.5   EVAPOTRANSPIRATION (ET)

The onset of stress may initially cause a loss of cell turgor, which in turn reduces gaseous exchange, and leaf elongation sinceboth are turgor-dependent processes. ET is known to positively correlate with yield of the crops, since it is a direct measure of crop water loss. Water stress causes a decrease in transpiration, an increase in foliage temperature and closure of stomata. Canopy temperature is dependent on climatic factors and internal plant water status. There seems to be a positive link between yield and transpiration rate. Important increases in crop yield might be possible if irrigation water is applied at the most appropriate time to prevent excessive and nutrient leaching. In order to improve irrigation efficiency, it is necessary to adjust the water application rate based on crop ET. Stomata regulated reduction in transpiration is a common response of plants to drought stress which also provides an opportunity to increase plant water-use efficiency. Bahadur et al. (2010) reported that mild or severe water stress in tomato significantly reduces the transpiration rate and increase the leaf temperature.

9.3.6 CELL MEMBRANE STABILITY (CMS)

A major impact of plant environmental stress is cellular membrane modification, which results in its perturbed function or total dysfunction. The cellular membrane dysfunction due to stress is well expressed in increased permeability and leakage of ions, which can be readily measured by the efflux of electrolytes, andmay be used as a tolerance index for drought stress (Sayar et al., 2008; Yang et al., 2008). The degree of cell membrane injury caused by stress can be assessed using this technique. Electric conductivity of solution containing the electrolytes leaking from leaf segment is used to assess the degree of drought tolerance. The tolerant genotypes show less electrolyte leakage due to maintenance of integrity of cell membrane.

9.3.7 RELATIVE WATER CONTENT (RWC)

The relative water content (RWC) of a tissue is the fraction of turgid weight that is remaining in the tissue. Tissue watercontent can vary greatly among organs, developmental stages, seasons, habitats, and species. However, a relative degree of tissue hydration can be calculated by comparing the current hydration of a tissue to its maximum potential hydration. RWC represents a useful indicator of the state of water balance of a plant, essentially because it expresses the absolute amount of water, which the plant requires to reach artificial full saturation. This method has gained favor over LWP as a very relevant physiological measure of plant water deficit. Its advantage is that it accounts for the effect of OA in affecting plant water status. Two plants with the same LWP can have different RWC if they differ for OA. Drought stress results in decreased RWC (e.g., Fu and Huang, 2001; Shaw et al., 2002). Bahadur et al. (2009 and 2010) noticed significant reduction in leaf RWC in okra and tomato, respectively with imposing drought tolerance. Higher RWC is a good indicator of drought stress tolerance (Shaw et al., 2002).

9.3.8 PLANT CANOPY TEMPERATURE OR TRANSPIRATIONAL COOLING

It has long been recognized that leaf or canopy temperature is highly dependent on the rate of transpiration and therefore, can be used as an indicator of stomatal opening. Accordingly, infrared thermometry has been developed as a means for irrigation scheduling. Plant canopy temperature is directly correlated to stomatal conductance and transpiration. As long as the plants continue to transpire through open stomata the canopy temperatures could be maintained at metabolically comfortable range otherwise higher temperature would destroy the vital enzyme activities. Stomatal closures for a considerable period of time are known to increase the leaf temperature. The thermal imagery system is a powerful tool as it can capture the temperature differences of plant canopies fairly quickly and instantly. The trend in canopy temperature and differences in temperatures between canopy and air (T_c-T_a) is an indicator of the plant water stress. The relationships between canopy temperature, air temperature and transpiration is not simple, involving atmospheric conditions (vapor pressure deficit, air temperature and wind velocity), soil (mainly available soil moisture) and plant (canopy size, canopy architecture and leaf adjustments to water deficit). These variables are considered when canopy temperature is used to develop the crop water stress index (CWSI), which is gaining importance in scheduling irrigation in crops. Relatively lower canopy temperature in drought stressed crop plants indicates a relatively better capacity for taking up soil moisture and for maintaining a relatively better plant water status by various plant constitutive or adaptive traits. Besides, it should be noted that canopy temperature is dependent on climatic parameters and internal plant water status. High crop canopy temperature in water-stressed plants may also be related to decreased transpiration rate and leaf water retention capacity values Tan (1993). Drought resistant genotypes show higher values for T_c-T_a. There seems to be a positive link between yield and transpiration rate.

9.3.9   SPECIFIC LEAF AREA (SLA)

It is a sign of leaf thickness, and usually decreased under drought stress (Marcelis et al., 1998). Decreasing pattern of SLA under drought situation is due to the different sensitivity of photosynthesis and leaf area expansion to soil drying. Research has indicated that WUE of a crop is related to the morphological characteristics of leaves. Wright et al. (1994) proposed that under field conditions at moderate temperatures there is a close negative correlation between WUE and SLA. Moisture stress affects leaf expansion earlier than photosynthesis (Jensen et al., 1996; Tardieu et al., 1999). Reduction of SLA is believed to be a mode to improve WUE (Craufurd et al., 1999; Wright et al., 1994). This is due to thicker leaves generally having a higher density of chlorophyll and proteins per unit leaf area and, thus, have a greater photosynthetic capability than thinner leaves. This correlation ship between WUE and SLA may be due to the reality that plants with low SLA (thicker leaves) have high nitrogen content and more mesophyll cells per unit area, both leading to higher rates of CO_2 assimilation, and consequently, higher biomass production (Thumma et al., 2001). Drought stress, that reduces SLA, may also raise WUE in leaves. This is possibly a part of an adaptive mechanism for reducing leaf area and transpiration rate (Craufurd et al., 1999).

9.4   CONCLUSION

The environmental uncertainties especially drought stress faced by crop, therefore, the primary objective will be to optimize confined management practices to reduce severe stress as far as possible and, in particular, to intensify the search for important physiological traits such as osmotic adjustment, CMS, photosynthetic and transpiration rate, stomatal conductance, WUE, chlorophyll fluorescence parameters, photosynthetic pigment content, plant canopy temperature simultaneously with biochemical attributes including level of antioxidant enzymes, nonenzymatic antioxidant and pyridine nucleotides involve in drought tolerance mechanism in respect of climate change scenario.

KEYWORDS

- **Climate change**
- **Drought stress**
- **Physiological traits**
- **Tolerance mechanism**
- **Vegetables**

REFERENCES

Anyia, A. O., & Herzog, H. (2004). Water-use efficiency, leaf area and leaf gas exchange of cowpeas under mid-season drought. Eur. J. Agron. 20, 327–339.

Bahadur, A., Chatterjee, A., Kumar, R., Singh, M., & Naik, P. S. (2011). Physiological and Biochemical Basis of drought tolerance in vegetables. Veg. Sci. 38, 1–16.

Bahadur, A., Kumar, R., Mishra, U., Rai, A., & Singh, M. (2010). Physiological approaches for screening of tomato genotypes for moisture stress tolerance. National Conference of Plant Physiology (NCPP-2010) BHU, Varanasi during Nov. 25–27, 2010. 142pp.

Bahadur, A.,Singh, K. P., Rai, A., Verma, A., & Rai, M. (2009). Physiological and yield response of okra (*Abelmoschus esculentus* Moench) to irrigation scheduling and organic mulching. Indian J. Agric. Sci. 79, 813–815.

Blum, A. (1988a). Plant Breeding for Stress Environments. CRC Press. Plant breeding for stress environments. Boca Raton. CRC Press Inc. 223p.

Blum, A. (1988b). Improving wheat grain filling under stress by stem reserve mobilisation. Euphytica. 100, 77–83.

Blum, A. (2005). Drought resistance, water-use efficiency, and yield potential-are they compatible, dissonant, or mutually exclusive? Aust. J. Agric. Res. 56, 1159–1168.

Blum, A., Mayer, J., & Golan, G. (1982). Infrared thermal sensing of plant canopies as a screening technique for dehydration avoidance in wheat. Field Crops Res. 57, 137–146.

Blum, A., Schertz, K. F., Toler, R. W., Welch, R. I., Rosenow, D. T., Johnson, J. W., & Clark, L. E. (1978). Selection for drought avoidance in sorghum using aerial infrared photography. Agron. J. 70, 472–477.

Bohm, W. (1974). Mini-rhizotrons for root observations under field conditions. Z. Acker-u. Pflanzenbau. J. Agron. Crop Sci. 140, 282–287.

Boyer, J. S. & Westgate, M. E. (2004). Grain yields with limited water. J. Exp. Bot. 55, 2385–2394.

Cattivelli, L., Baldi, P., Crosatti, C., Di Fonzo, N., Faccioli, P., Grossi, M., Mastrangelo, A. M., Pecchioni, N., & Stanca, A. M. (2002) Chromosome regions and stress-related sequences involved in resistance to abiotic stress in Triticeae. Plant Mol. Biol. 48, 649–665.

Cattivelli Luigi, Rizza Fulvia, Badeck Franz, W., Mazzucotelli Elisabetta, Mastrangelo Anna, M., Francia Enrico, Mare Caterina, Tondelli Alessandro & Stanca, A. Michele. (2008).

Drought tolerance improvement in crop plants. An integrated view from breeding to genomics. Field Crops Res. 105, 1–14.

Chaves, M. M., Pereira, J. S., Maroco, J., Rodrigues, M. L., Ricardo, C. P. P., Oserio, M. L., Carvalho, I., Faria, T., & Pinheiro, C. (2002). How do plants cope with water stress in the field? Photosynthesis and growth. Ann. Bot. 89, 907–916.

Condon, A. G., Richards, R. A., Rebetzke, G. J., & Farquhar, G. D. (2002). Improving intrinsic water-use efficiency and crop yield. Crop Sci. 42, 122–131.

Craufurd, P. Q., Wheeler, T. R., Ellis, R. H., Summerfield, R. J., & Williams, J. H. (1999). Effect of temperature and water deficit on water-use efficiency, carbon isotope discrimination, and specific leaf area in peanut. Crop Sci 39, 136–142.

Cushman, J. C. (2001). Crasulacean acid metabolism. A plastic photosynthetic adaptation to arid environments. Plant Physiol. 127, 1439–1448.

Duniway, J. M. (1971). Water relation of Fusarium wilt of tomato. Physiol. Plant 15, 10–21.

Edmeades, G. O., Bolanos, J., Elings, A., Ribaut, J. M., Banziger, M., & Westgate, M. E. (2000). The role and regulation of the anthesis silking interval in maize. In Westgate, M. E., & Boote, K. J. (Eds.), Physiology and Modeling Kernel Set in Maize. CSSA Special Publication No. 29. CSSA, Madison, WI. 43–73 pp.

Farquhar, G. D., & Richards, R. A. (1984). Isotopic composition of plant carbon correlates with water-use efficiency of wheat genotypes. Aust. J. Plant Physiol. 11, 539–552.

Flexas, J., Bota, J., Escalona, J. M., Sampol, B., & Medrano, H. (2002). Effects of drought on photosynthesis in grapevines under field conditions an evaluation of stomatal and mesophyll limitations. Funct. Plant Biol. 29, 461–471.

Foyer, C. H., & Noctor, G. (2000). Oxygen processing in photosynthesis: regulation and signalling. New Phytol. 146, 359–388.

Fu, J., & Huang, B. (2001). Involvement of antioxidants and lipid peroxidation in the adaptation of two cool season grasses to localized drought stress. Environmental and Experimental Botany 45, 105–114.

Gay, A. P. (1986). Variation in Selection for Leaf Water Conductance in Relation to Growth and Stomatal Dimensions in *Lolium perenne* L. Ann. Bot. 57, 361–369.

Govind, jee. (1999). On the requirement of minimum number of four versus eight quanta of light for the evolution of one molecule of oxygen in photosynthesis a historical note. Photosynthesis Res. 59, 249–254.

Hessini, K., Martínez, J. P., Gandour, M., Albouchi, A., Soltani, A., & Abdelly, C. (2009). Effect of water stress on growth, osmotic adjustment, cell wall elasticity and water use efficiency in Spartina alterniflora. Environ. Exp. Bot. 67, 312–319.

Jensen, C. R., Mogensen, V. O., Mortensen, G., Andersen, M. N., Schjoerring, J. K., Thage, J. H., & Koribidis, J. (1996).

Jones, H. G. (1999). Use of thermography for quantitative studies of spatial and temporal variation of stomatal conductance over leaf surfaces. Plant Cell Environ. 22, 1043–1055.

Keenan, T., Sabate, S., & Gracia, C. (2010). The importance of mesophyll conductance in regulating forest ecosystem productivity during drought periods. Global Change Biol. 16, 1019–1034.

Kerstiens, G. (1996). Cuticular water permeability and its physiological significance. J. Exp. Bot. 47, 1813–1832.

Kumar, A., & Elston, J. (1992). Genotypic differences in leaf water relations between *Brassica juncea* and *B. napus*. Ann. Bot. 70, 3–9.

Kumar, A., Singh, P., Singh, D. P., Singh, H., & Sharma, H. C. (1984). Differences in osmoregulation in *Brassica* species. Ann. Bot. 54, 537–541.

Kumar Rajesh, Solankey Shashank Shekhar, & Singh Major (2012). Breeding for drought tolerance in vegetables. Veg. Sci. 39(1), 1–15.

Lawlor, D. W. & Cornic, G. (2002). Photosynthetic carbon assimilation and associated metabolism in relation to water deficits in higher plants. Plant Cell Environ. 25, 275–294.

Leaf photosynthesis and drought adaptation in field-grown oilseed rape (*Brassica napus* L.). Aust. J. Plant Physiol. 23, 631–644.

Lopezcastaneda, C., & Richards, R. A. (1994). Variation in temperate cereals in rainfed environments. 3. Water use and water use efficiency. Field Crops Res. 39, 85–98.

Marcelis, L. F. M., Heuvelink, E., & Goudriaan J. (1998). Modeling biomass production and yield of horticultural crops a review. Sci. Hort. 74, 83–111.

Martin, B., & Ruiz Torres, N. A. (1992). Effects of water-deficit stress on photosynthesis, its components and component limitations, and on water use efficiency in wheat (*Triticumaestivum* L.). Plant Physiol. 100, 733–739.

Mitra Jiban (2001). Genetics and genetic improvement of drought resistance in crop plants. Current Science. 80(6), 758–763.

Morgan, J. M. (1980) Osmotic adjustment in the spikelet and leaves of wheat J. Exp. Bot. 31, 655–665.

Morgan, J. M. (1983). Osmo regulatiom as selection criterion for drought tolerance in wheat Aust. J. Agric. Res. 34, 607–614.

Ndunguru, B. J., Ntare, B. R., Williams, J. H. & Greenberg, D. C. (1995). Assessment of groundnut cultivars for end of season drought tolerance in a Sahelian environment J. Agric. Sci. of Cambridge. 125, 79–85.

Parry, M. A. J., Flexas, J., & Medrano, H. (2005). Prospects for crop production under drought: research priorities and future directions. Ann. Appl. Biol. 147, 211–226.

Pervez, M. A., Ayub, C. M., Khan, H. A., Shahid, M. A. & Ashraf, I. (2009). Effect of drought stress on growth, yield and seed quality of tomato (*Lycopersicon esculentum* L.) Pak. J. Agri. Sci., 46(3): 174–178.

Pastenes, C., Pimentel, P., & Lillo, J. (2005). Leaf movements and photo inhibition in relation to water stress in field-grown beans. J. Exp. Bot. 56, 425–433.

Rajcan, I. & Tollenaar, M. (1999). Source sink ratio and leaf senescence in maize. I. Dry matter accumulation and partitioning during the grain filling period. Field Crop Res. 90, 245–253.

Ramanjulu, S., & Bartels, D. (2002). Drought and desiccation induced modulation of gene expression in plants. Plant Cell Environ. 25, 141–151.

Reddy, A. R., Chaitanya, K. V. & Vivekanandan, M. (2004). Drought-induced responses of photosynthesis and antioxidant metabolism in higher plants. J. Plant Physiol. 161, 1189–1202.

Richards, R. A. (2006). Physiological traits used in the breeding of new cultivars for water-scarce environments. Agric. Water Manage. 80, 197–211.

Robertson, B. M., Hall, A. E. & Foster, K. W. (1985). A field technique for screening for genotypic differences in root growth. Crop Sci. 25, 1084–1090.

Rouhi, V., Samson, R., Lemeur, R., & Damme, P. V. (2007). Photosynthetic gas exchange characteristics in three different almond species during drought stress and subsequent recovery. Environ. Exp. Bot. 59, 117–129.

Sayar, R., Kemira, H., Kameli, A., & Mosbahi, M. (2008). Physiological tests as predictive appreciation for drought tolerance in durum wheat (*Triticum durum* Desf.). Agron. Res. 6, 79–90.

Serraj, R., & Sinclair, T. R. (2002). Osmolyte accumulation can it really increase crop yield under drought conditions? Plant Cell Environ. 25, 333–341.

Sharp, R. E., Poroyko, V., Hejlek, L. G., Spollen, W. G., Springer, G. K., Bohnert, H. J. & Nguyen, T. (2004). Root growth maintenance during water deficits physiology to functional genomics. J. Exp. Bot. 55, 2343–2351.

Shaw, B., Thomas, T. H., & Cooke, D. T. (2002). Responses of sugar beet (Beta vulgaris L.) to drought and nutrient deficiency stress. Plant Growth Regulation 37, 77–83.

Singh, S. & Reddy, K. Raja (2011). Regulation of photosynthesis, fluorescence, stomatal conductance and water use efficiency of cowpea (*Vigna unguiculata* [L.] Walp.) Under drought. J. Photochem. Photobio. B Bio. 105, 40–50.

Slafer, G. A., Araus, J. L., Royo, C. & Del Moral, L. F. G. (2005). Promising eco physiological traits for genetic improvement of cereal yields in Mediterranean environments. Ann. Appl. Biol. 146, 61–70.

Souza, R. P., Machado, E. C., Silva, J. A. B., Lagôa AMMA & Silveira, J. A. G. (2004). Photosynthetic gas exchange, chlorophyll fluorescence and some associated metabolic changes in cowpea (*Vigna unguiculata*) during water stress and recovery. Environ. Exp. Bot. 51, 45–56.

Srinivasa Rao, N. K. & Bhatt, R. M. (1988). Photosynthesis, transpiration, stomatal diffusive resistance, and relative water content of *Capsicum annum* L. Grossum (bell pepper) grown under water stress. Photosynthetica 22, 377–382.

Srinivasa Rao, N. K., & Bhatt, R. M. (1990). Response of photosynthesis to water stress in two eggplant cultivars (*Solanum melongena* L.). Photosynthetica 24, 506–513.

Srinivasa Rao, N. K., & Bhatt, R. M. (1992). Response of tomato to moisture stress: plant water balance and yield. Plant Physiol. Biotech. 19, 36–41.

Srinivasa Rao, N. K., Bhatt, R. M., Mascarenhas, J. B. D., & Naren, A. (1999). Influence of moisture stress on leaf water status, osmotic potential, chlorophyll fluorescence and solute accumulation in field grown tomato cultivars. Veg. Sci. 26, 129–132.

Tan, C. S. (1993). Tomato yield evapo-transpiration relationships, seasonal canopy temperature and stomatal conductance as affected by irrigation. Canadian J Plant Sci. 73, 257–264.

Thumma, B. R., Naidu, B. P., Chandra, A., Cameron, D. F., Bahnisch, L. M., & Liu, C. (2001). Identification of causal relationship among traits related to drought resistance in *Stylosanthes scabra* using QTL analysis J. Exp. Bot. 52, 203–214.

Turner, N. C., Wright Graeme, C., & Siddique, K. H. M. (2001). Adaptation of grain legumes (pulses) to water limited environments. Adv. Agron. 71, 193–231.

Tyerman, S. D., Niemietz, C. M., & Bramley, H. (2002). Plant aquaporins: multifunctional water and solute channels with expanding roles. Plant Cell Environ. 25, 173–194.

Ugherughe, P. O. (1986). Drought and Tropical Pasture Management. Z. Acker u. P, flanzenbau. J. Agron. Crop Sci., 157, 13–23.

Walter, A., & Shurr, U. (2005). Dynamics of leaf and root growth: endogenous control versus environmental impact. Ann. Bot. 95, 891–900.

Wilhelm, C., &Selmar, D. (2011). Energy dissipation is an essential mechanism to sustain the viability of plants. The physiological limits of improved photosynthesis. J. Plant Physiol. 168, 79–87.

Wright, G. C., Rao, R. C. N., & Farquhar, G. D. (1994). Water-use efficiency and carbon isotope discrimination in peanuts under water deficit conditions. Crop Sci. 34, 92–97.

Yang, Y., Han, C., Liu, Q., Lin, B. & Wang, J. (2008). Effect of drought and low light on growth and enzymatic antioxidant system of *Picea asperata* seedlings. Acta Physiol. Plant. 30, 433–440.

CHAPTER 10

HARNESSING HEAT STRESS IN VEGETABLE CROPS TOWARDS MITIGATING IMPACTS OF CLIMATE CHANGE

SHIRIN AKHTAR[1], ABHISHEK NAIK[2], and PRANAB HAZRA[2]

[1]Department of Horticulture (Vegetable and Floriculture) Bihar Agricultural University, Sabour, Bhagalpur, India.

[2]Department of Vegetable Crops Faculty of Horticulture Bidhan Chandra KrishiViswavidyalaya, Mohanpur, Nadia, India.

CONTENTS

ABSTRACT

Environmental stress is the primary cause of crop losses worldwide, reducing average yields for most major crops by more than 50%. High temperatures and limited soil moisture are the major causes of low yields of vegetables. Elevated temperature can disrupt the normal growth and development of plants, which ultimately affects crop productivity. Under heat stress seed germination, seedling and vegetative growth, flowering and fruit set, and fruit ripening are adversely affected. The plants respond to the heat stress by altering different morphological, anatomical, physiological and biochemical mechanisms. High temperature stress can be avoided by crop management practices, *viz.*, selection of proper sowing methods, choice of proper sowing date, cultivars, irrigation methods,etc., as well as by developing improved cultivars resistant to heat stress. The most recent trend in control of heat stress is by use of molecular tools.

10.1   INTRODUCTION

Stress is any factor of environment that interferes with the complete expression of the genotypic potential of the plant (Singh, 2000).Environmental stress is the primary cause of crop losses worldwide, reducing average yields for most major crops by more than 50% (Boyer, 1982; Bray et al., 2000). The response of plants to environmental stresses depends on the plant developmental stage and the length and severity of the stress (Bray, 2002). Plants may respond similarly to avoid one or more stresses through morphological or biochemical mechanisms (Capiati et al., 2006). Environmental interactions may make the stress response of plants more complex or influence the degree of impact of climate change. One of the most detrimental stresses among the ever-changing components of the environment is the constantly rising ambient temperature. The prediction of IPCC (2007) that global air temperature is to rise by 0.2 °C per decade, eventually leading to rise in temperatures up to 1.8–4.0°C higher than the current level by 2100 is a threat to the biosphere, as heat stress has known effects on the life processes of organisms, acting directly or through the modification of surrounding environmental components. Plants being sessile organisms, cannot move to more favorable environments as a result of which plant growth and developmental processes are substantially affected, often lethally, by high temperature stress (Lobel and Asner, 2003; Lobel and Fiend, 2007).

Heat stress is often defined as the rise in temperature beyond a threshold level for a period of time sufficient to cause irreversible damage to plant growth and development (Wahid et al., 2007). Generally heat shock or heat stress is considered as a transient elevation in temperature, usually 10–15 °Cabove the ambient. Heat stress involves intensity (temperature in degrees), duration, and rate of increase in temperature and the probability and period of high temperatures occurring during the day and/or the night determines its extent. Heat stress due to high ambient temperatures is a serious threat to crop production worldwide (Hall, 2001).The effects of heat stress on plant may be expressed as alterations in plant growth, development, physiological processes, and yield (Hasanuzzaman et al., 2012a, 2013a). The tolerance mechanism of plants against this stress is by physical as well as physiological changes as well as change in their metabolism. Plants alter their metabolism in various ways in response to heat stress, especially by producing compatible solutes that are able to organize proteins and cellular structures, maintain cell turgor by osmotic adjustment, and modify the antioxidant system to reestablish the cellular redox balance and homeostasis (Janska et al., 2010; Munns et al., 2008; Valliyodan et al., 2006). At the molecular level, heat stress causes alterations in expression of genes that are directly involved in protection from heat stress (Chinnusamy et al., 2007; Shinozaki et al., 2007). Genes responsible for the expression of osmoprotectants, detoxifying enzymes, transporters, and regulatory proteins are some worthy of mention in this context (Krasensky et al., 2012; Semenov et al., 2009).These modification of physiological and biochemical processes by gene expression changes gradually leads to the development of heat tolerance in the form of acclimation, and further leads to adaptation (Hasanuzzaman et al., 2010a; Moreno and Orellana, 2011).

Vegetables play a vital role in ensuring food and nutritional security. However, these are highly perishable crops and their prices rise fast under adverse climatic conditions reducing their availability, thus putting them out of reach of the poor. The small and marginal farmers are the more affected ones by the vagaries of climate (FAO, 2009). Most vegetables prefer mild temperatures and are generally sensitive to environmental extremes. Hence, high temperatures and limited soil moisture are the major causes of low yields in the tropics. Significant influence on different development phases like vegetative growth, flowering and fruiting has been observed by adversity of climate. Elevated temperature can disrupt the normal growth and development of plants, which ultimately affects crop productivity.

Measures to adapt to these climate change-induced stresses are critical for sustainable vegetable production. Until now, the scientific information on the effect of environmental stresses on vegetables is overwhelmingly on tomato. There is a need to do more research on how other vegetable crops are affected by increased abiotic stresses as a direct potential threat from climate change.

10.2   THRESHOLD LEVELS OF HEAT-STRESS

A value of daily mean temperature at which a detectable reduction in growth begins may be referred to as threshold temperature. Upper and lower developmental threshold temperatures specific for many plant species have been determined through controlled laboratory and field experiments. A lower developmental threshold or a base temperature is the temperature belowwhich plant growth and development cease. Similarly, an upper developmental threshold is the temperature abovewhich growth and development stop. Base threshold temperatures vary with plant species, for example, spinach (2 °C), lettuce (4.4 °C), pea (4.4 °C), French bean (10 °C), asparagus (5.5 °C), pumpkin (13 °C), tomato (15 °C), etc. The lower threshold temperature of cool season and temperate crops are often lower compared to tropical crops. Upper threshold temperatures also differ for different plant species and genotypes within species. However, determining a consistent upper threshold temperature is difficult because the plant behaviormay differ depending on other environmental conditions (Miller et al., 2001). In tomato, for example, when the ambient temperature exceeds 35 °C, its seed germination, seedling and vegetative growth, flowering and fruit set, and fruit ripening are adversely affected. For other plant species, the higher threshold temperature may be lower or higher than 35 °C. Legumes are particularly sensitive to heat stress at the bloom stage; only a few days of exposure to high temperatures (30–35 °C) can cause heavy yield losses through flower drop or pod abortion (Siddique et al., 1999). In general, base and upper threshold temperatures vary in plant species according to varying habitats. Hence, appraisal of threshold temperatures of new cultivars is very much desirable to prevent damages by unfavorable temperatures during the plant ontogeny.

10.3 PLANT ADAPTATION TO HEAT STRESS

Plants can be classified into three groups according to the optimum temperature for growth:

(a) Psychrophiles: plants which grow optimally at low temperature ranges between 0 °C and 10 °C.

(b) Mesophyles: plants whichfavor moderate temperature and grow well between 10 °C and 30 °C.

(c) Thermophyles: plants which grow well between 30 °C and 65 °C or even higher (Źróbek-Sokolnik, 2012). Within this group there are three types of plants: heat-sensitive, relatively heat resistant and heat tolerant.

Vegetables may be cool season (those tolerant to low temperature and even frost, for, for example, cabbage, onion, garlic, Brussels sprout, knolkhol, turnip, radish, etc.) and warm season (those suited for relatively high temperature, for example, cucurbits, tomato, brinjal, chili, beans, etc.). Cool season vegetables are heat sensitive crops, while among warm season crops tomato, brinjal, chili, pumpkin, cucumber, etc., are relatively heat resistant or warm loving crops and muskmelon, sweet potato, winged bean, cluster bean, etc., that can tolerate up to 40 °C temperature can be classed into heat resistant crops.

There is a great variation among the different plants in terms of their response and tolerance to high temperature. Plants survive in hot and dry environments by their adaptation mechanisms (Fitter and Hay, 2002). Avoidance and tolerance mechanisms are the strategies adapted by the plants to counter heat stress.

10.3.1 AVOIDANCE MECHANISMS

High temperature conditions result in changes in various mechanisms of plant that include long-term evolutionary phenological and morphological adaptations and short-term avoidance or acclimatization mechanisms such as changing leaf orientation, transpirational cooling, or alteration of membrane lipid compositions. Closure of stomata and reduced water loss, increased stomatal and trichomatous densities, and larger xylem vessels are common heat induced features in plant (Srivastava et al., 2011). Early maturation has been often closely correlated with smaller yield losses under high temperature, whichmay beat tributed to the engagement of an escape mechanism (Adams et al., 2001; Rodríguez et al., 2005). Plants growing in a hot climate avoid heat stress by reducing the absorption of solar radiation. The presence of small hairs (tomentose) that form a thick coat on the surface of the leaf as well as cuticles, protective waxy covering facilitate this. Besides, orientation of the leaf blades may also change and leaf blades often turn away from light and orient themselves parallel to sun rays, the phenomenon being termed paraheliotropism. Rolling of leaf blades is another physical mechanism of reduction of solar radiation. Leaf size may be reduced in order to avoid heat stress. Smaller leaves evacuate heat more quickly due to smaller resistance of the air boundary layer in comparison with large leaves. Transpiration is another adaptive mechanism to control heat stress. In well-hydrated plants, intensive transpiration prevents leaves from heat stress, and leaf temperature may be 6 °C or even 10–15 °C lower than ambient temperature. Avoidance mechanisms such as leaf abscission, leaving heat resistant buds have been found in plants. Desert annuals complete their entire reproductive cycle during the cooler months to avoid the heat stress (Fitter and Hay, 2002). Such morphological and phenological adaptations are commonly associated with biochemical adaptations favoring net photosynthesis at high temperature conditions (in particularC4 and CAM photosynthetic pathways), although C3 plants are also common in desert floras (Fitter and Hay, 2002). The degree of leaf rolling may be affected by high temperature. Leaf rolling has a physiological role in adaptation potential by increasing the efficiency of water metabolism under high temperature (Sarieva et al., 2010).

High temperature stress can also be avoided by crop management practices, *viz.*, selection of proper sowing methods, choice of proper sowing date, cultivars, irrigation methods,etc.For example, in sub tropical zones, cool-season vegetables such as lettuce when sown in the late summer may show incomplete germination and emergence due to high soil temperature (Hall, 2011). This incomplete emergence problem can be overcome by sowing the lettuce seed into dry beds during the day and then sprinkling water to the beds during the late afternoon. Seed priming can be another potential solution to this problem. It involves placing the seed in an osmotic solution for several days at moderate temperatures and then drying them. Due to high soil temperature, germination may be hampered leading to inadequate plant emergence and establishment, which can limit the productivity of several warm-season vegetables such as okra, tomato, brinjal and cucurbits. In such cases, deep placement can be done to overcome the problem. In temperate or subtropical climatic zones having seasonal variations in temperature, sowing date can be varied so that the crop can escape high temperature stress during the subsequent sensitive stages of crop development. Intense direct solar radiation and high temperature can cause damage to fruits as may be seen in tomato where sunscald is observed due to high temperature. Shading of the fruits by foliage can be the avoidance mechanism to this problem (Hall, 2011).

10.3.2   TOLERANCE MECHANISMS

Heat tolerance is generally defined as the ability of the plant to grow and produce economic yield under high temperature conditions. It is a highly specific trait and may vary significantly among closely related species, even different organs and tissues of the same plant. Plants have evolved various mechanisms for thriving under higher prevailing temperatures. These include short-term avoidance/acclimation mechanism or long-term evolutionary adaptations. Some major tolerance mechanisms are ion transporters, late embryogenesis abundant (LEA) proteins, osmoprotectants, antioxidant defense, and factors involved in signaling cascades and transcriptional control and these are essentially significant to counter act the stress effects (Rodríguez et al., 2005; Wang et al., 2004).

Leaf orientation, transpirational cooling and changes in membrane lipid composition are more important short-term response against heat stress (Radin et al., 1994, Rodríguez et al., 2005). Smaller yield losses due to early maturation in summer shows possible involvement of an escape mechanism in heat stress tolerance (Adams et al., 2001). Different tissues in plants show variations in terms of developmental complexity, exposure and responses towards the prevailing or applied stress types (Queitsch et al., 2000). The stress responsive mechanism is established by an initial stress signal that may be in the form of ionic and osmotic effector changes in the membrane fluidity. This helps to reestablish homeostasis and to protect and repair damaged proteins and membranes (Vinocur et al., 2005).

10.4   REACTIONS OF VEGETABLE CROPS TO HEAT STRESS

Heat stress due to increase in temperature is a major problem for vegetable crops. Most of the vegetables show marked reduction in growth above a temperature of 35 °C and above 50 °C thermal growth of the crop is said to occur (Hazra and Som, 2006). A constantly high temperature causes an array of morpho-anatomical changes in plant which affect the seed germination, plant growth, flower shedding, pollen viability, gametic fertilization, fruit setting, fruit size, fruit weight, fruit quality, etc.

The optimum soil temperature for most of the vegetable crops ranges from 20 to 30 °C. The maximum temperature for the warm season vegetables (tomato, brinjal, chili, cucurbits, beans, etc.) is 35 to 40 °Cwhile that of cool season vegetables is 32–35 °C. However, above 30 °C the germination of vegetable crops is impaired.

Vegetative and reproductive processes are strongly modified by temperature alone or in conjunction with other environmental factors (Abdalla and Verkerk, 1968). Heat stress above 35°C has become a major limiting factor for seed germination, seedling and vegetative growth, flowering and fruit setting, and ripening in most vegetable crops. The reproductive development is always more sensitive to high temperatures than vegetative development. The optimum temperatures for tomato cultivation are between 25 °C and 30 °C during the photoperiod and 20 °C during the dark period. However, only 2–4 °C increase in optimal temperature adversely affected gamete development and inhibited the ability of pollinated flowers into seeded fruits and thus, reduced crops yields (Firon et al., 2006; Peet et al., 1997). High temperatures also interfere with floral bud development due to flower abortion. High temperatures can cause significant losses in tomato productivity due to reduced fruit set, and smaller and lower quality fruits (Stevens and Rudich, 1978). In tomato bud drop, abnormal flower development, poor pollen production, dehiscence and viability, ovule abortion and poor viability, reduced carbohydrate availability, and other reproductive abnormalities have been reported due to high temperature by Hazra et al. (2007). In leafy vegetables like palak, high temperatures lead to bolting thus reducing the leaf yield. Likewise in all other vegetables, high temperatures leads to reduced fruit set thus lowering the yield as well as diminish the quality of the produce. Significant inhibition of photosynthesis occurs at temperatures above optimum, resulting in considerable loss of potential productivity. Brief exposure of plants to high temperatures during seed filling can accelerate senescence, diminish seed set and seed weight, and reduce yield (Siddique et al., 1999). This is because under such conditions plants tend to divert resources to cope with the heat stress and thus limited photosynthates would be available for reproductive development. Another effect of heat stress in many plant species is induced sterility when heat is imposed immediately before or during anthesis.

10.4.1 ANATOMICAL AND MORPHOLOGICAL REACTIONS

In tropical conditions, extra radiation and great temperatures are sometimes the major limiting factors that affect plant development and final yield. Greater temperatures can lead to remarkable pre and postharvest losses, burning of twigs and leaves, sunburns on stems and branches, senility of leaf and abscission, prohibition in the development of shoot and root, discoloration of fruit, and diminished production (Guilioni et al., 1997; Ismail and Hall, 1999; Vollenweider and Gunthardt-Goerg, 2005). In temperate regions, heat stress has been reported as one of the most important causes of reduction in yield and dry matter production in many crops (Giaveno and Ferrero, 2003). Reduction in number of fruits and flowers, percentage of fruit fresh weight has also been reported (Golam et al., 2012).

High temperature considerably affects the anatomical structures not only at the tissue and cellular levels but also at the subcellular level. The cumulative effects of all these changes under high temperature stress may result in poor plant growth and productivity. A general tendency of reduced cell size, closure of stomata and curtailed water loss, increased stomatal and trichomatous densities and greater xylem vessels of both root and shoot has been observed at the whole plant level (Anon et al., 2004). At the subcellular level, major modifications have been found to occur in chloroplasts, leading to significant reduction in photosynthesis by changing the structural organization of thylakoids (Karim et al., 1997). Studies have revealed that specific effects of high temperatures on photosynthetic membranes result in the loss of grana stacking or its swelling. Swelling of stroma, disruption of vacuoles and mitochondria has also been reported that result in reduced photosynthetic and respiratory activities (Zhang et al., 2005).

10.4.2   ALTERATIONS IN REPRODUCTIVE STAGE

The critical pre-anthesis high temperature stress is associated with developmental changes in anther, most strikingly, irregularities in the epidermis and endothesium, lack of opening of the stomium, and poor pollen formation (Sato et al., 2002). Bud drop before pollination, undeveloped flowers, persistence of flower and calyx for a long time without fruit set, splitting of antheridial cone, lack of anther dehiscence, poor pollen production, pollen sterility, degeneration of embryo sac, browning and drying of stigma, reduction in stigma receptivity, style elongation, under developed ovary, poor fertilization, slow pollen tube growth, poor pollen viability and germin ability, poor ovule viability, ovule abortion and embryo degeneration, disruption in meiosis and prevention of pollen formation, hindered sugar metabolism and failure of viable pollen production, reduced carbohydrate availability for the fruits, reduced total soluble protein content, developmental abnormalities and poor fruit set, reduction in fruit size and seeds/fruit and inhibition of pathogen induced resistance mechanisms are some of the reproductive manifestations of high temperature stress in vegetable crops (Abdalla and Verkerk, 1968; Akhtar et al., 2012; Ansary, 2006; Camejo and Torres, 2001; Charles and Harris 1972; El-Ahmadi and Stevens, 1979a; Hazra et al., 2007; Iwahori, 1967; Lohar and Peat, 1998; Peet et al., 2003; Rick and Dempsey, 1969). All these changes lead to a single result: nonsetting of fruits/pods and thus significant reduction in yield.

10.4.3   PHENOLOGICAL REACTIONS

The developmental stage at which the plant is exposed to the heat stress may determine the severity of possible damages experienced by the crop. However, it is still unknown whether damaging effects of heat spells occurring at different developmental stages are cumulative (Wollenweber et al., 2003). Vulnerability of species and cultivars to high temperatures may vary with the stage of plant development, but all vegetative and reproductive stages are affected by heat stress to some extent. During vegetative stage, for example, high day temperature can damage leaf gas exchange properties. During reproduction, a short spell of heat stress can cause significant increases in drop of floral buds and opened flowers; however, there variations in sensitivity within and among plant species are high (Guilioni et al., 1997; Young et al., 2004). Impairment of pollen and anther development by elevated temperatures is another important factor contributing to decreased fruit set in many crops at moderate to- high temperatures (Peet et al., 1998; Sato et al., 2006). The stage of pod filling in case of legumes is also very critical. High temperature during this period hampers yield and quality of the crop.

Thus, for vegetable production under high temperatures, it is important to know the developmental stages and plant processes that are most sensitive to heat stress, as well as whether high day or high night temperatures are more injurious. Such insights are important in determining heat-tolerance potential of crop plants.

10.4.4 PHYSIOLOGICAL REACTIONS

10.4.4.1 WATERS RELATIONS

Plant water status is the most important variable under changing ambient temperatures (Mazorra et al., 2002). Generally, plants tend to maintain stable tissue water status irrespective of temperature when adequate moisture is present. High temperatures, however, severely impair this tendency when water is limiting (Machado and Paulsen, 2001). Under field conditions, frequent association of high temperature stress with reduced water availability has been observed (Simoes-Araujo et al., 2003). In tomato heat stress disturbs the leaf water relations and root hydraulic conductivity (Morales et al., 2003). During daytime enhanced transpiration induces water deficiency in plants, leading to a decrease in water potential and hence altering many physiological processes (Tsukaguchi et al., 2003). High temperatures seem to cause water loss in plants more during daytime than nighttime.

10.4.4.2 COMPATIBLE OSMOLYTES ACCUMULATION

Accumulation of certain organic compounds of low molecular mass, known as compatible osmolytesis one of the key adaptive mechanisms in many plants grown under abiotic stresses, including salinity, water deficit and extreme temperatures (Hare et al., 1998; Sakamoto and Murata, 2002). Under stress conditions, different plant species may accumulate a variety of osmolytes such as sugars and sugar alcohols (polyols), proline, tertiary and quaternary ammonium compounds, and tertiary sulphonium compounds (Sairam and Tyagi, 2004). In tomato, fruit set failed due to the disruption of sugar metabolism and proline transport during pollen development under high temperature condition (Sato et al., 2006). Hexose sensing in transgenic plants engineered to produce trehalose, fructans or mannitol may be an important contributory factor to the stress-tolerant phenotypes (Hare et al., 1998). Due to significant roles of osmolytes in response to environmental stresses in plants, heat tolerance might be enhanced by increased accumulation of compatible solutes through traditional plant breeding, marker-assisted selection (MAS) or genetic engineering approaches (Ashraf and Foolad, 2007).

10.4.4.3   PHOTOSYNTHESIS

Any constraint in photosynthesis can limit plant growth at high temperatures. Alterations in various photosynthetic attributes under heat stress are indicative of thermo tolerance of the plant as they show correlations with growth. The primary sites of heat injury are photochemical reactions in thylakoid lamellae and carbon metabolism in the stroma of chloroplast (Wise et al., 2004). Chlorophyll fluorescence, the ratio of variable fluorescence to maximum fluorescence (Fv/Fm), and the base fluorescence ($F0$) are physiological parameters that have direct correlation with heat tolerance (Yamada et al., 1996). In tomato, an increased chlorophyll a: b ratio and a decreased chlorophyll: carotenoids ratio have been observed in the tolerant genotypes under high temperatures, indicating that these changes were related to heat tolerance of tomato (Camejo et al., 2005; Wahid and Ghazanfar, 2006). Furthermore, under high temperatures, degradation of chlorophyll a and b was more pronounced in developed compared to developing leaves (Karim et al., 1997, 1999). Being highly thermolabile, the activity of PSII is greatly reduced or even partially stopped under high temperatures (Bukhov et al., 1999; Camejo et al., 2005). Heckathorn et al. (1998) reported that photo system II electron transport of tomato plants was disturbed by stress of 42 °C for 6 h and exposure of continuous mild heat stress (32–34 °C/22–26 °C day/night) may not have been high enough to depress photosynthesis. Feller et al. (1998), however, reported that

Rubisco activity was inhibited via Rubisco activase inhibition under moderately high temperatures, 30 and 35 °C in cotton and wheat, respectively.

10.4.4.4   CELL MEMBRANE THERMO STABILITY

Cellular membranes play an important role for fundamental trends like respiration and photosynthesis under heat stress (Blum, 1988). Heat stress enhances the kinetic energy and motion of molecules in membranes lead to loss of chemical bonds in molecules of biological membranes. This causes the lipid bilayer of biological membranes to be rather liquid by either proteins de naturation or a rise in fatty acids that are unsaturated (Savchenko et al., 2002). The stability and roles of biological membranes are susceptible to high temperature, as heat stress changes the tertiary and quaternary structures of the membrane proteins. These changes increase the penetrance of membranes, as obvious from increased loss of electrolytes. The enhanced solute leakage, as a symptom of diminished cell membrane thermo stability (CMT), has long been applied as an indirect estimation of heat-stress resistance in different crop species, involving potato and tomato (Chen et al., 1982), soybean (Martineau et al., 1979), cotton (Ashraf et al., 1994), cowpea (Ismail and Hall, 1999), wheat (Blum et al., 2001), sorghum (Marcum, 1998), and barley (Wahid and Shabbir, 2005).

10.4.4.5   ALTERATIONS IN HORMONE

Crops have the capability to monitor and adjust to inappropriate environmental situations, although the adaptability or tolerance degree to special stresses varies between species and genotypes. The role of hormones in this regard is imperative. Under heat stress condition, hormonal homeostasis, stability, content, biosynthesis and compartmentalization are changed (Maestri et al., 2002). Stress hormones such as abscissic acid (ABA) and ethylene (C_2H_4), act as signal molecules and thus regulate various physiological mechanisms. Diverse environmental stresses, such as high temperature, leads to enhanced ABA levels (Larkindale and Huang, 2005). Induction of various HSPs (e.g., HSP70) by ABA can be another mechanism of developing thermo resistance (Pareek et al., 1998). Another hormone, brassinosteroidsis responsible for heat resistance in oilseed rape (*Brassicanapus*) and tomato. Different hormonal factors are also related to high temperature stress. Auxin and gibberellic acid content are reduced in plants in the glasshouse under high temperatures with an increase in proline content (Muthuvel al., 1999). Auxin production is reduced with a simultaneous high level of abscissic acid in the plant at high temperatures, whichfavors premature senescence and abscission of reproductive organs (El-Ahmadi and Stevens, 1979a; Iwahori, 1967; Levy et al., 1978). High temperature stress also suppresses ethylene production (Inaba et al., 1996; Johjima, 1995; Lurie et al., 1996) causing impaired ripening of fruits.

10.4.5 MOLECULAR REACTIONS

10.4.5.1 HEAT SHOCK PROTEINS

Production and accumulation of special proteins occur when heat stress is rapid and these proteins are identified as heat shock proteins (HSPs). Enhanced production of HSPs is found to incur when crops experience either sudden or slow rise in temperature (Nakamoto and Hiyama, 1999; Schoffl et al., 1999). HSPs expression is limited to certain steps of growth, including germination, embryogenesis, growth of pollen and maturation of fruit (Prasinos et al., 2005). Three sorts of proteins, as detected by molecular weight, account for most HSPs, viz., HSP90, HSP70 and less molecular weight proteins of 15–30 kDa. The ratios of these different types of protein vary between crop species (Feussner et al., 1997). The gene for a nuclear-encoded HSP, Hsa32, that encode a 32 kDa protein, has been cloned in tomato (Liu et al., 2006). In tomato crops suffering from heat stress, HSPs gather into a granular structure in the cytoplasm, probably preserving the protein bio production machinery (Miroshnichenko et al., 2005). Presence of HSPs can hinder denaturation of other proteins that can be influenced by high temperature. The dynamic nature and aggregate state of small HSPs may be vital for their roles in protection of crop cells from harmful influences of heat stress (Schöffl et al., 1999; Iba, 2002). The specific significance of small HSPs in crops is proposed by their abnormal diversity and abundance. The capability of small HSPs to gather into heat shock granules (HSGs) and their decomposition is a prerequisite for survival of crop cells under constant stress environments at sub-lethal temperatures (Miroshnichenko et al., 2005). Low molecular weight HSPs may have structural functions in stability of cell membrane. The presence of these low molecular weight HSPs in chloroplast membranes suggest that these proteins preserve the PSII from improper impacts of heat stress and have a major function in transport of photosynthetic electron (Barua et al., 2003).

10.5   DEVELOPMENT OF HEAT-STRESS TOLERANCE

Heat tolerance is generally defined as the ability of the plant to grow and produce economic yield under high temperatures. However, while some researchers believe that night temperatures are major limiting factors, while others have argued that day and night temperatures do not affect the plant independently and that the diurnal mean temperature is a better predictor of plant response to high temperature and day temperature plays a secondary role (Peet and Willits, 1998).

In general, the negative impacts of abiotic stresses on agricultural productivity can be reduced by a combination of genetic improvement and cultural practices. Genetic improvement entails development of cultivars, which can tolerate environmental stresses and produce economic yield. However, adjustment/modifications in cultural practices, such as planting time, plant density, and soil and irrigation managements, can minimize stress effects. The aim should be to synchronize the stress sensitive stage of the plant with the most favorable time period of the season. In recent times, exogenous applications of protectants in the form of osmoprotectants (proline, Pro; glycine betaine, GB; trehalose, Tre, etc.), phytohormones (abscisic acid, ABA; gibberellic acids, GA; jasmonic acids, JA; brassinosteroids, BR; salicylic acid, SA, etc.), signaling molecules (e.g., nitric oxide, NO), polyamines (putrescine, Put; spermidine, Spd and spermine, Spm), trace elements (selenium, Se; silicon, Si, etc.) and nutrients (nitrogen, N; phosphorus, P; potassium, K, calcium, Ca, etc.) have been found effective in mitigating high temperature stress induced damage in plants (Barnabás, 2008; Hasanuzzaman et al., 2010b, 2012b, 2012c, 2013b, 2013c; Waraich et al., 2012).

Agricultural productivity can be improved under stress environment by employing both genetic improvement and adjustment in cultural practices simultaneously. Agriculturists have long been aware of desirable cultural practices to minimize adverse effects of environmental stresses on crop production. However, genetic improvement of crops for stress tolerance is relatively a new endeavor and has been considered only during the past 2–3 decades. Traditionally, the majority of the plant breeding programs has focused on development of cultivars with high yield potential in favorable (i.e., stress-free) environments. Such efforts have been very successful in improving the efficiency of crop production per unit area and have resulted in significant increases in total agricultural production (Warren, 1998). However, genetic improvement of plants for stress tolerance can be an economically viable solution for crop production under stressful environments (Blum, 1988). The progress in breeding for stress tolerance depends upon an understanding of the physiological mechanisms and genetic bases of stress tolerance at the whole plant level, cellular level and molecular level. Considerable information regarding the physiological and metabolic aspects of plant heat-stress tolerance is available. However, information regarding the genetic basis of heat tolerance is generally scarce, though the use of traditional plant breeding protocols and contemporary molecular biological techniques, including molecular marker technology and genetic transformation, have resulted in genetic characterization and/ or development of plants with improved heat tolerance. In particular, the application of quantitative trait locus (QTL) mapping has contributed to a better understanding of the genetic relationship among tolerances to different stresses.

10.6   TRADITIONAL BREEDING STRATEGIES

Development of new vegetable cultivars tolerant to heat stress is a major challenge for vegetable breeders. The responses of crops to high temperature stress are dynamic depending upon the extremity and duration of the high temperature condition as well as the plant type and other environmental factors in the surroundings. However, the identification and confirmation of the traits that confer tolerance to heat stress still remain elusive (Rodríguez et al., 2005; Wahid et al., 2007). The aim of the scientists involved in research on high temperature stress is to discover the plant responses that lead to heat tolerance as well as management of the plants in high temperature stress environments.

Traditional breeding of heat resistant crops basically based on screening and selection. The common technique of selecting crops for heat stress resistance has been to grow breeding materials in a hot target production environment and detect individuals/lines with higher yield (Ehlers and Hall, 1998).

Several types of morphological traits help in heat tolerance in the conventional breeding approaches. Long root length has a good ability to uptake water and nutrients from the soil. Short life-span helps to minimize the temperature effect on plant. Hairiness provides partial shade to cell wall and cell membrane and repels sunrays. Small size of leaf resists evaporation due to reduction of stomata. Leaf orientation enhances the photosynthetic activity and produces tolerance against heat stress. Leaf glossiness and waxiness repel sunlight.

A proposed method has been detected in selection criteria during early steps of crop growth that can be linked to heat resistance during reproductive steps. In tomato, a potent positive correlation has been found between yield and fruit set under high temperature. Therefore, estimation of germ plasma to detect sources of heat resistance has regularly been performed by screening for fruit set under high temperature (Berry and Rafique-Uddin, 1988). Among various other characteristics that are influenced by high temperature, the nonreproductive trends involve efficiency of photosynthesis, assimilate translocation, mesophyll tolerance, and cellular membranes disorganization (Chen et al., 1982). Breeding to develop such characteristics under high temperatures can lead to improvement of varieties with heat resistance approaches.

Various other concerns when applying conventional breeding protocols to promote heat resistant crops may arise. Detection of genetic resources with heat resistance approaches is one such concern. In most of the vegetables, for instance tomatoes and legumes, the genetic differentiations within the cultivated species are restricted which necessitates the detection and use of wild accessions (Foolad, 2005). Sometimes heat resistance is linked to various unfavorable agronomical or horticultural traits. In tomato, for instance, two unfavorable traits generally observed in heat resistant lines are small fruit and limited foliar canopy (Scott et al., 1997). The small fruit production is mostly because of improper impacts of high temperature on the creation of auxins in the fruit and the poor canopy is for the sake of the highly reproductive nature of the heat resistant varieties (Scott et al., 1997).

Resistance breeding for heat stress is yet in its primitive step and less information on breeding for heat resistance in various vegetable crops is available. In spite of the complicatedness of heat resistance and hardships confronted during transfer of resistance, various heat resistant inbred lines and hybrid varieties with commercial acceptability have been developed and released in different vegetables, particularly in tomato (Scott et al., 1986, 1995).

AVRDC, Taiwan, has made significant contributions to the development of heat-tolerant tomato and Chinese cabbage lines (*Brassica rapa* subsp. *Pekinensis* and *chinenesis*) adapted to hot and humid climate. The key to achieving high yields with heat-tolerant cultivars is the broadening of their genetic base through crosses between heat-tolerant tropical lines and disease-resistant temperate or winter varieties (Opena and Lo, 1981). The heat-tolerant tomato lines were developed using heat-tolerant breeding lines and landraces from the Philippines (viz., VC11-3-1-8, VC 11-2-5, Divisoria-2) and the United States (viz., Tamu Chico III, PI289309) (Opena et al., 1992). However, lower yields in the heat-tolerant lines are still a concern.

More heat-tolerant varieties are required to meet the needs of a changing climate, and these must be able to match the yields of conventional, nonheat tolerant varieties under nonstress conditions. A wider range of genotypic variation must be explored to identify the additional sources of heat tolerance. For example AVRDC's tomato breeding line, CL5915, has demonstrated high levels of heat ranges from 15% to 30%, while there is complete absence of fruit set in heat-sensitive lines in mean field temperatures of 35 °C. Now, new breeding lines have been developed from CL5915 and other sources that exhibit increased heat tolerance. A CL5915 line is considered best combiners for percentage fruit set and total yield in hybrids developed for heat-tolerance (Metwally et al., 1996). Similarly, EC-520061 (*Solanumhabrochaites*) can set fruits under both high (40±2°C) and low (10±2°C) temperatures. Some heat tolerant breeding lines of tomato likeCLN-2413R, CLN 2116B and COML CR-7 have also been reported by Akhtar et al. (2012).

The Division of Vegetable Science, Indian Agricultural Research Institute, New Delhi has developed some varieties of vegetables to mitigate the harmful effect of heat. Tomato varieties Pusa Sadabahar and one hybrid Pusa Hybrid-1 have been developed which are tolerant to high temperatures. Radish variety, Pusa Chetaki has been developed having better root formation under high temperature regime, that is, April–August. Similarly carrot variety, Pusa Vrishti can form root at high temperature and high humidity, that is, March–August. Early cauliflower variety, Pusa Meghna has been developed which can form curd at high temperature.

One early cauliflower variety Sabour Agrim has been released at State level by Bihar Agricultural University that can produce compact white curd at high temperatures.

These varieties can be used directly for mitigating the effect of high temperatures as well as for future breeding programs. Since heat stress also causes higher infestation of disease-pest, there is a need to disease and pest resistance in heat tolerant lines through gene pyramiding using wild relatives for their wide adaptability.

10.7   MOLECULAR AND BIOTECHNOLOGICAL STRATEGIES

Recent genetic researches and endeavors to induce high-temperature resistance in vegetable crops with the use of conventional breeding approaches and transgenic attributes have vastly detected the polygenic nature of heat stress resistance. Various ingredients of resistance, handled by various sets of genes are vital for heat resistance at various steps of crop growth or in diverse tissues (Bohnert et al., 2006; Howarth, 2005). Therefore, the use of genetic stocks with diverse levels of heat resistance, cosegregation and correlation analyzes, molecular biology methods, molecular markers and quantitative trait loci (QTLs) are promising attributes to detect the genetic source of thermo-resistance (Maestri et al., 2002). Recently, biotechnology has assisted substantially to a proper understanding of the genetic source of heat resistance. Various genes, which are responsible for inducing the HSPs synthesis, have been detected and secluded in diverse crop species, involving maize and tomato (Liu et al., 2006; Momcilovic and Ristic, 2007; Sun et al., 2006). It has also been exhibited that tomato MT-sHSP has a molecular chaperone role in vitro (Liu and Shono, 1999). Recently it has been reported that MT-sHSP gene shows thermo-resistance in transformed tobacco with the tomato MT-sHSP gene (Sanmiya et al., 2004) at the crop level.

Recent widely studied molecular approaches have included omics techniques and the development of transgenic plants through manipulation of target genes (Kosová et al., 2011; Duque et al., 2013; Schöffl et al., 1999). Investigation of these underlying molecular processes may provide ways to develop stress tolerant varieties and to grow them under heat stress conditions.

10.8   CONCLUSIONS AND FUTURE PROSPECTS

In addition to genetic means of developing vegetable crops with improved heat tolerance, several new approaches have been used to induce heat tolerance in a range of plant species. These include preconditioning of plants to heat stress and exogenous applications of osmoprotectants or plant growth-regulating compounds on seeds or whole plants. Such approaches are promising and further research is worth undertaking. Also, while some notable progress has been reported as to the development of crop plants with improved heat tolerance via traditional breeding, the prospect for engineering plants with heat tolerance is also good considering accumulating molecular information on the mechanisms of tolerance and contributing factors. Further applications of genomics, proteomics and transcript omics approaches to a better understanding of the molecular basis of plant response to heat stress as well as plant heat tolerance are imperative. Little information is available for the use of markers in vegetable breeding especially for the development of complex characteristics like heat resistance. However, depending on the most recent discoveries and research progresses, it is clear that the future of routine application of markers in heat resistant vegetable breeding is prospective.

KEYWORDS

- **Breeding Strategies**
- **Climate Change**
- **Heat-Stress**
- **Osmolytes Accumulation**
- **Phenological Reactions**
- **Tolerance Mechanisms**

REFERENCES

Anon, S., Fernandez, J. A., Franco, J. A., Torrecillas, A., Alarcon, J. J., & Sanchez-Blanco, M. J. (2004). Effects of water stress and night temperature preconditioning on water relations and morphological and anatomical changes of *Lotus creticus* plants. *Sci. Hortic.* 101, 333–342.

Abdalla, A. A., & Verderk, K. (1968). Growth, flowering and fruit set of tomato at high tempera- ture. *The Neth J Agric Sci* 16, 71–76.

Adams, S. R., Cockshull, K. E., & Cave, C. R. J. (2001). Effect of temperature on the growth and development of tomato fruits. *Ann. Bot.* 88, 869–877.

Akhtar, S., Ansary, S. H., Dutta, A. K., Karak, C., & Hazra, P. (2012). Crucial reproductive char- acters as screening indices for tomato (*Lycopersiconesculentum* Mill.) under high temperature stress. *Journal of Crop and Weed.* 8(1), 114–117.

Ansary, S. H. (2006). Breeding tomato (*Lycopersiconesculentum*Mill.) tolerant to high tempera- ture stress. PhD Thesis, Bidhan Chandra Krishi Viswavidyalaya, West Bengal, India, 147 pp.

Ashraf, M. & Hafeez, M. (2004). Thermo tolerance of pearl millet and maize at early growth stages: growth and nutrient relations. *Biol. Plant.* 48, 81–86.

Ashraf, M., Saeed, M. M., & Qureshi, M. J. (1994). Tolerance to high temperature in cotton (*Gossypiumhirsutum* L.) at initial growth stages. *Environ. Exp. Bot.* 34, 275–283.

Barnabás, B., Jäger, K., & Fehér, A. (2008). The effect of drought and heat stress on reproductive processes in cereals. *Plant Cell Environ.* 31, 11–38.

Barua, D., Downs, C. A., & Heckathorn, S. A. (2003). Variation in chloroplast small heat-shock protein function is a major determinant of variation in thermo-tolerance of photosynthetic electron transport among ecotypes of *Chenopodium album. Funct Plant Biol*, 30, 1071–1079.

Bell, G. D., Halpert, M. S., Schnell, R. C., Higgins, R. W., Lowrimore, J., Kousky, V. E., Tinker, R., Thiaw, W., Chelliah. M., & Artusa, A. (2000). Climate Assessment for 1999. *Supplement June 2000 Bull Am Meteorol Soc* Vol 81.

Berry, S. Z., & Rafique-Uddin, M. (1988). Effect of high temperature on fruit set in tomato cultivars and selected germ plasm *Hort Science*23, 606–608.

Blum, A. (1988). Plant Breeding for Stress Environments. CRC Press Inc., Boca Raton, Florida, 223.

Blum, A., Klueva, N., & Nguyen, H. T. (2001). Wheat cellular thermo tolerance is related to yield under heat stress. *Euphytica* 117, 117–123.

Bohnert, H. J., Gong, Q., Li, P., & Ma, S. (2006). Unraveling abiotic stress tolerance mecha- nisms getting genomics going. *Curr. Opin. Plant Biol.* 9, 180–188.

Boyer, J. S. (1982). Plant productivity and environment. *Science.* 218, 443–448.

Bray, E. A. (2002). Abscisic acid regulation of gene expression during water deficit stress in the era of the *Arabidopsis* genome. *Plant Cell Environ.* 25, 153–161.

Bukhov, N. G., Wiese, C., Neimanis, S., & Heber, U. (1999). Heat sensitivity of chloroplasts and leaves leakage of protons from thylakoids and reversible activation of cyclic electron transport. *Photosyn. Res.* 59, 81–93.

Camejo, D., & Torres, W. (2001). High temperature effect on tomato (*Lycopersiconesculentum*) pigment and protein content and cellular viability. *Cultivos Tropicales* 22, 13–17.

Camejo, D., Rodr'ıguez, P., Morales, M. A., Dell'amico, J. M., Torrecillas, A., & Alarc'on, J. J. (2005). High temperature effects on photosynthetic activity of two tomato cultivars with dif- ferent heat susceptibility. *J. Plant Physiol.* 162, 281–289.

Capiati, D. A., País, S. M., & Téllez-Iñón, M. T. (2006). Wounding increases salt tolerance in tomato plants: evidence on the participation of calmodulin like activities in cross-tolerance signaling. *J Exp Bot* 57, 2391–2400.

Charles, W. B., & Harris, R. E. (1972). Tomato fruit set at high and low temperature. *Canadian Journal of Plant Science.* 52, 497.

Chen, T. H. H., Shen, Z. Y., & Lee, P. H. (1982). Adaptability of crop plants to high temperature stress. *Crop Sci.* 22, 719–725.

Chinnusamy, V., Zhu, J., Zhou, T., & Zhu, J. K. (2007). Small RNAs Big Role in Abiotic Stress Tolerance of Plants. In *Advances in Molecular Breeding toward Drought and Salt Tolerant Crops*; Jenks, M. A., Hasegawa, P. M., Jain, S. M. (Eds.) Springer: Dordrecht, the Netherland, 223–260.

Duque, A. S., de Almeida, A. M., da Silva, A. B., da Silva, J. M., Farinha, A. P., Santos, D., Fevereiro, P., & de Sousa Araújo, S. (2013). Abiotic stress responses in plants unraveling the complexity of genes and networks to survive. In *Abiotic Stress Plant Responses and Applications in Agriculture* Vahdati, K., Leslie, C. (Eds.) In Tech Rijeka, Croatia, 3–23.

Ehlers, J. D., & Hall, A. E. (1998). Heat tolerance of contrasting cowpea lines in short and long days. *Field Crops Res.* 55, 11–21.

El Ahmadi, A. B., & Stevens, M. A. (1979a). Responses of heat tolerant tomatoes to high temperature. *Journal of the American Society for Horticultural Science* 104, 686–691.

Erickson, A. N., & Markhart, A. H. (2002). Flower developmental stage and organ sensitivity of bell pepper (*Capsicum annuum* L.) to elevated temperature. *Plant Cell Environ* 25, 123–130.

FAO (2009). *Global Agriculture towards 2050. Issues Brief.* High level expert forum. Rome, 12–13 October. www.fao.org/wsfs/forum2050/wsfs_background_documents/hlef issues-briefs/en/ Accessed March 2010.

Feller, U., Crafts Brandner, S. J., & Salvucci, M. E. (1998). Moderately high temperatures inhibitribulose–1, 5–bisphosphate carboxylase/oxygenase (Rubisco) activase mediated activation of Rubisco. *Plant Physiology* 116, 539–546.

Feussner, K., Feussner, I., Leopold, I., & Waster nack, C. (1997). Isolation of a cDNA coding for an ubi quit in conjugating enzyme UBC1 of tomato-the first stress induced UBC of higher plants. *FEBS Lett,* 409, 211–215.

Firon, N., Shaked, R., Peet, M. M., Phari, D. M., Zamsk, E., Rosenfeld, K., Althan, L., & Pressman, N. E. (2006). Pollen grains of heat tolerant tomato cultivars retain higher carbohydrate concentration under heat stress conditions. *Scient. Hort.,* 109, 212–217.

Fitter, A. H., & Hay, R. K. M. (2002). *Environmental Physiology of Plants*, 3rd ed. Academic Press London, UK.

Foolad, M. R. (2005). Breeding for abiotic stress tolerances in tomato. In:Ashraf,M., Harris, P. J. C. (Eds.), Abiotic Stresses: Plant Resistance Through Breeding and Molecular Approaches. The Haworth Press Inc., New York, USA. 613–684.

Giaveno, C., & Ferrero, J. (2003). Introduction of tropical maize genotypes to increase silage production in the central area of Santa Fe, Argentina. Crop Breed. *Appl. Biotechnol.* 3, 89–94.

Golam., F., Hossain., Z., Arash., P., Nezhadahmadi., & Rahman, M. (2012). Heat Tolerance in Tomato. *Life Science Journal*, 9(4), 1936–1950.

Guilioni, L., Wery, J., & Tardieu, F. (1997). Heat stress induced abortion of buds and flowers in pea: is sensitivity linked to organ age or to relations between reproductive organs. *Ann Bot* 80, 159–168.

Hall, A. E. (2001). Crop Responses to Environment. CRC Press LLC, Boca Raton, Florida.

Hall, A. E. (2011). Accessed on 3rd June. The mitigation of heat stress. Available online: http://www.planstress.com.

Hare, P. D., Cress, W. A., & Staden, J. V. (1998). Dissecting the roles of osmolyte accumulation during stress. *Plant Cell Environ.* 21, 535–553.

Hasanuzzaman, M., Gill, S. S., & Fujita, M. (2013a). Physiological role of nitric oxide in plants grown under adverse environmental conditions. In *Plant Acclimation to Environmental Stress*; Tuteja, N., Gill, S. S., Eds. ; Springer: New York, NY, USA, 269–322.

Hasanuzzaman, M., Hossain, M. A., & Fujita, M. (2010a). Selenium in higher plants Physiological role, antioxidant metabolism and abiotic stress tolerance. *J. Plant Sci.,* 5, 354–375.

Hasanuzzaman, M., Hossain, M. A. & Fujita, M. (2010b). Physiological and biochemical mechanisms of nitric oxide induced abiotic stress tolerance in plants. *Am. J. Plant Physiol.,* 5, 295–324.

Hasanuzzaman, M., Hossain, M. A., & Fujita, M. (2012a). Exogenous selenium pretreatment protects rapeseed seedlings from cadmium induced oxidative stress by up regulating the antioxidant defense and methylglyoxal detoxification systems. *Biol. Trace Elem. Res.,* 149, 248–261.

Hasanuzzaman, M., Hossain, M. A., da Silva, J. A. T., & Fujita, M. (2012b). Plant Responses and Tolerance to Abiotic Oxidative Stress: Antioxidant Defenses is a Key Factor. In *Crop Stress and ItsManagement: Perspectives and Strategies*; Bandi, V., Shanker, A. K., Shanker, C., Mandapaka, M. (Eds.) Springer: Berlin, Germany, 261–316.

Hasanuzzaman, M., Nahar, K., & Fujita, M. (2013a). Extreme Temperatures, Oxidative Stress and Antioxidant Defense in Plants. In *Abiotic Stress Plant Responses and Applications in Agriculture*; Vahdati, K., Leslie, C. (Eds.). In Tech Rijeka, Croatia, 169–205.

Hasanuzzaman, M., Nahar, K., & Fujita, M. (2013b). Plant response to salt stress and role of exogenous protestants to mitigate salt-induced damages. In *Ecophysiology and Responses of Plants under Salt Stress* Ahmad, P., Azooz, M. M., Prasad, M. N. V. (Eds.) Springer: New York, NY, USA, 25–87.

Hasanuzzaman, M., Nahar, K., Alam, M. M., & Fujita, M. (2012c). Exogenous nitric oxide alleviates high temperature induced oxidative stress in wheat (*Triticum aestivum*) seedlings by modulating the antioxidant defense and glyoxalase system. *Aust. J. Crop Sci.,* 6, 1314–1323.

Hazra, P. & Som, M. G. (2006). Environmental influences on growth, development and yield of vegetable crops. *In*: Vegetable Science. Kalyani Publishers. 37–108.

Hazra, P., Ansary, S. H., Sikder, D., & Peter, K. V. (2007). Breeding tomato (*LycopersiconEsculentum*Mill) resistant to high temperature stress. *Int. J. Plant Breed* 1, 1.

Heckathorn, S. A., Downs, C. A., Sharkey, T. D., & Coleman, J. S. (1998). The small, methionine-rich chloroplast heat-shock protein protects photo system II electron transport during heat stress. *Plant Physiology* 116, 439–444.

Hopf, N., Plesofskv Vig, N., & Brambl, R. (1992). The heat response of pollen and other tissues of maize. *Plant Mol Biol,* 9, 623–630.

Howarth, C. J. (2005). Genetic improvements of tolerance to high temperature. In Ashraf, M., Harris, P. J. C. (Eds.). Abiotic stresses: Plant Resistance Thriugh Breeding and Molecular Approaches. Howarth Press Inc., New York.

Iba, K. (2002). Acclimative response to temperature stress in higher plants: approaches of gene engineering for temperature tolerance. *Annu Rev Plant Biol,* 53, 225–245.

Inaba, M., Hamauzu, Y., & Chachin, K. (1996). Influence of temperature stress on color development, respiration rate, and physiological injury in harvested tomato. *Bulletin of the University of Osaka, Prefecture Series B, Agriculture and Life Sciences* 48, 1–11.

Intergovernmental Panel on Climate Change (IPCC). Climate change (2007). The physical science basis. In *Contribution of Working Group I to the Fourth Assessment Report of the Intergovernmental Panel on Climate Change* Cambridge University Press Cambridge, UK, (2007).

Ismail, A. M., & Hall, A. E. (1999). Reproductive stage heat tolerance, leaf membrane thermo-stability and plant morphology in cowpea. *Crop Sci.* 39, 1762–1768.

Iwahori, S. (1967). Auxin of tomato fruit at different stages of its development with special reference to high temperature injuries. *Plant Cell Physiology* 8, 15–22.

Janska, A., Marsik, P., Zelenkova, S., & Ovesna, J. (2012). Cold stress and acclimation: What is important for metabolic adjustment? *Plant Biol.,* 12, 395–405.

Johjima, T. (1995). Inheritance of heat tolerance of fruit coloring in tomato. *Acta Horticulturae* 412, 64–70.

Karim, M. A., Fracheboud, Y., & Stamp, P. (1997). Heat tolerance of maize with reference of some physiological characteristics. *Ann. Bangladesh Agri.* 7, 27–33.

Kosova, K., Vítámvás, P., Prášil, I. T., & Renaut, J. (2011). Plant proteome changes under abiotic stress Contribution of proteomics studies to understanding plant stress response. *J. Proteom.* 74, 1301–1322.

Krasensky, J., & Jonak, C. (2012). Drought, salt, and temperature stress induced metabolic rear-rangements and regulatory networks. *J. Exp. Bot.,* doi:10. 1093/jxb/err460.

Larkindale, J., & Huang, B. (2005). Effects of abscisic acid, salicylic acid, ethylene and hydro-gen peroxide in thermo tolerance and recovery for creeping bentgrass. *Plant Growth Regul.* 47, 17–28.

Levy, A., Rabinowitch, H. D., & Kedar, M. (1978). Morphological and physiological characters. 218.

Liu, J., & Shono, M. (1999). Characterization of mitochondria located small heat shock protein from tomato (*Lycopersiconesculentum*). *Plant Cell Physiol.* 40, 1297–1304.

Liu, N., Ko, S., Yeh, K. C., & Charng, Y. (2006). Isolation and characterization of tomato Hsa32 encoding a novel heat shock protein. *Plant Sci.* 170, 976–985.

Lobell, D. B., & Asner, G. P. (2003). Climate and management contributions to recent trends in U.S. agricultural yields. *Science* 299, doi: 10.1126/science.1078475.

Lobell, D. B., & Field, C. B. (2007). Global scale climate Crop yield relationships and the im-pacts of recent warming. *Environ. Res. Lett.* 2, doi: 10.1088/1748-9326/2/1/014002.

Lohar, D. P., & Peat, W. E. (1998). Floral characteristics of heat tolerant and heat sensitive to-mato cvs. at high temperature. *Scientia Horticulturae* 73, 53–60.

Lurie, S., Handros, A., Fallik, E., & Shapira, R. (1996). Reversible inhibition of tomato fruit gene expression at high temperature Effects on tomato fruit ripening. *Plant Physiology* 110, 1207–1214.

Machado, S., & Paulsen, G. M. (2001). Combined effects of drought and high temperature on water relations of wheat and sorghum. *Plant Soil,* 233.

Maestri, E., Klueva, N., Perrotta, C., Gulli, M., Nguyen, H. T., & Marmiroli, N. (2002). Molecu-lar genetics of heat tolerance and heat shock proteins in cereals. *Plant Mol. Biol.* 48, 667–681.

Marcum, K. B. (1998). Cell membrane thermostability and whole plant heat tolerance of Ken-tucky bluegrass. *Crop Sci.* 38, 1214–1218.

Martineau, J. R., Specht, J. E., Williams, J. H., & Sullivan, C. Y. (1979). Temperature tolerance in soybean. I. Evaluation of technique for assessing cellular membrane thermostability. *Crop Sci.* 19, 75–78.

Mazorra, L. M., Nunez, M., Echerarria, E., Coll, F., & S´anchez-Blanco, M. J. (2002). Influence of brassinosteriods and antioxidant enzymes activity in tomato under different temperatures. *Plant Biol.* 45, 593–596.

Metwally, E., El Zawily, A., Hassan, N., & Zanata, O. (1996). Inheritance of fruit set and yields of tomato under high temperature conditions in Egypt. *First Egypt-Hung. Hort Conf.*, 1, 112–122.

Miller, P., Lanier, W., & Brandt, S. (2001). Using Growing Degree Days to Predict Plant Stages. Ag/Extension Communications Coordinator, Communications Services, Montana State University-Bozeman, Bozeman, MO.

Miroshnichenko, S., Tripp, J., Nieden, U., Neumann, D., Conrad, U., & Manteuffel, R. (2005). Immuno modulation of function of small heat shock proteins prevents their assembly into heat stress granules and results in cell death at sub lethal temperatures. *Plant J*, 41, 269–281.

Momcilovic, I., & Ristic, Z. (2007). Expression of chloroplast protein synthesis elongation factor, EF Tu, in two lines of maize with contrasting tolerance to heat stress during early stages of plant development. *J. Plant Physiol.* 164, 90–99.

Morales, D., Rodr´ıguez, P., Dell'amico, J., Nicol´as, E., Torrecillas, A., & S´anchez Blanco, M. J. (2003). High-temperature preconditioning and thermal shock imposition affects water relations, gas exchange and root hydraulic conductivity in tomato. *Biol. Plant.* 47, 203–208.

Moreno, A. A., & Orellana, A. (2011). The physiological role of the unfolded protein response in plants. *Biol. Res.,* 44, 75–80.

Munns, R., & Tester, M. (2008). Mechanisms of salinity tolerance. *Ann. Rev. Plant Biol.,* 59, 651–681.

Muthuvel, I., Thamburaj, S., Veeraragavathatham, D., & Kanthaswamy, V. (1999). Screening of tomato (*Lycopersiconesculentum* Mill.) genotypes for high temperature. *South Indian Horticulture* 47, 231–233.

Nakamoto, H., & Hiyama, T. (1999). Heat-shock proteins and temperature stress. In Pessarakli, M, (ed). Handbook of Plant and Crop Stress. Marcel Dekker, New York, 399–416.

Neumann, D. M., Emmermann, M., Thierfelder, J. M., Zur, N. U., Clericus, M., Braun, H. P., Nover, L., & Schmitz, U. K. (1993). HSP68–a DNA K-like heat-stress protein of plant mitochondria. *Planta*, 190, 32–43.

Nieto-Sotelo, J., Mart´ınez, L. M., Ponce, G., Cassab, G. I., Alag´on, A., Meeley, R. B., Ribaut, J. M., & Yang, R. (2002). Maize HSP101 plays important roles in both induced and basal thermo-tolerance and primary root growth. *Plant Cell*, 14, 1621–1633.

Opena, R. T., Chen, J. T., Kuo, C. G., & Chen, H. M. (1992). Genetic and physiological aspects of tropical adaptation in tomato. In *Adaptation of Food Crops to Temperature and Water Stress*. Eds Kuo, C. G. AVRDC, Shanhua, Taiwan 321–334.

Pareek, A., Singla, S. L., & Grover, A. (1998). Proteins alterations associated with salinity, desiccation, high and low temperature stresses and abscissic acid application in seedlings of Pusa 169, high yielding rice (*OryzasativaL.*) cultivar. *Curr. Sci.* 75, 1023–1035.

Peet, M. M., Sato, S., & Gardner, R. G. (1998). Comparing heat stress effects on male-fertile and male-sterile tomatoes. *Plant Cell Environ.* 21, 225–231.

Peet, M. M., Willits, D. H., & Gardner, R. (1997). Response of ovule development and post pollen production processes in male-sterile tomatoes to chronic, sub-acute high temperature stress. *J. Exp. Bot.,* 48(306) 101–111.

Peet, M. M., & Willits, D. H. (1998). The effect of night temperature on greenhouse grown tomato yields in warm climate. *Agric. Forest Meteorol.* 92, 191–202.

Prasinos, C., Krampis, K., Samakovli, D., & Hatzopoulos, P. (2005). Tight regulation of expression of two Arabidopsis cytosolic *Hsp90* genes during embryo development. *J Exp Bot*, 56, 633–644.

Queitsch, C., Hong, S. W., Vierling, E., & Lindquist, S. (2000). Hsp101 plays a crucial role in thermotolerance in *Arabidopsis*. *Plant Cell*, 12, 479–492.

Radin, J. W., Lu, Z., Percy, R. G., & Zeiger, E. (1994). Genetic variability for stomatal conductance in Pima cotton and its relation to improvements of heat adaptation. *Proc. Natl. Acad. Sci. USA*, 91, 7217–7221.

Rick, C. M., & Dempsey, W. H. (1969). Position of the stigma in relation to fruit setting of the tomato. *Botanical Gazette* 130, 180–186.

Rodríguez, M., Canales, E., & Borrás-Hidalgo, O. (2005). Molecular aspects of abiotic stress in plants. *Biotechnol. Appl.* 22, 1–10.

Sairam, R. K., & Tyagi, A. (2004). Physiology and molecular biology of salinity stress tolerance in plants. *Curr. Sci.* 86, 407–421.

Sakamoto, A., & Murata, N. (2002). The role of glycine betaine in the protection of plants from stress clues from transgenic plants. *Plant Cell Environ.* 25, 163–171.

Sanmiya, K., Suzuki, K., Egawa, Y., & Shono, M. (2004). Mitochondrial small Heatshock protein enhances thermo tolerance in tobacco plants. *FEBS Lett.* 557:265–268.

Sarieva, G. E., Kenzhebaeva, S. S., & Lichtenthaler, H. K. (2010). Adaptation potential of photosynthesis in wheat cultivars with a capability of leaf rolling under high temperature conditions. *Russ. J. PlantPhysiol.* 57, 28–36.

Sato, S., Kamiyama, M., Iwata, T., Makita, N., Furukawa, H., & Ikeda, H. (2006). Moderate increase of mean daily temperature adversely affects fruit set of *Lycopersiconesculentum* by disrupting specific physiological processes in male reproductive development. *Ann. Bot.* 97, 731–738.

Sato, S., Peet, M. M., & Gardner, R. G. (2001). Formation of partenocarpic fruit, undeveloped flowers and aborted flowers in tomato under moderately elevated temperatures. *Sci Horti*, 90, 243–254.

Savchenko, G. E., Klyuchareva, E. A., Abrabchik, L. M., & Serdyuchenko, E. V. (2002). Effect of periodic heat shock on the membrane system of etioplasts. Russ. *J. Plant Physiol.* 49, 349–359.

Sch"offl, F., Prandl, R., & Reindl, A. (1999). Molecular responses to heat stress. In: Shinozaki, K., Yamaguchi-Shinozaki, K. (Eds.), Molecular Responses to Cold, Drought, Heat and Salt Stress in Higher Plants. R. G. Landes Co., Austin, Texas, 81–98.

Scott, J. W., Bryan, H. H., & Ramos, L. J. (1997). High temperature fruit setting ability of large-fruited, joint less pedicel tomato hybrids with various combinations of heat-tolerance. *Proc. Fla. State Hortic. Soc.* 110, 281–284.

Scott, J. W., Olson, S. M., Howe, T. K., Stoffella, P. J., Bartz, J. A., & Bryan, H. H. (1995). 'Equinox' heat-tolerant hybrid tomato. *HortScience*30, 647–648.

Scott, J. W., Volin, R. B., Bryan, H. H., & Olson, S. M. (1986). Use of hybrids to develop heat tolerant tomato cultivars. *Proc. Fla. State Hortic. Soc.* 99, 311–315.

Semenov, M. A., & Halford, N. G. (2009). Identifying target traits and molecular mechanisms for wheat breeding under a changing climate. *J. Exp. Bot.*,60, 2791–2804.

Shinozaki, K., & Yamaguchi-Shinozaki, K. (2007). Gene networks involved in drought stress response and tolerance. *J. Exp. Bot.*,58, 221–227.

Siddique, K. H. M., Loss, S. P., Regan, K. L., & Jettner, R. L. (1999). Adaptation and seed yield of cool season grain legumes in Mediterranean environments of south-western Australia. *Aust. J. Agric. Res.* 50, 375–387.

Simoes-Araujo, J. L., Rumjanek, N. G., & Margis-Pinheiro, M. (2003). Small heat shock proteins genes are differentially expressed in distinct varieties of common bean. *Braz. J. Plant
Physiol.* 15, 33–41.

Singh, B. D. (2000). Breeding for resistance to abiotic stresses. I Drought resistance. In: Plant
breeing: Principles and practices. Kalyani publishers. 381–409.

Srivastava, S., Pathak, A. D., Gupta, P. S., Shrivastava, A. K., & Srivastava, A. K. (2012). Hydrogen peroxide scavenging enzymes impart tolerance to high temperature induced oxidative
stress in sugarcane. *J. Environ. Biol.,*33, 657–661.

Stevens, M. A., & Rudich, J. (1978). Genetic potential for overcoming physiological limitations
on adaptability, yield, and quality in tomato. *HortScience*13, 673–678.

Tsukaguchi, T., Kawamitsu, Y., Takeda, H., Suzuki, K., & Egawa, Y. (2003). Water status of
flower buds and leaves as affected by high temperature in heat tolerant and heat-sensitive
cultivars of snap bean (*Phaseolus vulgaris* L.). *Plant Prod. Sci.* 6, 4–27.

Valliyodan, B., & Nguyen, H. T. (2006). Understanding regulatory networks and engineering for
enhanced drought tolerance in plants. *Curr. Opin. Plant Biol.,* 9, 189–195.

Vierling, E. (1991). The role of heat shock proteins in plants. In: Annu Rev Plant Physiol Plant
MolBiol, 42, 579–620.

Vinocur, B., & Altman, A. (2005). Recent advances in engineering plant tolerance to abiotic
stress: achievements and limitations. *Curr. Opin. Biotechnol.* 16, 123–132.

Vollenweider, P., & Gunthardt-Goerg, M. S. (2005). Diagnosis of abiotic and biotic stress factors
using the visible symptoms in foliage. *Environ Pollut,* 137, 455–465.

Wahid, A., & Shabbir, A. (2005). Induction of heat stress tolerance in barley seedlings by pre-
sowing seed treatment with glycine betaine. *Plant Growth Reg.* 46, 133–141.

Wahid, A. & Ghazanfar, A. (2006). Possible involvement of some secondary metabolites in salt
tolerance of sugarcane. *J. Plant Physiol.* 163, 723–730.

Wahid, A., Gelani, S., Ashraf, M., & Foolad, M. R. (2007). Heat tolerance in plants: An overview. *Environ. Exp. Bot.* 61, 199–223.

Wang, W., Vinocur, B., Shoseyov, O., & Altman, A. (2004). Role of plant heat-shock proteins
and molecular chaperones in the abiotic stress response. *Trends Plant Sci.* 9, 244–252.

Waraich, E. A., Ahmad, R., Halim, A., & Aziz, T. (2012). Alleviation of temperature stress by
nutrient management in crop plants: A review. *J. Soil Sci. Plant Nutr.* 12, 221–244.

Warren, G. F. (1998). Spectacular increases in crop yields in the twentieth century. *Weed Technol.* 12, 752–760.

Weis, E., & Berry, J. A. (1988). Plants and high temperature stress. *Soc of Expt Biol,* 329–346.

Wise, R. R., Olson, A. J., Schrader, S. M., & Sharkey, T. D. (2004). Electron transport is the
functional limitation of photosynthesis in field grown Pima cotton plants at high temperature.
Plant Cell Environ. 27, 717–724.

Wollenweber, B., Porter, J. R., & Schellberg, J. (2003). Lack of interaction between extreme
high temperature events at vegetative and reproductive growth stages in wheat. *J. Agron. Crop
Sci.* 189, 142–150.

Yamada, M., Hidaka, T., & Fukamachi, H. (1996). Heat tolerance in leaves of tropical fruit crops
as measured by chlorophyll fluorescence. *Sci. Hortic.* 67, 39–48.

Yang, K. A., Lim, C. J., Hong, J. K., Park, C. Y., Cheong, Y. H., Chung, W. S., Lee, K. O., Lee, S.
Y., Cho, M. J., & Lim, C. O. (2006). Identification of cell wall genes modified by a permissive
high temperature in Chinese cabbage. *Plant Sci.,* 171, 175–182.

Young, L. W., Wilen, R. W., & Bonham-Smith, P. C. (2004). High temperature stress of *Brassica napus* during flowering reduces micro and mega gametophyte fertility, induces fruit abortion, and disrupts seed production. *J. Exp. Bot.* 55, 485–495.

Zhang, J. H., Huang, W. D., Liu, Y. P., & Pan, Q. H. (2005). Effects of temperature acclimation pretreatment on the ultra structure of mesophyll cells in young grape plants (*Vitisvinifera*L. cv. Jingxiu) under cross-temperature stresses. *J. Integr. Plant Biol.* 47, 959–970.

Zhang, Y., Mian, M. A. R., & Bouton, J. H. (2006). Recent molecular and genomic studies on stress tolerance of forage and turf grasses. *Crop Sci.* 46, 497–511.

Źróbek-Sokolnik, A. (2012). Temperature stress and responses of plants. In *Environmental Adaptations and Stress Tolerance of Plants in the Era of Climate Change*; Ahmad, P., Prasad, M. N. V., Eds. Springer New York, NY, USA, 113–134.

CHAPTER 11

CLIMATE CHANGE IMPACT ON BLACK PEPPER AND CARDAMOM

K. S. KRISHNAMURTHY[1], K. KANDIANNAN, B. CHEMPAKAM, S. J. ANKEGOWDA, and M. ANANDARAJ

Indian Institute of Spices Research, Kozhikode 673 012, Kerala, India
[1]E-mail: kskrishnamurthy@gmail.com

CONTENTS

ABSTRACT

India is the land of spices. Black pepper and small cardamom are popularly known as king and queen of spices, respectively. Both the crops are generally grown as rainfed crops and are grown in the temperature range of 10–35 °C. Both are climate sensitive crops and under the present scenario of climate change, both the crops may suffer. India has witnessed 0.6 °C increase in temperature so far and it is expected to increase further. Decrease in winter precipitation with a lower number of rainy days and more frequent drought and floods are also predicted. Agricultural production is likely to be affected because of increase in temperature, decrease in rainfall, increases in the frequency of drought, floods, heat and cold waves, etc. and associated risks such as emergence of newer pests and pathogens. Climate change in terms of increased temperature may reduce the yield of black pepper in plains but increase in minimum temperature in higher elevations may enhance pepper production. Cardamom production is likely to increase with temperature rise. But decrease in rainfall, rainy days and enhanced frequency of drought and heat waves are likely to bring down the production of these crops. So, the challenge before the scientists is to look for effective climate resilient agriculture strategies to mitigate the ill effects of climate change.

11.1 INTRODUCTION

India is the land of spices. India grows more than 50 spices, though all 50 are not cultivated extensively. This is because of the gifted climate of the country, varying from tropical to subtropical to temperate in different parts of the country, which allows each state to grow one or more suitable spices. Table 11.1 shows area and production of spices in India and percent share of each state to total spices production in India. Area under spices in India is 2.6 million ha with a production of 4.1 million tons. Spices share 13% of the area and 2.0% of production of the total horticultural crops of India. Spices exports have registered substantial growth during the last five years, registering an annual average growth rate of 21% in value and 8% in volume. During the year 2010–2011, spices export from India has registered an all time high both in terms of quantity (525,750 tons) and value (Rs.6840.71 crores). Though grown under irrigated conditions in low rainfall areas, most of the spices are grown under rainfed conditions in general. Yield (productivity) of most of the spices is low in India compared to other growing countries, which is mainly attributed to the prevailing climatic conditions. In India, black pepper and cardamom growing regions of southern states receive on an average 1500–4000 mm rainfall of which around 80% is distributed between June and October, and very less during November to May. This uneven distribution affects growth, flowering and also nutrient absorption from soil. Low soil moisture coupled with high radiation load and high temperature leads to wilting of plants, thus affecting productivity. Basin irrigation of pepper vines at 40–50 L/vine at fortnightly intervals during this period can enhance productivity substantially (Ankegowda et al., 2011). Now the climate change is becoming a reality with shift in monsoon pattern, frequent drought, flood, heat and cold waves, sun stroke, etc. These climatic variables affect the productivity of crops to a great extent.

11.2 BLACK PEPPER (*PIPER NIGRUM* L.)

Black pepper, often referred to as black gold is one of the important foreign exchange earners among spices. Presently, it is cultivated in more than 25 countries. India, Brazil, Indonesia, Malaysia, Thailand, Sri Lanka, Vietnam, People Republic of China, Madagascar and Mexico are the important producers. In India, it is generally grown in southern states viz., Kerala, Karnataka, Tamilnadu, Maharashtra, Goa and is slowly spreading to nontraditional areas such as East and West Godavari districts of Andhra Pradesh, Orissa, West Bengal, Andaman and Nicobar islands and north eastern states. The optimum temperature for crop growth is in the range of 23–32 °C, though it tolerates a temperature range of 10–40 °C. It can be grown from sea level to 1500 m above MSL (Radhakrishnan et al., 2002). Tropical temperature and high relative humidity with little variation in day length throughout the year is relished by the crop. The crop requires a well-distributed rainfall of 2000–3000 mm for better productivity. It is susceptible to excessive heat and dryness (Sivaraman et al., 1999). Hao et al. (2012) reported that the minimum temperature of the coldest month, the mean monthly temperature range, and the precipitation of the wettest month were identified as highly effective factors in the distribution of black pepper and could possibly account for the crop's distribution pattern. Such climatic requirements inhibited this species from dispersing and gaining a larger geographical range. The rise in temperature would affect the native genotypes or wild types of black pepper that are endemic to Western Ghats region (Utpala Parthasarathy et al., 2008). This indicates the climate sensitivity of the crop.

TABLE 11.1   Area and Production of Spices in India (2007–2008)

STATE/UT'S	Spices		% Share of spices to Horticultural Crops of the State		% Share of spices to total spices in India	
	Area	Production	Area	Production	Area	Produc-tion
ANDAMAN & NICOBAR	1.6	3.1	4.7	2.4	0.06	0.08
ANDHRA PRADESH	317.8	1235.2	17.9	6.9	12.21	30.11
ARUNACHAL PRADESH	8.2	47.5	8.8	17.9	0.31	1.16
ASSAM	27.2	18.5	4.7	0.3	1.04	0.45
BIHAR	11.1	12.3	1.0	0.1	0.43	0.30
CHHATTISGARH	11.9	7.8	2.8	0.2	0.46	0.19
GOA	0.7	0.2	0.7	0.1	0.03	0.00
GUJARAT	299.8	356.8	28.6	2.6	11.52	8.70
HARYANA	4.9	30.9	1.5	0.9	0.19	0.75
HIMACHAL PRADESH	6.3	23.4	2.2	1.2	0.24	0.57
JAMMU & KASHMIR	3.6	0.6	1.0	0.0	0.14	0.01
KARNATAKA	235.2	344.9	14.5	3.0	9.04	8.41
KERALA	312.3	159.3	16.6	1.5	12.00	3.88
MADHYA PRADESH	194.6	249.8	41.2	5.5	7.48	6.09
MAHARASHTRA	114.3	100.2	5.2	0.6	4.39	2.44
MANIPUR	8.7	7.7	14.5	1.9	0.33	0.19
MEGHALAYA	18.4	80.9	18.1	11.8	0.71	1.97
MIZORAM	9.0	38.3	26.0	12.7	0.35	0.93
NAGALAND	4.5	26.2	16.2	18.2	0.17	0.64

TABLE 11.1   *(Continued)*

STATE/UT'S	Spices		% Share of spices to Horticultural Crops of the State		% Share of spices to total spices in India	
	Area	**Production**	**Area**	**Production**	**Area**	**Production**
ORISSA	147.0	199.2	11.7	2.0	5.65	4.86
PONDICHERRY	0.1	0.5	0.8	0.4	0.00	0.01
PUNJAB	5.2	23.7	2.2	0.6	0.20	0.58
RAJASTHAN	556.4	520.6	60.4	28.3	21.37	12.69
SIKKIM	34.0	42.4	53.7	27.8	1.31	1.03
TAMILNADU	126.8	279.2	10.4	1.4	4.87	6.81
TRIPURA	4.5	9.4	5.7	1.0	0.17	0.23
UTTAR PRADESH	56.8	166.9	3.8	0.7	2.18	4.07
UTTRANCHAL	3.3	2.9	1.2	0.2	0.13	0.07
WEST BENGAL	79.0	114.4	4.8	0.4	3.03	2.79
TOTAL (ALL INDIA)	2603.2	4102.8	13.0	2.0	100.00	100.00

Area (**'000** ha); Production ('000 tons); Source: National Horticultural Board, Gurgaon.

11.2.1   CLIMATE CHANGE AND PEPPER PRODUCTION W.R.T. RAINFALL

Black pepper is mainly grown as rainfed crop in India. Hence, total rainfall and its distribution play an important role in pepper cultivation and its productivity. In pepper growing areas of Indonesia and Malaysia (Sarawak) the average annual rainfall is 2300 mm (Wahid and Sitepu, 1987) and 3950 mm, respectively (De Waard, 1969). In India, pepper-growing areas receive 1500 to more than 4000 mm rainfall. Annual rainfall of 2000 mm with uniform distribution is ideal for pepper. Premonsoon and early monsoon seems to be very important for pepper from the point of new flush, flower initiation as well as yield. It is reported that rainfall during premonsoon period (March–April) is positively correlated with pepper productivity and that of December rainfall is negatively correlated (Krishnamurthy et al., 2011). Rainfall of 70 mm received in 20 days during May–June may be sufficient for triggering off flushing and flowering process in the plant, but once the process is set off there should be continuous shower until fruit ripening. Any dry spell even for a few days within this critical period of 16 weeks (flowering to fruit ripening) will result in low yield. Rainfall after stress induced profuse flowering (Pillay et al., 1988; Ridley, 1912). Heavy rains during flowering reduce the rate of pollination and continuous heavy rainfall promotes vegetative development and limits flowering (Pillay et al., 1988). Growth of fruit bearing lateral shoots (plagiotropes) and photosynthetic rate were maximum during peak monsoon (June–July) in India (Mathai, 1983). A relative humidity of 60–95% is conducive for optimum growth at various stages of growth.

Rainfall beyond normal during initial period of annual cycle (i.e., 5–11 March to 25 June–1 July) was harmful or would reduce the yield in major black pepper growing regions of India (Kandiannan et al., 2011a). Rainfall of 244.5 mm received in 26 days during March–April and 144.1 mm in 14 days during May resulted in very low yield, whereas, no rainfall in January–February and 40 mm in March and good rainfall from third week of April–August resulted in good yield (Kannan et al., 1988). Increased rainfall during December and January tend to decrease productivity while rains during premonsoon season (March–May) increase the productivity. A study on 140 years of climatic data of Kerala indicated cyclical rainfall pattern with a declining trend of annual as well assouthwest monsoon in the last 60 years and an increasing trend in post monsoon rains (Rao et al., 2009). Climatic data of two decades (1984–2004) revealed a declining trend in rainfall and rainy days in major black pepper growing areas of the country (Krishnamurthy et al., 2011). The impact of climate change in the form of climate variability like floods and droughts adversely affected food and plantation crops to a large extent (Rao et al., 2009). In Idukki (a predominant black pepper growing region), the change in rainfall pattern during 1999–2000 crop season affected the flowering and yield (John et al., 1999). Suparman (1998) reported similar observation in Bangka, Indonesia. Increasing trend in rainfall during summer months was observed in black pepper growing regions of India (Parthasarathy et al., 2010 and Kandiannan et al., 2011b) that could affect the flowering pattern. Temporal and spatial variation in rainfall is also noticed (Kandiannan et al., 2008). Comparison of the rainfall pattern and pepper yields during two extremely adverse years (1980–1981 and 1986–1987) with that of a favorable year (1981–1982) revealed that during both adverse years, there was a distinct break in the rainfall during critical period following flower initiation. Though the break was experienced during different crop growth stages, the pepper yields were only 24.3% of the normal years yield. But in the favorable year, rainfall was steady without any break and the pepper yield was also high (Pillay et al., 1988). This gives an indication that climate change in terms of drought period during the normal monsoon season affects black pepper productivity. All these studies point to the fact that there is a climate change in recent years, which is affecting flowering pattern and yield of black pepper.

11.2.2   RAINFALL DEFICIT AND PEPPER PRODUCTIVITY

Government of India had declared 1987 and 2002 as drought years as the rainfall during those years was less than 80% of long-term average throughout India. The rainfall deficit during 1987 was around 20% for Wynad and Coimbatore, 15% for Kannur and only around 7% for Idukki while the deficit during 2002 was more than 30% for black pepper growing regions except Coimbatore (28.5%). The black pepper productivity also was below normal (mean of 1984–04) for all the places (except Coimbatore where the black pepper growing area is very less) indicating the negative influence of rainfall deficit on black pepper productivity. Interestingly, though the rainfall deficit was more pronounced during 2002, productivity reduction was more during 1987 compared to 2002 for Wynad and Idukki. When we analyzed the reason, we found that good premonsoon (January to May) and early monsoon (June–July) rains and well distributed rainfall during 2002 compared to 1987 contributed for better productivity.

11.2.3   CLIMATE CHANGE AND PEPPER PRODUCTION W.R.T. TEMPERATURE

Though black pepper is a crop of humid tropics and tolerates temperatures between 10 and 40°C, the ideal temperature for its growth is 23–32 °C with an average of 28 °C. Optimum soil temperature for root growth is 26–28 °C (De Waard 1969; Wahid and Sitepu 1987). A study on 140 years of climatic data of Kerala indicated increase in day maximum by 0.64 °C and night minimum temperature by 0.23 °C (Rao et al., 2009). Correlation between pepper productivity and climatic parameters at Panniyur based on climatic data for 11 years showed that maximum temperature and number of sunshine hours in the first fortnight of March had positive impact on productivity, while mean relative humidity of the July first fortnight, number of sunshine hours received during the February first fortnight and April second fortnight, and mean maximum temperature during the June second fortnight had negative influence on productivity (Pradeepkumar et al., 1999). Most of the pepper growing areas of the country shows an increasing trend in temperature and decreasing trend in productivity (Table 11.2).

TABLE 11.2   Trend Analysis of Climatic Variables and Black Pepper Productivity

Place	Rainfall	T_{max}	T_{min}	Productivity
Kerala				
Ambalavayal (Wynad) 1979–2004	Decreasing	Increasing	Increasing	No change
Pampadumpara (Idukki) 1986–2004	Decreasing	-	Increasing	Increasing
Panniyur (Cannanore) 1974–2004	Decreasing	Increasing	Increasing	Decreasing
Trichur 1980–2004	Decreasing	Decreasing	No trend	Decreasing
Tamil Nadu				
Valparai (Coimbatore) 1976–2004	No trend	No trend	Decreasing	No change
Nilgiris 1980–1992	Increasing	Increasing	Decreasing	No change

Pepper productivity is generally higher in higher elevations such as Wynad and Idukki. Relatively cool climate of these regions may have influence on productivity. Table 11.3 reveals that the temperature (both T_{max} and T_{min}) Idukki and Wynad is about 6–7 degree lesser than that of plains viz. Cannanore or Trichur. The total rainfall of these two places is much lesser when compared to that of Cannanore or Trichur (Table 11.3). This justifies the point that prevailing temperature of high elevations may have a role in higher productivity, as rainfall by itself does not seem to have much role as it is less compared to plains. Also, T_{min} of high elevations had positive and both T_{max} and T_{min} in plains had negative correlation with pepper productivity indicating that climate change in terms of increase in temperature may negatively influence black pepper productivity especially in plains whereas increase in T_{min} may have positive influence in high elevations.

TABLE 11.3 Mean Temperature, Rainfall and Yield for Two Decades (1985–2004) in a Few Black Pepper Growing Regions

Place	Temperature (°C)			Rainfall (mm)	Mean yield (kg/ha)
	Mean Tmax	**Mean Tmin**	**Difference**		
Wynad	27.3	17.6	9.7	1931	402
Idukki	27.5	15.6	11.9	1902	327
Cannanore	33.1	22.6	10.5	3348	241
Trichur	32.1	23.4	8.7	2752	239

Analysis of data of a decade on the extent of relationship between climatic variables and productivity in six black pepper plantations in Coorg indicated that one degree increase in maximum temperature can result in up to 20.9 units (kgs) reduction in yield and one degree increase in minimum temperature can result in up to 20.1 units increase in yield in different plantations. One mm of rainfall can result in up to 0.05 units increase in yield in different plantations. Tmin and rainfall showed positive influence on yield in 5 and rainy days showed positive influence in 4 out of 6 plantations (Krishnamurthy et al., 2011).

11.2.4 ELEVATED TEMPERATURE AND GROWTH OF BLACK PEPPER

Plant height, leaf area and Photosynthetic rate of black pepper varieties grown under elevated temperatures (2.7 degrees higher than ambient) was less compared to those grown under ambient temperature during the initial growth period. There was significant difference among varieties for leaf area and diurnal variation in photosynthesis. Varieties such as Panniyur 5 and Panniyur 3 showed better adaptability at higher temperature compared to other varieties. This shows that climate change in terms of increased temperature may affect growth and yield of black pepper and that it may be possible to nullify it to a certain extent by growing varieties which are least sensitive to temperature fluctuations.

11.2.5   CLIMATE CHANGE AND BLACK PEPPER DISEASES

Phytophthora foot rot is the major disease of black pepper, which affects productivity. Nowadays, viral diseases are also becoming serious. Nematodes damage the root system of the plant. Phytophthora in association with nematodes destroy the plants totally. Pollu beetle is the main insect pest that feeds on pepper berries, thus bringing down the yield levels. A three years study on the relationship between weather parameters and Phytophthora foot rot at Horticultural Research Station, Pechiparai revealed that foot rot was positively correlated with rainfall, number of rainy days and relative humidity, and negatively correlated with maximum and minimum temperatures (Jayasekhar and Muthusamy, 1999). In Karnataka also, studies revealed that high RH and well-distributed rainfall supported the disease development. Significant negative regression between disease severity and rainy days, RH and significant negative correlation between maximum temperature and disease development was recorded (Shamarao and Siddaramaiah, 2002). Similar results were also reported by Arasumallaiah et al. (2008) and Ramachandran et al. (1988). A daily rainfall of 15.8–23 mm, RH of 81–99%, temperature range of 22.7 °C–29.6 °C and sunshine hours of 2.8–3.5/day during the peak monsoon period of the year are conducive for the maximum development of the disease (Anandaraj and Sarma, 1994). This implies that climate change (decrease in rainfall and increase in temperature) is likely to reduce the severity of the disease. Climate change in cardamom hills of Kerala has resulted in increased incidence of anthracnose disease of black pepper (Murugan et al., 2012). Highest population densities of *M. incognita* on black pepper root were observed during the first half of the dry season (Thuy et al., 2012) indicating that frequent drought may increase the population density of *M. incognita* on black pepper roots. Black pepper berry damage by pollu beetle was highest in the plains, and it was very low at higher elevations (300–900 m a.s.l.) and absent at >900 m a.s.l. (Kumar and Nair, 1987). All these studies indicate that climate change in terms of increase in temperature, decrease in rainfall, frequent drought and floods, etc., may bring about changes in disease incidence, population dynamics of pests, minor pests becoming major pests,etc.which ultimately affects productivity.

11.2.6 *CLIMATE CHANGE AND QUALITY OF BLACK PEPPER*

Quality in black pepper refers to physical quality constituents viz., grade of berries, bulk density, test weight, fiber, starch and protein content of the berries and intrinsic quality constituents viz., oil, oleoresin, piperine and oil constituents. Dry black pepper berries collected from low (10–200 m above MSL) and high elevation (400–1200 m above MSL) were analyzed for both physical and intrinsic quality to test if the elevation groups differ in quality constituents. But none of the physical quality constituents showed variation between high and low elevation. Intrinsic quality parameters viz., piperine, oleoresin and oil also did not show variation between elevation groups. But oil components limonene and sabinene + myrcene showed positive correlation while β-caryophyllene showed negative correlation with elevation. This indicates that climate change in terms elevation (differences in temperature and associated climatic parameters) may alter the proportion of quality constituents, thus affecting quality.

Higher β-caryophyllene and lower limonene and sabinene + myrcene were observed under low elevation (warmer climate) in black pepper. Telci et al. (2010) also observed higher transβ-caryophyllene and germacrene D, in warmer climate and higher d-limonene and β-phellandrene in temperate climate in spearmint. Llusià et al. (2006) reported that maximum concentration of foliar volatile terpenes in four Mediterranean woody species were found in the coldest periods and minimum concentrations in the summer. In general, concentrations increased when soil moisture increased and decreased when air temperature increased. These studies reveal that volatile oil constituents are vulnerable to change in climate, could be light, temperature, moisture stress,etc.

11.3 SMALL CARDAMOM (*ELETTERIA CARDAMOMUM MATON*)

Small cardamom, commonly known as queen of spices has a prominent place in global market. It is grown for its spicy seeds. This is one of the oldest spices and is a native to the evergreen forests of Western Ghats of India. The crop is grown in a well distributed rainfall of 1500–2500 mm, temperature range of 15–25 °C and an altitude of 600–1200 m above MSL. Cardamom is a shade loving plant. Based mainly on inflorescence type, cardamom is classified in to Malabar, Mysore and Vazhukka types.

11.3.1 CLIMATE CHANGE AND CARDAMOM

Temperature (1978 to 1997) and precipitation (1957–1996) recorded in Cardamom Hills of Kerala and their effect on cardamom production revealed that the minimum temperature exhibited drastic variation over the years. The difference between the warmest and coolest month had narrowed considerably and the days had become warmer markedly. The total annual average precipitation received was more or less equal (except during 1967–1976). The total number of rainy days had increased. The rainfall parameters had positive correlation with production of cardamom with significant relationship for number of rainy days (Murugan et al., 2000). An experiment conducted during 1991 to 2001 at the Regional Agricultural Research Station, Ambalavayal, Kerala revealed that maximum temperature had strong positive correlation with capsule yield in the first year of planting (still in vegetative stage) while during flowering stage, in the second year after planting, maximum temperature negatively influenced the final capsule yield (Sunil et al., 2010). The mean air temperature increased significantly during the last 30 years; the greatest increase and the largest significant upward trend was observed in the daily temperature. The highest increase in minimum temperature was registered for June (0.37 °C/18 years) at the Myladumpara station. December and January showed greater warming across the stations. Rainfall during the main monsoon months (June–September) showed a downward trend. Relative humidity showed an increasing trend in the cardamom growing tracts.

Rao et al. (2008) reported that South-west monsoon and annual rainfall showed declining trends from 1951 onwards at rates of 5.2 and 5.6 mm/year, respectively in the humid tropics. However, the occurrence of floods and droughts, as evident in 2007 (floods due to a 41% excess in monsoon rainfall) and the summer of 2004 (drought due to no significant rainfall from November 2003 to April 2004), is likely to increase and crop losses are expected. Climate change in addition to deforestation will affect these thermo sensitive crops (cardamom, tea, black pepper, etc.) as these are grown under the influence of typical forest and agricultural ecosystems. Deforestation, shift in cropping systems, decline in wetlands, and depletion of surface and groundwater resources may aggravate the adverse effects of floods and drought on crops.

Many studies have shown that the primary climatic elements (temperature and rainfall) have a profound influence on the phenology of crop plants. In the case of cardamom, the seasonal flowering habit has changed. Up to early1990s, June to December was the main flowering season and now cardamom has been flowering throughout the year. Accordingly, the number of harvests also has been increased from five to nine per year. This has resulted in manifold increase in productivity of cardamom. More harvests in recent years (since late 1990s) indicate the impact of climate change on cardamom phenology (Murugan et al., 2012). Cardamom productivity increased in the cardamom hills irrespective of the variety during the study period (1987 to 2007) indicating that warming may have positive influence on cardamom productivity (Murugan et al., 2012). But apart from warming, crop management practices may also have influence on productivity.

The spatial and temporal distribution and proliferation of pests is determined, largely, by climate, because temperature, light, and water are the major factors controlling the growth and development of pests (Rosenzweig et al., 2001). In cardamom hills, since 2000, the number of pesticide sprays has been significantly increased, and at present, 15–18 rounds of pesticide sprays are given (as against 7–8 rounds until 1990). But there was no great increase in the frequency of cardamom damage by major insect pests like thrips and borers indicating the involvement of more number of insect pests and diseases in damaging the crops. The incidence of many minor pest insects and disease pathogens has increased in the recent years along with warming. Increased frequency of break period during monsoon seasons (wet and dry spells) as observed in Pampadumpara station might favor the development of dry rot during dry spell and wet rot during wet spell (Murugan et al., 2012). Such situations necessitated the use of pesticides to manage these rot diseases in cardamom, otherwise the crop loss would reach up to 70% (DaMatta et al., 2006).

11.4   CONCLUSIONS

It is predicted that the global temperature will rise between 1 °C and 3.5 °C and that the rainfall is likely to decrease from the present levels by 2100. Agricultural production in temperate countries is likely to go up while that of tropical countries is likely to be affected because of increase in temperature and associated risks such as emergence of newer pests and pathogens. Climate change in terms of increased temperature may sustain/decrease total black pepper production as increased temperature in plains may decrease productivity but the increased minimum temperature in high elevation may enhance productivity, thus nullifying the negative effect. Studies have shown that warming has a positive influence on cardamom production. But decreased premonsoon rainfall will have negative impacts on these crops and hence, rainwater harvesting and providing it during premonsoon season should get top priority, apart from other soil and water conservation measures. Studies have shown that climate change (especially temperature increase) is likely to alter the flavor profile of aromatic crops and it is very essential that under changing climate scenario, original flavor of spices and aromatic crops is retained. Increased frequency of droughts, floods, heat stroke, etc. as predicted would affect the productivity of these crops. Spatial and temporal variation in weather particularly rainfall and temperature are of great concern in augmenting the productivity of these rainfed crops. Climate change is evident and it is a great challenge for scientific community to find solutions to mitigate the ill-effect. Studies should be directed towards climate resilient agriculture to mitigate the ill effects of climate change.

KEYWORDS

- **Climate Change**
- ***Eletteria cardamomum***
- **Growth and development**
- **Piper nigrum L**
- **Quality**
- **Rainfall Deficit**

REFERENCES

Anandaraj, M. & Sarma, Y. R. (1994). Biological control of black pepper diseases. *Indian Co-coa, Arecanut Spices J.,* 18(1), 22–23.

Ankegowda, S. J., Venugopal, M. N., Krishnamurthy, K. S., & Anandaraj, M. (2011). Impact of basin irrigation on black pepper production in coffee based cropping system in Kodagu District, Karnataka. *Ind. J. Hortic.,* 68(1), 71–74.

Arasumallaiah, L., Krishnamurthy, Y. L. & Krishnappa, M. (2008). Role of epidemiology on the incidence, development and spread of *Phytophthora capsici* Beon. Causing foot rot disease of black pepper in malnad regions of Karnataka. *Environ. Ecol.,* 26(3A) 1427–1431.

DaMatta, F. M., Cochicho, J. D., & Ramalho, G. (2006). Impacts of drought and temperature stress on coffee physiology and production a review. *Braz J Plant Physiol.,* 18(1), 55–81.

De Waard, P. W. F. (1969). Foliar diagnosis, Nutrition and yield stability of black pepper (*Piper nigrum* L.) in Sarawak. Communication No. 58. Department of Agricultural Research, Koninklijk Instituut voor de Tropen, Amsterdam.

Jayasekhar, M. & Muthusamy, M. (1999). Influence of weather parameters on the incidence of foot rot of black pepper. *Madras Agric. J.,* 86(4), 344–346.

John Koshy, Shankar, M., & Sudhakaran, K. V. (1999). Seasonal climatic influence in pepper production Idukki district. *Spice India,* 12(12), 2–3.

Hao ChaoYun, Fan Rui, Ribeiro, M. C., Tan LeHe, Wu HuaSong, Yang JianFeng, Zheng Wei-Quan & Yu Huan. (2012). Modeling the potential geographic distribution of black pepper (*Piper nigrum* L.) in Asia using GIS tools. *J. Integrative Agric.,* 11(4), 593–599.

Kandiannan, K., Utpala parthasarthy, Krishnamurthy, K. S., Thankmani, C. K., Srinivasan, V. & Aipe, K. C. (2011). Modeling the association of weather and black pepper yield. *Ind. J. Hortic.,* 68(1), 96–102.

Kandiannan, K., Thankamani, C. K., Krishnamurthy, K. S. & Mathew, P. A. (2011). Monthly rainfall trend at high rainfall tract of northern agro climatic zone in Kerala. In **National Seminar on Recent Trends in Climate and Impact of Climate Change on South-West India.** 11October 2011. Department of Physics, St Joseph's College, Devagiri, Calicut.

Kandiannan, K., Thankamani, C. K., & Mathew, P. A. (2008). Analysis of rainfall of the high rainfall tract of northern agro climatic zone of Kerala. *J. Spices Aromatic Crops,* 17, 16−20.

Kannan, K., Devadas, V. S., & George Thomas, C. (1988). Effect of weather parameters on the productivity of coffee and pepper yield in Wynad 147–151. In Agro meteorology of Plantation Crops, ((Eds.) GSLVP Rao and RR Nair). Kerala Agricultural University, Thrissur.

Krishnamurthy, K. S., Kandiannan, K., Sibin, C., Chempakam, B., & Ankegowda, S. J. (2011). Trends in climate and productivity and relationship between climatic variables and productivity in black pepper (*Piper nigrum* L.). *Ind. J. Agric., Sci.,* 81(8), 729–33.

Kumar, T. P., & Nair, M. R. G. K. (1987). Effect of some planting conditions on infestation of black pepper by Longitarsus nigripennis Mots. *Indian Cocoa, Arecanut Spices J.,* 10(4), 83–84.

Llusia, J., Peñuelas, J., Alessio, G. A., & Estiarte, M. (2006). Seasonal contrasting changes of foliar concentrations of terpenes and other volatile organic compounds in four dominant species of a Mediterranean shrub land submitted to a field experimental drought and warming. *Physiol. Plantarum,* 127(4), 632–649.

Mathai, C. K. (1983). Growth and yield analysis in black pepper varieties (*Piper nigrum* L.) under different light conditions. Ph. D Thesis, University of Agricultural Sciences, Bangalore.

Murugan, M., Raj, N. M. & Joseph, C. R. (2000). Changes in climatic elements and their impact on production of cardamom (*Elettaria cardamomum* Maton) in the Cardamom Hills of Kerala, India. *J. Spices Aromatic Crops*, 9(2), 157–160.

Murugan, M., Shetty, P. K., Raju Ravi, Aavudai Anandhi & Rajkumar, A. J. (2012). Climate change and crop yields in the Indian Cardamom Hills, (1978–2007) CE. *Climatic Change* 110 (3), 737–753.

Parthasarathy, V. A., Kandiannan, K., Utpala Parthasarathy & Ankegowda, S. J. (2010). Climate change and spices production. In 19th Biennial Symposium on Plantation Crops (Placrosym XIX), 7–10, December (2010). Rubber Research Institute of India, Kottayam.

Pillay, V. S., Sasikumaran, S. & Ibrahim, K. K. (1988). Effect of rainfall pattern on the yield of black pepper. In Agrometeorology of Plantation Crops. pp. 152–159. Kerala Agricultural University, Trichur.

Pradeepkumar, T., Vasanthkumar, Aipe, K. C., Kumaran, K., Susamma, P. George, Manmohandas, T. P., & Anith, K. N. (1999). Studies on yielding behaviour of Black pepper CV Panniyur-I. *Indian J. Arecanut Spices Medicinal Plants*, 1(3), 88–90.

Pradeepkumar, T., Kumaran, K., Aipe, K. C. & Manmohandas, T. P. (1999). Influence of weather on the yield of pepper cv Panniyur-I (*Piper nigrum* L.). *J. Tropical Agric.*, 37, 56–59.

Radhakrishnan, V. V., Madhusoodanan, K. J., Kuruvilla, K. M. & Vadivel, V. (2002). Production technology for black pepper. *Indian J. Arecanut, Spices Medicinal Plants*, 4(2), 76–80.

Ramachandran, N., Sarma, Y. R. & Anandaraj, M. (1998). Effect of climatic factors on Phytophthora leaf infection in black pepper grown in Areca nut Black pepper mixed cropping system. *J. Plantation Crops*, 16(2), 110–118.

Ridley, H. N. (1912). Pepper. In Spices 239–312. Macmillan and Co. Ltd. London.

Rao, G. S. L. H. V. P., Mohan, H. S. R., Gopakumar, C. S. & Rishnakumar, K. N. (2008). Climate change and cropping systems over Kerala in the humid tropics. *J. Agrometeorol.*, 10, (Special Issue 2) 286–291.

Rao, G. S. L. H. V. P., Kesava Rao, A. V. R., Krishnakumar, K. N. & Gopakumar, C. S. (2009). Impact of climate change on food and plantation crops in the humid tropics of India. In ISPRS Archives XXXVIII-8/W3 Workshop Proceedings: Impact of Climate Change on Agriculture. Space Applications Centre (ISRO), Ahmedabad, India.

Rosenzweig, C., Iglisias, A., Yang, X. B., Epstein, P. R., & Chivian, C. (2001). Climate change and extreme weather events implications for food production, plant diseases, and pests. *Glo Change Hum Health*, 2(2), 90–104.

Shamarao, J., & Siddaramiah, A. L. (2002). Influence of weather forecasters on the epidemiology of foot rot of black pepper in Karnataka. *Ind. J. Agric. Res.*, 36(1), 49–52.

Sivaraman, K., Kandiannan, K., Peter, K. V. & Thankamani, C. K. (1999). Agronomy of black pepper (*Piper nigrum* L.)—A review. *Journal of Spices and Aromatic Crops*, 8, 1–18.

Sunil, K. M., Devadas, V. S., Sreelatha, A. K. & George, S. P. (2010). Effect of weather parameters on the yield of small cardamom (*Elettaria Cardamomum* Maton.). *Indian J. Arecanut, Spices Medicinal Plants*, 12(3), 6–8.

Suparman, U. (1998). The effect of El-Nino & La-Nina on the production of white pepper in Bangka, Indonesia. *International Pepper News Bulletin*, XXII (3&4) 44–45.

Telci, I., Demirtas, I., Bayram, E., Arabaci, O. & Kacar, O. (2010). Environmental variation on aroma components of pulegone/piperitone rich spearmint (*Mentha Spicata* L.). *Indu. l Crops Prod.* 32(3), 588–592.

Thuy, T. T. T., Yen, N. T., Tuyet, N. T. A., Te, L. L. & Waele, D. de. (2012). Population dynamics of *Meloidogyne incognita* on black pepper plants in two agro-ecological regions in Vietnam. *Archives Phytopathol. Plant Protection*, 45(13), 1527–1537.

Utpala Parthasarathy, Parthasarathy, V. A., & Jayarajan, K. (2008). A temperature sensitivity analysis on plantation crops A GIS approach. *J. Plantation Crops*, 36, 372–374.

Wahid, P., & Sitepu, D. (1987). Current status and future prospect of pepper development in Indonesia. Food and Agricultural Organization, Regional Office for Asia and Pacific, Bangkok, Thailand.

CHAPTER 12

CLIMATE CHANGE: THREAT TO FLORICULTURE

ANIL K. SINGH and ANJANA SISODIA

Department of Horticulture, Institute of Agricultural Sciences, Banaras Hindu University, Varanasi-221 005, India.

CONTENTS

ABSTRACT

Climate change is one of the most important global environmental challenges in the history of mankind. It is mainly caused by increasing concentration of Green House Gases (GHGs) in the atmosphere. In 1980s, scientific evidences linking GHGs emission due to human activities causing global climate change, started to concern everybody. Climate change has become increasingly recognized as one of the greatest challenges to humankind and all other life on Earth. Worldwide changes in seasonal patterns, weather events, temperature ranges and other related phenomena have all been reported and attributed to global climate change. Numerous experts in a wide range of scientific disciplines have warned that the negative impacts of climate change will become much more intense and frequent in the future particularly if environmentally destructive human activities continue unabated. Climate change is causing noticeable effects on the life cycles and distributions of the world's vegetation. Future effects of climate change are largely uncertain, but current evidence suggests that these phenomena are having an impact on various flowering plants and that there are some potential threats worthy of concern and discussion.

12.1 ADVERSE EFFECT OF CLIMATE CHANGE

Climate change produces warmer temperatures and increases CO_2 gases, rainfall and drought that enhance disease, pests and weeds. Experts predict that these events will cause pests and diseases to spread further, covering more areas that increasingly become suitable for them and to multiply faster in current areas. Variability in global temperatures and rainfall amount, intensity and frequency patterns will alter agricultural productivity and may necessitate a change or limit in the species and cultivars traditionally planted in specific regions. Increased temperatures also have an effect on the growing season of many plant species requiring a change in the agricultural calendar. Increased temperatures and reduced rainfall will increase the rate of evapo-transpiration and increase water demand. An increase in climate change variability hinders the ability to predict, forecast, manage and prepare for various agricultural activities such as the sowing of seeds, harvesting of produce as well as measures to respond to any extreme weather events.

- An increase in the frequency and intensity of extreme weather events will damage flowers in farms.
- Changes in climate may foster ideal conditions for agricultural pests and diseases including emergence of new traits as well as reemergence of those that may have been eradicated.
- Increased evaporation rates as well as increased erosion due to extreme winds or rainfall will lead to soil degradation and nutrient loss.
- Changes in temperature, rainfall and seasonality may have an impact on produce that is pollinated by insects due to changes in the distribution and species compliment.
- There is also a likelihood of change in the quality flower features such as color, fragrance, stem length and thickness and vase life. New cultivar that can adapt to the new growing conditions will have to be sought and that at significant cost.
- To maintain production, as an example, farms may have to look into better quality greenhouse plastic sheeting which can withstand higher temperatures and solar radiations, high wind velocity and rainstorms, whichever the scenario in a particular growing area will be, and with increased cost.
- Power transmission lines affected by storms and lightning will drive farmers to depend on expensive stand-by power generation to sustain production.
- Since fresh water supply for farms will be reduced due to poor rains or due to pressure on preservation of water bodies, farms will be expected to invest in water recycling and water efficiency measures to adapt to new water regime. In some instances, excessive rains will impact on water supply via destruction of water supply infrastructure.
- An invasion by introduced or migrated alien species of plants or animal pests.
- A reduction of crops' tolerance and resistance to pests and disease an increase in food toxins (mycotoxins) such as Aflatoxin and the appearance of new strains of toxin-producing fungi.
- The loss of some wild relatives of crops that could be used to introduce desired traits in classical and modern crop resistance breeding programs.
- A decrease in soil fertility and an increase in soil erosion that reduces the natural capacity of soils to control soil-borne pests and diseases.

- A reduction of beneficial organisms for pest and disease control.
- A reduction in the effectiveness of safe pesticides and herbicides negative effects on plant resistance.

12.2   WIDE SPREAD EFFECTS OF CLIMATE CHANGE

Some effects of climate change appear to be impacting plants worldwide. For instance, evidence has shown that climate change has been affecting vegetation patterns such as phenology (the timing of life cycle events in plants and animals, especially in relation to climate) and distribution. Some wild plants have begun to flower earlier and shift their ranges in response to changing temperatures and weather patterns. Shifting phenologies and ranges may seem of little importance at first glance, but they have the potential to cause great challenges to species survival. They further serve as harbingers of future environmental conditions from climate change. Increased weather extremes are also predicted to accompany climate changeand plant species resilience in the face of these weather events may also factor into their abilities to adapt and survive.

12.2.1   SHIFTS IN PHENOLOGY

As the climate warms, many plants are flowering 8.5 times sooner than experiments had predicted, raising questions for the world's future food and water supply, a new international study concludes. Higher carbon dioxide emissions from burning fossil fuels can affect how plants produce oxygen, and higher temperatures can alter their behavior. Shifts in natural events such as flowering or leafing, which biologists call phenology, are obvious responses to climate change. They can impact human water supply, pollination of crops, the onset of spring (and allergy season), chances of wildfires and the overall health of ecosystems.

12.2.2   EVIDENCES OF EFFECT OF CLIMATE CHANGE

The life cycles of plants correspond to seasonal cues, so shifts in the timing of such cycles provide some of the most compelling evidence that global climate change is affecting species and ecosystems (Cleland et al.,

2007 and Walther et al., 2002). Available evidence indicates that spring emergence has generally been occurring progressively earlier since the 1960s. Such accelerated spring onset has generated noticeable changes in the phenolgical events of many plant species, such as the timing of plants' bud bursts, first leafings, first flowerings or first seed, etc. Records indicate that many plants have started blooming earlier in response to the earlier occurrences of spring temperatures and weather. There is a concern made by environmentalists as flowers are losing their scent due to climate change and air pollution, also their fragrance may be lost forever.

In recent news of Times of India, the total export from Hosur (District Krishnagiri, Tamil Nadu), last season was more than 12 million rose flowers but now it has reduced to 8 million as the nights are extremely cold while the day is steaming hot thus climate change has had a negative impact on rose production (Shaji and Kumaran, 2013).

According to recent research review it was found that plants have already begun emitting more smelly chemicals known as biogenic volatile organic compounds (BVOCs). Higher temperatures not only cause plants to produce more BVOCs, they also often lengthen their growing seasons, prolonging the period over which they generate the fragrant compounds (Hattam, 2010). Penuelas and Staudt (2010) finds that these changed emissions can lead to unforeseeable consequences for the biosphere structure and functioning, and can disturb biosphere feedback on atmospheric chemistry and climate with a direction and intensity that warrants in-depth investigation.

Absence of required chilled weather during winter and severe heat during the summer has adversely hit production of roses in Pune district. The climate has affected quality of the flowers, thus leading to 50% decline in their export. The rose producers, which were already facing crisis due to recession, are now facing additional problem due to the unfavorable climate. A large number of farmers from Maval taluka in the district produce export quality flowers, especially roses, in green houses. There are over 50 commercial firms, which are involved in flower production at the 250 hectares Floriculture Park in Talegaon Dabhade (Anon, 2009). The temperature fluctuation has not only affected early production but stunted the growth of the stems. The roses were blooming before they could grow to their full size and the stems were shorter. These aberrations in production found no takers in the export markets and are now being dumped in the domestic market. The flower is known to last only for 10 days after harvesting. Their longevity can be stretched a bit with cold storage but the export market will be cold to preserved products, which are known to wilt fast once out in the open. Whereas in other side In Pune and across Bangalore, the winter temperature has increased by 4 °C. Both these centers are major areas for rose cultivation. The higher than normal temperature has resulted in an earlier bloom of roses. This, in turn, has advanced the arrival of the roses in the market. Planters projected that at least 25% of the blooms are set to arrive in the market much ahead of Valentine's Day this year. The ongoing concern of floriculturists is that the sharp rise in temperature could lead to damage and consequent rejection of export consignments. In the domestic market, most planters are expecting a glut that would result in prices plunging.

Deshmukh (2013) reported that under naturally ventilated polyhouse technology for rose cultivation, it was found that the warmer climate in December advanced the maturity cycle of the roses. The buds bloomed before attaining maturity, thereby reducing the flower size. Fitter and Fitter (2002) report that the average first flowering date of 385 British plant species has advanced by 4.5 days during the past decade compared with the previous four decades: 16% of species flowered significantly earlier in the 1990s than previously, with an average advancement of 15 days in a decade. Ten species (3%) flowered significantly later in the 1990s than previously. These data reveal the strongest biological signal yet of climatic change. Flowering is especially sensitive to the temperature in the previous month and spring-flowering species are most responsive. However, large inter specific differences in this response will affect both the structure of plant communities and gene flow between species as climate warms. Annuals are more likely to flower early than congeneric perennials, and insect-pollinated species more than wind-pollinated ones.

12.3   FLORICULTURE SCENARIO IN INDIA

India is becoming a strong center of commercial floriculture in the international market. During the last 5–7 years, there was a great surge in the floricultural activity in the production of flowers (cut and loose), ornamental plants (potted and cut-greens) and dry flowers (value added products), besides marketing. The horticultural sector contributed around 28% of the GDP annually from 13.08% of the area and 37% of the total exports of agricultural commodities (2004–2005). Albeit, India's present contribution in the global floricultural export market is negligible (about 0.4%) as compared to the Netherlands (58%), Columbia (14%), Ecuador (7%), Kenya (5%), Israel (2%), Italy (2%), Spain (2%) and others 10%, it is not far when India will come up as a major grower/exporter by virtue of well planned policies formulated by the Government of India backed with foreign technologies for greenhouse production.

12.3.1 IMPACT OF CLIMATE CHANGE ON FLORICULTURE

The impact of climate change on flowering plants and crops will be more pronounced. Melting of ice cap in the Himalayan regions will reduce chilling required for the flowering of many of the ornamental plants like Rhododendron, Orchid, Tulip, Alstromerea, Magnolia, Saussurea, Impatiens, Narcissus, etc. Some of them will fail to bloom or flower with less abundance while others will be threatened. Indigenous species in the natural habitat will be under threat for not getting favorable agro climatic conditions for their proliferation. Western Ghats and surrounding regions may be deprived of normal precipitation due to abnormal monsoon. Plant species requiring high humidity and water may find them under difficult conditions for survival. Plains of India will also have similar kind of problems and will be affected either by drought or excessive rains, floods and seasonal variations. Commercial production of flowers particularly grown under open field conditions will be severely affected leading to poor flowering, improper floral development and color besides reduction in flower size and short blooming period.

12.3.2 SOME OF THE ABIOTIC EFFECTS ON ORNAMENTAL FLOWER CROPS

12.3.2.1 CHRYSANTHEMUM

Photoperiod: Chrysanthemum is a short day plant. So flowering round the year in open field condition is not possible.

12.3.2.2   JASMINE

Low temperatures shut down flowering in jasmine (<19 °C) and lead to reduction in flower size.

12.3.2.3   CARNATION

Whenever night temperature is less than 13 °C the deterioration of flower quality due to calayx splitting.

12.3.2.4   JASMINE

In *Jasminum sambac*day time temperatures of 27–32 °C and night time temperatures of 21°–27 °C are ideal. If night temperatures fall below 19 °C, flower production and size are reduced. Mechanical flower forcing by advancing pruning to September resulted in off-season flowering with increased yield and quality (by offsetting abiotic stress caused by low temperature). Among the *Jasminum sambac* varieties, Single mogra performed best with respect to winter season flowering and peak season flowering in terms of yield and quality.

12.3.2.5   CARNATION

Optimum temperature range of 13.3–23.8 °C has been reported for producing quality with least calyx splitting.

12.3.2.6   TROPICAL ORCHIDS

Flowers do not open up fully in tropical orchids wherever temperatures below 15 °C. High temperature leads to flower bud drop and unmarketable spikes in tropical orchids when temperature is >35 °C. In Dendrobium cv. Sonia 17, the flower yield and quality have found to drastically reduce in peak summer and winter months. Nutrition management has found to alleviate temperature stresses. Biweekly foliar spray of 1:6:1 NPK at 0.2% to plants at flowering stage resulted in round the year flowering and increase in number of spikes per plant per year, number of florets per spike, maximum average yield of spikes per plant, spike length and a B:C ratio of 2.38 which were significantly higher compared to control.

Gardeners know how color in the flower garden depends on changes in the weather. Many, but not all, species are triggered into flower by higher temperatures. Miller and Primack, (2008) in a recent study compared flowering times and found that there had been an average rise of 2.4 °C and also found that plants were flowering about a week earlier than 150 years ago. Naturally this figure masked huge variations between species and even between closely related species. Sweet birch, Betula lenta, for example, now flowers 3 days earlier for every increase of 1 °C, while the Gray birch catkins, *Betula populifolia* have been advanced by 11 days for each degree. In a study on effect of higher temperature on flower colorand changes in the content of the main anthocyanins. Results revealed that poor coloration of flowers was observed at 30 °C in all genotypes except 'Chatoo.' All genotypes showed lower contents of the two anthocyanins tested [cyanidin 3-O-(6"-O-monomalonyl-beta-glucopyranoside) and cyanidin 3-O-(3",6"-O-dimalonyl-beta-glucopyranoside)] at higher temperatures (Nozaki et al., 2006). It found that flowering was delayed in all cultivars of spray chrysanthemum as day length increased from 13 to 14 h and abnormal flower formation was induced at over 13 h day length. These results suggest that there were formation of abnormal flowers and delay in flowering under high temperature. Delayed flowering of chrysanthemum under hightemperature conditions is a serious obstacle for all-year-round cut chrysanthemumflower production in southern temperate and subtropical zones. In a study of Nozaki and Fukai (2008) two different genotypes of spray chrysanthemum (*Dendranthema grandiflorum* syn. *Chrysanthemum morifolium*) were grown under high-temperature conditions: summer-to-autumn flowering type (SA type, hightemperature tolerant) and autumn flowering type (A type, hightemperature sensitive). Results showed that high temperatures slowed floral development in inflorescence, thereby increasing the number of florets in both SA and A chrysanthemum genotypes. Secondly, high temperatures slowed the developmental speed of inflorescence after the budding stage, and the time to reach the bud break stage was prolonged, thereby delaying flowering. Eun Joo et al. (2004) found that chrysanthemums that were exposed to high temperature developed two types of abnormal flowers, that is, formation of ray florets only and formation of ray florets with bracts at the center of the floral disk. The most susceptible developmental stage was 14–20 days after short-day (SD) period. Flowers exposed to 20 °C night temperature formed only ray florets, whereas those exposed to 30 °C night temperature developed both ray florets and bracts. Due to climate change a number of threats were noticed to orchid community in the Yachang Orchid Nature Reserve in Guangxi a good representative of the region. Decreased soil moisture is likely to have a negative effect on growth and survival of orchids, especially terrestrial and saprophytic ones. The greater majority of the orchids in Yachang Reserve (72%) have populations on or close to the limestone mountain tops.

These populations are likely to shrink or even become extinct as the warming continues because they have no higher places to which they are able to migrate. Extreme rainfall events are projected to occur more frequently, which can exacerbate erosion. These mayimpactorchid populations that grow on steep cliffs. The majority of orchid species have specialized insect pollination systems. It is unknown whether the change or lack of change in plant phenology will be in synchrony with the potential phonological shifts of their pollinators (Liu et al., 2010). Natural and manmade changes have influenced the climate and in turn the ecosystem through several of its parameters such as rainfall and temperature. One of the direct effects of this on orchid is, change in the plant structure and composition. The changes in the climate have also affected sensitive events such as flowering and pollination. This has forced the plant species to shift to more congenial climate for its survival and eventually affecting the coexisting biota (Shashidhar and Kumar, 2009).

12.4   FUTURE STRATEGIES

In view of these problems, horticulturists will have to play a significant role in the climate change scenario and proper strategies have to be envisaged for saving horticulture/floriculture from future turmoil (Sharma and Roy, 2010).

- The most effective way to address climate change is to adopt a sustainable development pathway, besides using renewable energy, forest and water conservation, reforestation, etc.
- Awareness and educational programs for the growers, modification of present horticultural practices and greater use of green house tech-

nology are some of the solutions to minimize the effect of climate change. Hi-tech horticulture is to be adopted in an intensive way.

• It is necessary that selection of plant species/cultivars is to be considered keeping in view the effects of climate change. The performance of different seasonal may not be satisfactory due to shorter and warmer winter.

• Judicious water utilization in the form of drip, mist and sprinkler will be a key factor to deal with the drought conditions.

• Development of new cultivars of floricultural crops tolerant to high temperature, resistant to pests and diseases, short duration and producing good yield under stress conditions, will be the main strategies to meet this challenge.

KEYWORDS

- **Abiotic Effects**
- **Adverse Effect**
- **Climate Change**
- **Floriculture**
- **Flower Crops**
- **Strategies**

REFERENCES

Anonymous (2013). Climate change hit production of roses in Pune district. *Punekar* Newspaper, May 5.

Cleland, E. E., Chuine, I., Menzel, A., Mooney, H. A., & Schwartz, M. D. (2007). Shifting plant penology in response to global change. *Trends Ecol. Evol,* 72(7), 357–364.

Deshmukh, N. (2013). Climate variations in December force early export of rose. *Times ofIndia* News Paper, Pune, Feb. 12.

EunJoo, H., HakKi, S., KwangJin, K. & Youl, C. S. (2004). High temperature-induced flower abnormalities at bud development in chrysanthemum. *J. Korean Soc. Hortic. Sci.,* 45(6), 345–348.

Fitter, A. H., & Fitter, R. S. R. (2002).*Rapid changes in flowering time in British plants.Science,* 1689–1691.

Hattam, J. (2010). Will Roses smell sweeter in a warming world. Science/Natural Sciences. Treehungger, Feb. 10.

Liu, H., Feng, Chang Lin., Luo, YiBo, Chen, BaoShan., Wang, ZhongSheng & Gu, HongYa. (2010). Potential challenges of climate change to orchid conservation in a wild orchid hotspot in southwestern China. *Bot. Rev.,* 76(2), 174–192.

Miller-Rushing, A. J. & Primack, R. B. (2008). Global warming and flowering times in Thoreau's Concord: a community perspective. *Ecology,* 89(2), 332–341.

Nozaki, K., Takamura, T. & Fukai, S. (2006). Effects of high temperature on flower colour and anthocyanin content in pink flower genotypes of greenhouse chrysanthemum (*Chrysanthemum morifolium* Ramat.). *J. Hortic. Sci. Biotechnol.,* 81(4), 728–734.

Nozaki, K., &Fukai, S. (2008). Effects of hightemperature on floral development and flowering in spray chrysanthemum. *J. Appl. Hortic.,* 10(1), 8–14.

Penuelas, J., & Staudt, M. (2010). Induced biogenic volatile organic compounds from plants. *Trends Plant Sci.,* 15(3), 133–144.

Shaji, K. A., & Kumaran, V. S. (2013). Not all roses in Hosur this Valentine's day. *Times of India* News Paper, Coimbatore. Feb. 14.

Sharma, S. C. & Roy, R. K. (2010). Impact of global climate change on floriculture in India *Environews* newsletter, 16(1), 7–8.

Shashidhar, K. S. & Kumar, A. N. A. (2009). Effect of climate change on orchids and their conservation. *Indian Forester,* 135(8), 1039–1049.

Walther, G. R., Post, E., & Convey, P. (2002). Ecological responses to recent climate change. *Nature,* 16, 389–395.

CHAPTER 13

CLIMATE CHANGE: BREEDING STRATEGIES TO MITIGATE ABIOTIC STRESS IN ORNAMENTAL CROPS

K. V. PRASAD[1], T. JANAKIRAM, SAPNA PANWAR, BHARAT SING HADA, and PRATIVA LAKHOTIA

[1]Division of Floriculture and Landscaping Indian Agricultural Research Institute, New Delhi, India; [1]Email: kvprasad66@gmail.com

CONTENTS

ABSTRACT

The global floriculture trade is estimated to be at US$ 70 billion. The floriculture industry is growing at the rate of 8–10% per annum. There are nearly 120 countries that are actively involved in floriculture business. India ranks second in flower cultivation next to China producing 1,651,000 MT of loose flowers and 75,065 lakh numbers of cut flowers annually (2011–2012). In the last decade, there was a great surge in the production as well as consumption of flowers, ornamental plants and value added products. Though, India's present contribution in the global floricultural export market is negligible (~0.4%) as compared to the Netherlands (58%), Columbia (14%), Equador (7%), Kenya (5%), Israel (2%), Italy (2%), Spain (2%) and Others 10%. Europe continues to be the largest destination for Indian floriculture exports. However, in the recent years, India has been exporting floriculture products to the Japanese, Australian and the Middle East markets.

The main global challenge facing the floriculture industry today is the impact of climate change and the issue of carbon footprint. IPCC-GCM projections for India indicate that the annual temperature will increase by 2 to 3.5 °C. Similarly annual precipitation is expected to increase by 10–20% but, the seasonal variations would range from deficits to excesses of the annual precipitation.

The higher ambient temperature has a direct impact on: (1) Volatile fragrances the flowers emit; (2) Deterioration of pigments leading to dull shades; (3) Reduced production and productivity under open and protected environment; (4) Shift in insect pest and disease outbreaks; (5) Absence of winter chilling will reduce flowering; (6) reduced post harvest life; (7) poor pollination and seed set due to changes in insect behavior. Similarly higher rainfall would increase anaerobic stress at the root zone leading to yellowing, poor growth and even mortality.

It is expected that the hi-tech floriculture units are likely to be buffered to an extent from direct effects of climate change as most of the flower production is carried out in greenhouses. Small-scale players who depend on rain-fed floriculture will be extremely vulnerable to climate change because of their direct impact. Hi-tech growers are also not entirely immune as the industry depends heavily on natural resources like water availability and its quality besides the vagaries like drought and floods that are likely to cause havoc in flower production.

A number of flowers like chrysanthemum, poinsettia and carnation are either photo sensitive or thermo sensitive or both. Changing pattern in photo periodism and thermo periodism would greatly alter the blooming pattern in such flower crops. Climate change is expected to enhance the global temperatures by 2–3 °C by 2050. This would alter the chilling requirement of some of the temperate flower crops. The insect pest and disease dynamics would bring about a change in use of pesticides in open and protected environments. Increase in temperature would alter the relative humidity levels that would have profound impact on disease incidence and its spread.

Breeding specific varieties that are tolerant to high/low temperature by incorporating the suitable species like*Rosa macarantha* and *Rosa spinosissima* in roses, selecting lines that are capable of selective salt uptake or exclusion or make osmotic adjustments to regulate turgor pressure to select salt tolerant lines in salt sensitive crops.

The paper outlines the breeding strategies to be adopted to mitigate the challenges enlisted in some of the important floricultural crops.

13.1   INTRODUCTION

Climate change is one of the most important global environmental challenges facing humanity with implications for food production, natural ecosystems, freshwater supply, health, etc. According to the latest scientific assessment, the earth's climate system has demonstrably changed on both global and regional scales since the preindustrial era. The average global temperature has increased by 0.8 °C in the past 100 years. The Intergovernmental Panel on Climate Change (IPCC) projects that the global mean temperature may increase between 1.4 and 5.8 °C by 2100. It is now well known that warming will also be associated with changes in rainfall patterns, increased frequency of extreme events of drought, frost and flooding. Collectively all these factors will affect output of agriculture and allied sectors. India is one of the 27 countries likely to be most affected. This unprecedented change is expected to have severe impacts on the global hydrological system, ecosystems, sea level, crop production and related processes. The impact would be particularly severe in the tropical areas, which mainly consist of developing countries, including India. The UN Conference on Environment and Development (UNCED) in

1992 at Rio de Janeiro led to FCCC (Framework Convention on Climate Change), which laid the framework for the eventual stabilization of green-house gases in the atmosphere, recognizing the common but differenti-ated responsibilities and respective capabilities, and social and economic conditions. The Convention came into force in 1994. Subsequently, the 1997 Kyoto protocol, which came into force in 2005, reasserted the im-portance of stabilizing greenhouse gas concentrations in the atmosphere and adhering to sustainable development principles. The Protocol laid out guidelines and rules regarding the extent to which a participating industri-alized country should reduce its emissions of six greenhouse gases carbon dioxide, methane, nitrous oxide, chlorofluorocarbon, hydro fluorocarbons and per fluorocarbons. It requires industrialized countries to reduce their greenhouse gas emissions by a weighted average of 5.2%, based on the 1990 greenhouse gas emissions. The reduction is to be achieved by the end of the five-year period, 2008 to 2012. The Kyoto Protocol does not require the developing countries to reduce their greenhouse gas emissions. Horticulture, in India with approximately 30% contribution in agricultural GDP from only 8% of cultivated land is threatened with serious conse-quences in production, quality and processing and increased cost of plant protection.

Climate is a phenomenon, which is a truth, and it continues. The change could be for betterment or it could have a disastrous impact. Technological changes have provided immense comfort, and also caused the imbalances in climatic parameters, threatening the sustainability. For the last one year, various institutions, Government organizations and also the Indian Coun-cil of Agricultural Research (ICAR) have been deliberating upon the likely change in climate and its adverse impact on agriculture, and there have been clear conclusion that climate is changing, which needs public inter-vention and also preparedness to face the challenges.

13.2   WHAT IS CLIMATE CHANGE?

Climate change is a natural part of the Earth's dynamic system. However, there is very strong evidence that human activity, particularly the emission of greenhouse gases, is increasing the rate and degree of change. Climate change in IPCC usage refers to a change in the state of the climate that can be identified (e.g., using statistical tests) by changes in the mean and/

or the variability of its properties and that persists for an extended period, typically decades or longer. It refers to any change in climate over time, whether due to natural variability or as a result of human activity. This usage differs from that in the United Nations Framework Convention on Climate Change (UNFCCC), where climate change refers to a change of climate that is attributed directly or indirectly to human activity that alters the composition of the global atmosphere and that is in addition to natural climate variability observed over comparable time periods. The earth is only the planet in our solar system that supports life, because of unique environmental conditions that are present- water, an oxygen- rich atmosphere, and a suitable surface temperature. About 30% of incoming energy from the sun is reflected back to space while the rest reaches the earth, warming the air, oceans, and land, maintaining an average surface temperature of about 15 °C. The chemical concentration in atmosphere for nitrogen is 78%; about 21% is oxygen, and only a small (0.036%) is made up of carbon dioxide, which plants require for photosynthesis. In atmosphere, energy is absorbed by land, seas, mountains, etc., and simultaneously released in the form of infrared waves. All this released heat is not lost to space, but is partly absorbed by some gases present in very small quantities in atmosphere, called greenhouse gases (GHG) consisting of carbon dioxide, methane, Nitrous dioxide, chloro fluorocarbon, hydro fluorocarbons and per fluorocarbons, Ozone, etc. Among GHGs, nitrous oxide 6%, methane 13%, fluorocarbons 5%, and carbon dioxide 76%. These gases are also called greenhouse gases because they act as a blanket and trap heat radiating from the earth and make the atmosphere warm in absence of emission of heat imbalances are created. Thus, increased concentration of GHG leads to increased temperature, which in turn has impact on the world climate, leading to phenomena known as climate change.

The earth's climate change system constantly adjusts so as to maintain the balance between the energy that reaches it from the sun and the energy goes from earth back to space. This means that, even a small rise in temperature could mean accompanying changes in cloud cover and wind patterns. Some of these changes may enhance the warming, while others may counteract. Cooling effect may result from an increase in the levels of aerosols. Climate change per se is not necessarily harmful, but the problems arise from extreme events that are difficult to predict, like more erratic rainfall pattern and unpredicted warm spells shall affect productivity.

At the same time, more availability of CO_2 would help in improved yield of root crops and increased temperature may shorten the period.

Conventional approaches to understanding climate change were limited to identifying and quantifying the potential long-term climate impacts on different ecosystems and economic sectors. While useful in depicting general trends and dynamic interactions between the atmosphere, biosphere, land, oceans and ice, this top-down, science-driven approach failed to address the regional and local impacts of climate change and the local abilities to adapt to climate-induced changes. This impact-driven approach (Adger et al., 1999) gave way to a new generation of scholarship, whichutilized bottom-up or vulnerability-driven approaches that assessed past and present current vulnerability, existing adaptation strategies, and how these might be modified with climate change. Vulnerability in this context is defined as, "the degree to which a system issusceptible to, or unable to cope with, the adverse effects of climate change, including climate variability and extremes" and adaptation as, "adjustmentsin ecological, social or economic systems in response to actual or expectedstimuli and their effects or impacts. This term refers to changes in processes, practices and structures to moderate potential damages or to benefit fromopportunities associated with climate change" (IPCC, 2001).

The international community is continuing to grapple with the likely socioeconomic and environmental impacts that shall result from climate change. Adaptation to climate change is a new process for both developed and developing nations, and concrete experience in applying an integrated approach to adaptation is limited. The adaptation line of inquiry reflects the international community's escalating need to prepare for and adapt to climate change and to ensure that any future climate change regime will bestow on the issue its legitimate recognition. It also recognizes growing international awareness of the need to integrate adaptation issues into core policy and decision-making processes. The question that needs to be address is how adaptation to climate variability and change can be more fully integrated into development policies and what are the funding instruments for adaptation?

The rationale for integrating adaptation into development strategies and practices is underlined by the fact that interventions required to increase resilience to climate variability and change generally further development objectives. Adaptation calls for natural resource management, buttressing food security, development of social and human capital and strengthening of institutional systems (Adger et al., 2003). Such processes, besides building the resilience of communities, regions and countries to all shocks and stresses, including climate variability and change, are good development practice in themselves. Hence the inclusion of climatic risks in the design and implementation of development initiatives is vital to reduce vulnerability and enhance sustainability.

13.3   GLOBAL CLIMATE CHANGE CURRENT SCENARIO

Global average sea level rose at an average rate of 1.8 mm per year over 1961 to 2003. This rate was faster over 1993 to 2003, about 3.1 mm per year. Analyzes done by the Indian Meteorology Department and the Indian Institute of Tropical Meteorology generally show the same trends for temperature, heat waves, glaciers, droughts and floods, and sea level rise as by the Inter-Governmental Panel on Climate Change (IPCC) of United Nations. Magnitude of the change varies in some cases. There are evidences that glaciers in Himalayas are receding. The rainfall is also likely to become more uncertain. The projected global, mean annual temperature increase by the end of this century is likely to be in the range 2 to 4.5 °C. Values substantially higher than 4.5 °C cannot be excluded. For south Asia (Indian region), the IPCC has projected 0.5 to 1.2 °C rise in temperature by 2020, 0.88 to 3.16 °C by 2050 and 1.56 to 5.44 °C by 2080, depending on the scenario of future development (Table 13.1; IPCC 2007b). Overall, the temperature increases are likely to be much higher in winter (rabi) season than in rainy season (kharif). Precipitation is likely to increase in all time slices in all months, except during December–February when it is likely to decrease. It is likely that future tropical cyclones will become more intense, with larger peak wind speeds and heavier precipitation. Himalayan glaciers and snow cover are projected to contract. It is very likely that hot extremes, heat waves, and heavy precipitation events will continue to become more frequent. For the next two decades, a warming of about 0.2 °C per decade is projected. Even if all future emissions were stopped now, a further warming of about 0.1 °C per decade would be expected. The projected sea level rise by the end of this century is likely to be 0.18 to 0.59 meters.

Projected changes in surface air temperature and precipitation for South Asia under SRES A1FI (highest future emission trajectory) and B1 (lowest future emission trajectory) pathways for three time slices, viz. 2020s, 2050s and 2080s. (Source: IPCC, 2007).

13.4   INTER-GOVERNMENTAL PANEL ON CLIMATE CHANGE (IPCC)

Climate Change is a very complex issue: policymakers need an objective sourceof information about the causes of Climate Change, its potential environmental andsocioeconomic consequences and the adaptation and mitigation options to respondto it. This is why WMO and UNEP have established IPCC in 1988. Ministry of Environment and Forests is the nodal agencyfrom India for the IPCC. Membership of IPCC is open to all member countries of WMO & UNEP. The IPCC does not conduct any research nor does it monitor climaterelated data or parameters. Its role is to assess on a comprehensive, objective, openand transparent basis the latest scientific, technical and socioeconomic literatureproduced worldwide relevant to the understanding of the risk of human-inducedClimate Change, its observed and projected impacts and options for adaptation andmitigation. IPCC reports should be neutral with respect to policy, although they needto deal objectively with policy relevant scientific, technical and socio economic factors.

13.5   CLIMATE CHANGE: INDIAN SCENARIO

13.5.1   VARIATIONS IN SURFACE TEMPERATURE

Surface air temperature for the period 1901–2000 indicated a significant warming of 0.4 °C for 100 years. The spatial distribution of temperature changes indicated a significant warming trend has been observed along the west coast, central India, and interior peninsula and over north-east India. However, cooling trend has been observed in north-west and some parts in southern India. Season wise temperature trends indicated that maximum increase in temperature was observed in post monsoon (0.7 °C) followed by winter (0.67 °C) and pre monsoon (0.5 °C) and monsoon (0.3 °C).

13.5.2 VARIATIONS IN ANNUAL RAINFALL

At all India level, there are some regional patterns in rainfall over the last 100 years. Increasing trend in monsoon rainfall is found along the west coast, north Andhra Pradesh and north-west India, and those of decreasing trend over east Madhya Pradesh and adjoining areas, north-east India and parts of Gujarat and Kerala (6 to 8% of normal over 100 years). Rainfall analysis of data of 1140 stations in the country indicated that greater than 70% of the stations showed short-term fluctuations in annual rainfall for less than 10 years period. A 20% rise in all India summer monsoon rainfall and further rise in rainfall is projected over all states except Punjab, Rajasthan and Tamil Nadu, which show a slight decrease.

13.5.3 IMPACT OF CLIMATE CHANGE

13.5.3.1 POSITIVE IMPACT

Global climate change is expected to affect floricultural crops through its direct and indirect effects. Scientific evidence suggests a positive effect of increase in atmosphere CO_2 in C3 photosynthetic pathway promoting their growth and productivity. Increased CO_2 will reduce evapo-transpiration and thus increase in water-use efficiency.

TABLE 13.1 List of C3 and C4 Ornamental Plants

C3 Ornamental Plant Species	C4 Ornamental Plant Species
Carnation	Amaranth
Chrysanthemum	Bermuda Grass
Day lilies	Cacti
Ferns	Orchids
Geranium	Succulents
Kentucky blue grass	
Lavender	
Marigold	
Moon flower	
Morning glory	
Potted azalea	
Roses	
Tulips	

The C3 plants are likely to gain from the increasing level of CO_2 in the atmosphere. Elevated carbon dioxide has positive effect by increasing the productivity ranging from 24–51%. CO_2 enrichment in greenhouses (800–2000 ppm depending on the crop) to increase the production of flower crops is popular. It would be possible to spend less on energy to generate CO_2 for carbon fertilization.

13.5.3.2   NEGATIVE IMPACT

The above positive effects will be counteracted by increase in temperature, reduced humidity levels, reduced water levels, increased salt levels in water and soil. Rise in temperature will reduce crop duration, increase respiration rate, alter photosynthate partitioning to economic product, alter phenology, particularly flowering, fruiting and reduce chilling unit accumulation, hasten senescence, fruit ripening and maturity. However, the overall impact of climate change and global warming will depend on interaction effect of elevated carbon dioxide and temperature rise.

Besides these the major impact would be on the following:
1. Drastic reduction in volatile fragrances the flowers emit;
2. deterioration of pigments leading to dull shades;
3. reduced production and productivity under open and protected environment;
4. shift in insect pest and disease outbreaks;
5. absence of winter chilling will reduce flowering in bulbous ornamental crops;
6. reduced post harvest life;
7. poor pollination and seed set due to changes in insect behavior;
8. Similarly higher rainfall would increase anaerobic stress at the root zone leading to yellowing, poor growth and even mortality.

13.5.4   ABIOTIC STRESS ENCOUNTERED

A number of factors induce stress in plants when the plants are grown in adverse growing conditions. Plant encounters stress when grown under high temperature, low temperature, high salt and heavy metal concentrations. Such stresses are known as abiotic stress in comparison to biotic stresses that the plant encounters due to pathogens. With the advent of modern tools, a number of strategies are now available to develop abiotic stress tolerant varieties.

13.5.5 BREEDING STRATEGIES

13.5.5.1 DROUGHT TOLERANCE

Insufficient availability of water, that is, drought, is presumably the most common stress experienced by terrestrial plants. On the cellular level drought stress will affect vital metabolic functions and maintenance of turgor pressure. Cell expansion and cell wall formation are therefore especially sensitive to water limitation. In order to minimize water loss, plants respond to lower water availability with the closure of stomata. However, this protective measure is not without drawbacks for the plant as this will also decrease the CO_2 supply within the plant leaves and finally affect photosynthesis (Tippmann et al., 2009).

13.5.5.2 CONVENTIONAL APPROACHES

Using the available sources of resistance one can incorporate such species in the breeding program to transfer the genes of interest. For instance in case of roses *Rosa canina* hardy species adaptable to drought and alkaline conditions (Hartmann and Kester, 1972) can be used as one of the parents to cross with cultivated varieties belonging to *Rosa hybrid*a. Similarly *Rosa indica* (*var, odorata*) can be incorporated in the inter specific hybridization program as it is well adapted to both excessively dry or wet soil conditions and can withstand high soil pH. In Carnations, *Dianthus aydogduii* is salt tolerant species, which can be used as one of the parents to introduce the trait in to cultivated Dianthus.

13.5.5.3 BIOTECH APPROACHES

The major signal during drought induced stress comprises of the formation of Reactive Oxygen Species (ROS) likesinglet oxygen ($^1O^{2-}$), superoxide anions (O_2⁻), hydrogen peroxide (H_2O_2) and hydroxyl radicals (OH) which are highly reactive and damage cells by ROS mediated oxidative processes such as membrane lipid peroxidation, protein oxidation, enzyme inhibition and damage of nucleic acids (Grene, 2002) The biotech tools available aim at expression of ROS scavenging enzymes like superoxide peroxidase, ascorbate peroxidase, catalase, glutathione peroxidase and peroxiredoxin. In cooperation with the antioxidants like ascorbic acid and glutathione, these scavenging enzymes detoxify the ROS and prevent serious damage to the cells (Scheller and Haldrup, 2005).

Another important messenger during drought stress adaptation is the phytohormone abscisic acid (ABA). ABA biosynthesis is initiated by decreasing water potential, and seems to be essential for the activation of many protective measures towards abiotic stresses. In drying soil, an ABA signal can be produced early during stress in the roots and be transported via the phloem to the shoot. There, ABA accumulation represents an important signal for the closure of stomata in response to drought stress. ABA induces changes in the turgor of guard cells and the changed guard cell expansion leads to stomatal closure. Biosynthesis of proline for instance increases the concentration of compatible osmoprotectants in the cells, while aquaporins can facilitate water permeability of cellular membranes and maximize water uptake potential of the plant, and ROS scavenging proteins can limit damage by secondary oxidative stress (Chaves et al., 2003). The specific function of the majority of drought-induced genes is, however, still unknown and revelation of their function should give new insights into the plant protective mechanisms against drought. Strategies to mitigate the drought stress aim at expression of ABA and proline in the cells.

13.5.6   SALT STRESS

High concentrations of salt (i.e., ions, mostly Na+) in the soil solution can impair water and nutrient uptake, reduce growth and photosynthetic activity. Furthermore high salinity can lead to an unfavorable Ca2+ or K+ to Na+ ratio, toxic intracellular Na+ concentrations and peroxidation of membrane lipids (Levitt, 1980).

13.5.6.1   CONVENTIONAL APPROACHES

Using the available sources of resistance one can incorporate tolerant/resistant species in the breeding program to transfer the genes of interest. In

case of roses, *Rosa indica* (var Odorata) has the abilities to withstand high levels of pH an indication of excessive soil salinity. Such species can be included in the breeding program to transfer the genes of interest. Similarly resistance to high level of salts is found in carnations species *Dianthus aydogduii*which can be incorporated in the breeding program.

13.5.6.2   BIOTECH APPROACHES

The reduced ability of the plant to take up water induces water deficit effects comparable to drought stress. It is therefore not surprising to find similarities in the signaling and response of drought and salt stressed plants. This includes ABA biosynthesis and accumulation, which regulate measures against water loss such as closure of stomata and increased production of compatible osmoprotectants and antioxidants.

Limitations in photosynthesis and changes in redox status of the mitochondrion promote increased production of ROS and downstream signaling pathways. More than half of the known drought-responsive genes show also induced gene expression by salt stress. This includes ABA responsive pathways, which are active under both types of stress indicating significant signalingcross-talk (Shinozaki et al., 2003).

Tolerance to the resulting secondary oxidative stress of salt stress is apparently highly important for stress tolerance: thus, a number of salt tolerant species increase the activity of antioxidant enzymes and accumulate antioxidants in response to salt stress, while salt-sensitive species fail to do so. Biotech tools can therefore be used to selectively express the anti oxidant enzymes to mitigate the stress induced by high salt concentrations.

13.5.7   TEMPERATURE STRESS

Ornamental plant species perform best only in a characteristic temperature range, which depending on the species might be very narrow. Extremely high or low temperatures affect vital cell functions such as enzyme activity, cell division and membrane integrity. Nevertheless, heat and cold acclimation is possible, as mild stress pretreatment can significantly enhance the thermo-tolerance of plants (Thomashow, 1999).

13.5.7.1   HIGH TEMPERATURE STRESS

Effects of high temperature stress can range from moderate effects such as oxidative stress and enhanced transpiration to fatal consequences for the plant, leading to tissue collapse and plant death. To cope with the high temperature stress, plants usually react with enhanced transpiration rates to achieve evaporative cooling. Since high temperatures are often accompanied with limited water availability, the water potential in the plant cell can decrease, leading to drought conditions and initiate responses similar to drought stress. Under these conditions, different events take place: small heat shock proteins (HSPs), acting as molecular chaperones, are synthesized in abundance (Vierling 1997), Ca^{2+} influx is enhanced and ROS accumulate rapidly in different parts of the cell (Doke et al., 1996; Foyer et al., 1997).

Although the function of many HSPs during heat stress is still unknown, they have been found to accumulate in all prokaryotic and eukaryotic organisms under high temperature stress, and are thought to contribute to the stabilization of cellular structures and prevent the thermal aggregation of proteins (Lee et al., 1997).

13.5.7.2   CONVENTIONAL APPROACHES

Conventional hybridization followed by repeated back crosses with the resistant parent have great potential to introduce the tolerance/resistance genes present in the wild species. For instance in case of roses, *Rosa arkansana* (Collicut, 1992) is tolerant to hot, dry summer as well as cold which can be utilized in the breeding program to develop high temperature tolerant rose varieties.

13.5.7.3   BIOTECH APPROACHES

Responses to heat stress show overlapping signal events to other nonrelated stresses, for example the accumulation of HSPs, which is seen in different stresses. The expression of some HSPs is ABA dependent, and ABA treatment can induce HSP accumulation at ambient temperatures. Apart from activating local and systemic stress responses under pathogen attack, salicylic acid (SA) has an additional role in the acclimation process to heat stress. SA accumulation has been observed for instance in heat stressed mustard (Dat et al., 1998b) and SA application can improve plant tolerance to subsequent heat stress (Dat et al., 1998a). The role of SA in this process is not fully understood and is apparently not linked to the induction of HSPs biosynthesis, but to the induction of oxidative responses. The application of exogenous SA can mimic temperature acclimation, similar to heat shock generated oxidative stress. However, there might be a stabilizing role for SA in HSP gene activating transcription factor complexes (Jurivich et al., 1992). ROS on the other hand, serve as fast messengers inside the cell, activating multiple downstream responses and increased ROS generation has been observed during heat shock in plants (Dat et al., 1998b).

With the advent of genetic engineering it is now possible to clone and transfer the genes of interest including HSP to induce tolerance to high temperature.

13.5.8 COLD AND FREEZING STRESS

Chilling stress is a suboptimal temperature, where the plant faces reduced enzyme activity and maybe water availability, however, the temperature is above the freezing point of water. Cold stress affects mainly metabolic processes, impairing enzyme reactions, substrate diffusion rates and membrane transport properties. Thereby some reactions are more affected by the cold then others. In particular, the dark reaction of photosynthesis and oxidative phosphorylation seem to be sensitive to chilling. The discrepancy between the speeds of biochemical reactions can cause ROS accumulation in the chloroplast and mitochondrion (Scheller and Haldrup, 2005). One early response to cold and osmotic stress is the rapid Ca^{2+} influx into the cell. Physical alterations in the cellular structure may cause this Ca^{2+} influx by activation of Ca^{2+} channels and initiate downstream Ca^{2+} dependent signaling pathways (Xiong et al., 2002).

When temperature drops below zero degrees, plants can experience freezing stress. Intracellular freezing of water can damage the protoplast membrane structure, mechanically injure and finally kill the cells by the expanding ice crystals. During extracellular freezing, the protoplasm of the plant becomes severely dehydrated when water is transferred to the ice crystals in the intercellular spaces. Freezing acclimation can be achieved by the expression of a class of cold-induced "cold-regulated genes" (COR), which encode hydrophilic polypeptides and stabilize membranes against freeze-induced injury (Artus et al., 1996). The exact role in cold acclimation of these COR genes remain unknown, and several of them are also induced by drought and ABA, as well as by low temperature stress.

13.5.8.1 CONVENTIONAL APPROACHES

Sources of resistance for cold tolerance are well documented in some of the flower crops. For instance in roses *Rosa * odorata* ahardy species is used for budding or grafting other roses (Hartmann and Kester, 1972). Similarly *Rosa macrantha* has been used for breeding hardy cultivars. Similarly in case of florist chrysanthemum, *Chrysanthemum bubellum* is exceptionally sturdy and used by breeders for its hardiness earlier. In lilium, *Lilium hansonii* is extremely winter hardy and thrives well even in lightly shaded location. Another important species, *Lilium martagon* is winter hardy with wide adaptability for soil.

13.5.8.2 BIOTECH APPROACHES

C-binding factors (CBF), proteins with a conserved 60 amino acid AP2 domain, bind to C-repeat/DRE (drought responsive) elements in the promoter region of different stress response genes such as responsive to desiccation 29a (*RD29a*) and are crucial for the activation of *COR* genes. The over expression of these activators (e.g., CBF1) induces *COR* gene activity and can enhance also the freezing tolerance on nonacclimated plants (Jaglo-Ottesen et al., 1998). The isolation of highly freeze tolerant mutant *eskimo1* (*esk1*) in *Arabidopsis*, revealed another way how plants can cope with freezing stress: *esk1* accumulates high levels of a compatible osmolyte, proline, yet does not depend on increased expression of several (COR) (Xin and Browse, 1998).

Protective anti freeze proteins (AFP) accumulate after cold and drought treatment, presumably by ethylene induction (Griffith et al., 2005; Yu et al., 2001; Yu and Griffith, 2001). These AFPs show similarities to pathogenesis-related (PR) proteins (Hon et al., 1995) and might confer additional resistance against cold tolerant pathogens. Finally, SA accumulates in cold-stressed plants and is apparently involved in low-temperature growth inhibition (Scott et al., 2004). The application of exogenous SA (Janda et al., 1999), but also ethylene (Yu et al., 2001) can actually enhance cold tolerance of plants.

In contrast to drought and salt stress, the role of the phytohormone ABA in cold stress and acclimation is highly controversial. Initial experiments in *Solanum commersonii* showed a correlation between ABA and cold acclimation (Chen et al., 1983). Similar ABA accumulation has been reported from other cold acclimated plants (Lång et al., 1994; Mäntylä et al., 1995) and accordingly ABA deficient mutants exhibit lower cold tolerance (Heino et al., 1990). However, as Thomashow (1999) points out, the observed ABA accumulation in different plants is only transient and is not correlated to the enduring effects of cold acclimation. As for ABA deficient mutants, the reduced cold stress tolerance might be rather a side effect of the general reduced plants' health, caused by impaired ABA biosynthesis. Thus there are at least two signaling pathways that activate cold-regulated gene expression, an ABA dependent pathway (Lång et al., 1994) and an ABA independent pathway (Capel et al., 1997).

13.6   POLLUTANTS

Many natural occurring substances, termed xenobiotics, are toxic for plants at high concentrations. Human activity contributed substantially to the accumulation of heavy metals in soils and ground water. Most plants do not possess specific mechanisms to prevent the excessive uptake of heavy metals from the soil. Accumulation of heavy metal ions in the plant can impair membrane integrity, affect enzyme activity and hinder nutrient uptake. Metal ions can generate ROS by auto-oxidation, Haber-Weiss cycle or Fenton reaction and disturb the redox status of cells.

Metals without redox capacity such as cadmium, mercury and lead can disturb the antioxidative glutathione pool, activate Ca2+ dependent systems and iron mediated processes (Pinto et al., 2003). Apart from soil pollution human activity also increased concentration of air pollutants such as O_3, SO_2, NO, NO_2, NH_3, HNO_3 and HF. Through stomata they can enter the leaves and affect the plant metabolism. Entry of O_3 into the cell forces the creation of ROS and induces oxidative stress. It is assumed that the O_3 derived ROS production mimics the signaling pathways following oxidative burst during a virulent pathogen attack (Rao and Davies, 2001).

Similar to the pathogen response, O_3 stress also activates the production of SA, JA and ethylene.

13.7 IN VITRO SELECTION PRESSURE TECHNIQUE

In vitro culture of plant cells, tissues or organs on a medium containing selective agents offers the opportunity to select and regenerate plants with desirable characteristics. The technique has also been effectively used to induce tolerance, which includes the use of some selective agents that permit the preferential survival and growth of desired phenotypes (Purohit et al., 1998). The selecting agents usually employed for in vitro selection include NaCl (for salt-tolerance), PEG or mannitol (for drought-tolerance), heavy metal, specific fungal culture filtrate (FCF) or phytotoxin such as fusaric acid or the pathogen itself (for disease-resistance). The explants are exposed to a broad range of these selective agents added to the culture medium.

Only the explaints capable of sustaining such environments survive in the long run and are selected. Two types of selection methods has been suggested: (a) stepwise long-term treatment, in which cultures are exposed to stress with gradual increase in concentrations of selecting agent and (b) shock treatment, in which cultures are directly subjected to a shock of high concentration and only those which would tolerate that level will survive (Purohit et al., 1998). These methods are based on the induction of genetic variation among cells, tissues and/or organs in cultured and regenerated plants (Mohamed et al., 2000). The tissue culture induces variation in regenerated plants, called somaclonal variation (Larkin and Scowcroft, 1981), can result in a range of genetically stable variations, useful in crop improvement. In vitro selection can considerably shorten the time for the selection of desirable traits under selection pressure with minimal environmental interaction, and can complement field selection (Jain, 2001). Despite many advantages, development of stress tolerant plants through in vitro selection has some limitations like loss of regeneration ability during selection, lack of correlation between the mechanisms of tolerance operating in cultured cell, tissue or organ and those of the whole plants, and phenomenon of epigenetic adaptation (Tal, 1994).

13.8   CONCLUSIONS

Response of plant to stress induced by abiotic factors is complex and is regulated by a number of genes acting directly or indirectly. The advent of molecular tools has helped in understanding the basic mechanism of the tolerance or resistance to various factors paving the way for developing the mitigation strategies to be adopted in flower crops.

KEYWORDS

- **Abiotic Stress**
- **Breeding Strategies**
- **Climate Change**
- **Negative Impact**
- **Pollutants**
- **Positive Impact**

REFERENCES

Adger, W. N., Saleemul Huq, Katrina Brown, Declan Conway & Mike Hulmea (2003). Adaptation to climate change in the developing world. Progress in Development Studies 33, 179–195.

Capel, J, Jarillo, J. A., Salinas, J., & Martinez-Zapater, J. M. (1997). Two homologous low-temperature-inducible genes from *Arabidopsis* encode highly hydrophobic proteins. *Plant Physiology* 115, 569–576.

Chaves, M. M., Maroco, J. O., & Pereira, J. S. (2003) Understanding plant responses to drought from genes to the whole plant. *Functional Plant Biology* 30, 239–264.

Chen, H. H., Li, P., & Brenner, M. (1983). Involvement of abscisic acid in potato cold acclimation. *Plant Physiology* 71, 362–365.

Collicut, L. (1992). Hardy-rose Breeding at the Morden Research Station *HortScience*, 27, 1070–1147.

Dat, J., Foyer, C., & Scott, I. (1998a). Changes in salicylic acid and antioxidants during induced thermo tolerance in mustard seedlings. *Plant Physiology* 118, 1455–1461.

Dat, J., Lopez-Delgado, H., Foyer, C., & Scott, I. (1998b). Parallel changes in H2O2 and catalase during thermo tolerance induced by salicylic acid or heat acclimation in mustard seedlings. *Plant Physiology* 116, 1351–1357.

Doke, N., Miura, Y., Sanchez, L. M., Park, H. J., Noritake, T., Yoshioka, H., & Kawakita, K. (1996). The oxidative burst protects plants against pathogen attack mechanism and role as an emergency signal for plant bio-defence a review. *Gene* 179, 45–51.

Foyer, C., Lopez-Delgado, H., Dat, J., & Scott, I. (1997). Hydrogen peroxide and glutathione associated mechanisms of acclamatory stress tolerance and signaling. *Physiologia Plantarum* 100, 241–254.

Griffith, M., Lumb, C., Wiseman, S. B., Wisniewski, M., Johnson, R. W., & Marangoni, A. G. (2005). Antifreeze proteins modify the freezing process in planta. *Plant Physiology* 138, 330–340.

Hartmann, H. T., & Kester, D. E. (1972). Plant Propagation: Principles and Practices. Chapman and Hall, London. 283 p.

Heino, P., Sandman, G., Lång, V., Nordin, K., & Palva, E. (1990). Abscisic acid deficiency prevents development of freezing tolerance in *Arabidopsis thaliana* (L.) Heynh. *Theoretical and Applied Genetics* 79, 801–806.

Hon, W. C., Griffith, M., Mlynarz, A., Kwok, Y. C., & Yang, D. (1995). Antifreeze proteins in winter rye are similar to pathogenesis related proteins. *Plant Physiology* 109, 879–889.

IPCC (2001). Climate Change 2001 Synthesis Report Summary for Policymakers http://www. ipcc.ch/pdf/climatechanges2001/synthesisspm/synthesisspmen.pdf.

IPCC (2007). Climate Change 2007 Synthesis Report http://www.ipcc.ch/publications_and_data/publications_ipcc_fourth_assessment_report_synthesis_report.htm

Jaglo Ottosen, K., Gilmour, S., Zarka, D., Schabenberger, O., & Thomashow, M. (1998).*Arabidopsis* CBF1 over expression induces COR genes and enhances freezing tolerance. *Science* 280, 104–106.

Jain, M. (2001). Tissue culture-derived variation in crop improvement. Euphytica 118, 153–166

Janda, T., Szalai, G., Tari, I., & Paldi, E. (1999). Hydroponic treatment with salicylic acid decreases the effects of chilling injury in maize (*Zea mays* L.) plants. *Planta* 208, 175–180.

Jurivich, D. A., Sistonen, L., Kroes, R. A., & Morimoto, R. I. (1992). Effect of sodium salicylate on the human HS response.*Science* 255, 1243–1245.

Lång, V., Mäntylä, E., Welin, B., Sundberg, B., & Palva, E. (1994). Alterations in water status, endogenous abscisic acid content, and expression of *rab*18 gene during the development of freezing tolerance in *Arabidopsis thaliana*.*Plant Physiology* 104, 1341–49.

Lång, V., Mäntylä, E., Welin, B., Sundberg, B., & Palva, E. (1994). Alterations in water status, endogenous abscisic acid content, and expression of *rab*18 gene during the development of freezing tolerance in *Arabidopsis thaliana*. *Plant Physiology* 104, 1341–49.

Larkin, P. J., & Scowcroft, S. C. (1981). Somaclonal variation a novel source of variability from cell culture for plant improvement. Theor Appl. Genet. 60, 197–214.

Lee, G. J., Roseman, A. M., Saibil, H. R., & Vierling, E. (1997). A small heat shock protein stably binds heat denatured model substrates and can maintain a substrate in a folding competent state. *EMBO Journal* 16, 659–671.

Levitt, J. (1980). Water, radiation, salt and other stresses. In Kozlowski, T. T. (ed).*Responses of Plants to Environmental Stresses* (Vol 2), Academic Press NY, 365–488.

Mäntylä, E., Leng, V., & Palva, E. (1995). Role of abscisic acid in drought induced freezing tolerance, cold acclimation, and accumulation of LT178 and RAB18 proteins in *Arabidopsis* thaliana. *Plant Physiology* 107, 141–148.

Mohamed, M. A. H., Harris, P. J. C., & Henderson, J. (2000). In vitro selection and characterization of a drought tolerant clone of *Tagetes minuta*. Plant Sci. 159, 213–222.

Pinto, E., Sigand-Kutner, T. C. S., Leitão, M. A. S., Okamoto, O. K., Morse, D., & Colepicolo, P. (2003). Heavy metal induced oxidative stress in algae. *Journal of Phycology* 39, 1008–1018.

Purohit, M., Srivastava, S., & Srivastava, P. S. (1998). Stress tolerant plants through tissue culture. In: Srivastava, P. S. (Ed.), Plant Tissue Culture and Molecular Biology: Application and Prospects. Narosa Publishing House, New Delhi, 554–578.

Rao, M. V., & Davis, K. R. (2001). The physiology of ozone induced cell death. *Planta* 213, 682–690.

Scheller, H. V., & Haldrup, A. (2005). Photo inhibition of photo system I. Planta 2005, 221, 5–8.

Scott, I. M., Clarke, S. M., Wood, J. E., & Mur, L. A. (2004). Salicylate accumulation inhibits growth at chilling temperature in *Arabidopsis. Plant Physiology* 135, 1040–1049.

Shinozaki, K., Yamaguchi-Shinozaki, K., & Seki, M. (2003). Regulatory network of gene expression in the drought and cold stress responses. *Current Opinion in Plant Biology* 6, 410–417.

Tal, M. (1994). In vitro selection for salt tolerance in crop plants. Theoretical and practical considerations. In Vitro Cell Dev. Biol. Plant 30, 175–180.

Thomashow, M. (1999). Plant cold acclimation. Freezing tolerance genes and regulatory mechanisms. *Annual Review in Plant Physiology and Plant Molecular Biology* 50, 571–599.

Thomashow, M. (1999). Plant cold acclimation: Freezing tolerance genes and regulatory mechanisms. *Annual Review in Plant Physiology and Plant Molecular Biology* 50, 571–599.

Tippmann, H. F., Urte Schlüter, David, B., & Collinge (2006). Common Themes in Biotic and Abiotic Stress Signalling in Plants in Floriculture, Ornamental and Plant Biotechnology Advances and Topical Issues, Jaime A. Teixeira da Silva (ed) Global Science Book Ltd 52–67.

Vierling, E. (1997). The small heat shock proteins in plants are members of an ancient family of heat induced proteins. *ACTA Physiologia Planta* 19, 539–547.

Xin, Z., & Browse, J. (1998). Eskimo1 mutants of *Arabidopsis* are constitutively freezing-tolerant. *Proceedings of the National Academy of Sciences USA* 95, 7799–7804.

Xiong. L., Schumaker, K. S., & Zhu, J. (2002). Cell signaling during cold, drought, and salt stress. *Plant Cell* 14, 165–183.

Yu, X. M., & Griffith, M. (2001). Winter rye antifreeze activity increases in response to cold and drought, but not abscisic acid. *Physiologia Plantarum* 112, 78–86.

Yu, X. M., Griffith, M., & Wiseman, S. B. (2001). Ethylene induces antifreeze activity in winter rye leaves. *Plant Physiology* 126, 1232–1240.

CHAPTER 14

BAMBOO AND SUSTAINABLE DEVELOPMENT WITH CLIMATE CHANGE: OPPORTUNITIES AND CHALLENGES

KAMESH SALAM

World Bamboo Organization (WBO) and South Asia Bamboo Foundation (SABF) E-mail: kameshsalam@gmail.com

CONTENTS

14.1 INTRODUCTION

Bamboo, perhaps the fastest growing plant on the planet, has a very important role to play in restoring balance to the Earth's climate system. Currently, the 30 billion tons of carbon dioxide equivalent produced each year by human activity are wreaking havoc on the global environment. Efforts to curb our CO_2 emission are essential but much more needs to be done. Soon! Global efforts are underway to reduce our planetary carbon emissions below 1990 levels. That still leaves a lot of CO_2 being put into the atmosphere each year by human activities. Bamboo offers perhaps the quickest way to remove vast amounts of that carbon dioxide from the atmosphere. Each acre of bamboo sequesters up to 40 tons of CO_2. The bamboo plant eats carbon dioxide, takes CO_2 from the atmosphere and through the process of photosynthesis turns it into sugars. The bamboo plant transforms these sugars into the compounds that make up bamboo fiber. The carbon from the atmosphere is thus locked up in the bamboo fiber itself. When that bamboo fiber is used to construct buildings the carbon in it is sequestered for the 100-year lifetime of the building. Bamboo is only effective for long-term carbon sequestration if the bamboo plant is being regularly harvested and that harvest turned into durable goods or biochar. Left unharvested the sequestration rate of the bamboo plant levels off. By harvesting 20% of the biomass of the plant each year as 3+ year old mature bamboo culms, the high rates of carbon sequestration are maintain for the 50–75 year life of the bamboo plant. Unlike most trees you are not killing the bamboo plant when you harvest. Each year the mat of primitive roots called rhizomes is expanding, sequestering additional carbon for the life of the bamboo plant. Also unlike trees the bamboo plant produces microscopic plant stones that encapsulated carbon in silica and sequester an additional half-ton per acre of carbon for possibly thousands of years. Bamboos are among the fastest-growing plants, growing at up to a meter per day. Unlike trees, bamboos form extensive rhizome and root systems, which can extend up to 100 km/ha and live for a hundred years. Culms that emerge from the rhizomes die naturally after about 10 years if not harvested before. The rhizome system survives the harvesting of individual culms, so the bamboo ecosystem can be productive while continuing to store carbon, as new culms will replace the harvested ones. The lost biomass is usually replaced within a year. Bamboo can be an efficient tool for both climate change mitigation and adaptation, but there is a lack of scientific knowledge and awareness of its potential Bamboos versatility and unique characteristics provide communities with options to diversify their economies and decrease their sensitivity to climate change. Increasing the cultivation and use of bamboos will help enable rural and urban populations adapt to the effects of climate change. Bamboos are relatively easy to grow and can provide additional food, energy and income security to the rural poor, as well as a range of environmental services and uses in their growing and harvested forms. Bamboo products such as houses and charcoal, can contribute to the livelihood resilience of rural and urban dwellers.

14.2   AREA AND PRODUCTION OF BAMBOO

When we look at 'Bamboo' also known as 'Green Gold' in the Asian culture, it is a symbol of friendship in India; while bamboo's long life makes it a Chinese symbol of longevity.

Several diseases can be cured using bamboo as traditional medicine. Ayurveda describes that bamboo manna; scientifically the siliceous concretion can be used as a healthy tonic to cure diseases. In addition to its health benefits, the bamboo growing could be a prosperous business generating good income for several parts of the world including India. The bamboo covers about 20 million hectares of the world and the bamboo market is to be jumped to US $ 20 billion by 2015 from US $ 10 billion at present. In India, it covers about 12.8% of the total forest of India consisting of 130 MT having an annual harvest of 13.47 million tons. The current market of bamboo/ bamboo products in India is estimated to be Rs.4500 crores, which is expected to increase to Rs.20,000 crores by 2015 with major contribution from wood substitute, processed bamboo shoots, industrial products (activated charcoal, etc.) and Structural applications segments. The employment potential of bamboo is very high and the major work force constitutes of the rural poor, especially women and 432 million workdays per annum are provided by the bamboo sector in India. Rapid increase in the demand of bamboos in the industrial sector coupled with increase in domestic demand due torising population have caused depletion of the natural bamboo resources which calls for concerted efforts for the awareness to raise bamboo plantations in land hitherto barren, degraded or in association with agriculture crops. With the trend of decrease in production and rise in human population, the gap between supply and demand is going to be larger stressed that in India the demand for bamboo planting stocks are 90–120 million per annum, which is expected to increase to up to 300 million seedlings per annum. Large-scale cultivation is the only way to prevent further depletion of bamboo resource, and to ensure a regular and sustained supply of raw material for growing industrial uses. This situation elucidates the need for increase in bamboo production. Due attention on raising bamboo plantation under various programs has not been paid so far. Now farmers and villagers need to be involved in bamboo cultivation/production. Apart from protecting natural vegetation of bamboos, the activity has to be brought to the non-forestlands.

However, there is a low awareness regarding the potential of bamboo and associated products among users and even other stakeholders including the government. The locals in the region use it for everyday uses, but they are not aware of the economical, commercial and industrial applications of bamboo. The database available with the Indian Council of Forestry Research and Education (ICFRE) reflects that 18 million hectares of world is covered by Bamboo. In India, approximately 10 million hectares of area is covered by Bamboo out of which 28% is in the NER. There are more than 125 species belonging to 23 genera of bamboos found in India. Out of this, only 30 species are commercially important. They grow naturally up to 3500 m above mean sea level. Around 66% of India's bamboo resources exist in NER and hence, the potentialities.

14.3 INDUSTRIAL OUTLOOK OF BAMBOO

The requirement of "bamboo wood" for multiple uses by the industries and the common man will definitely increase in far greater dimensions. In India, the total demand of various bamboo-consuming sectors is estimated at 26.9 million tons. The estimated supply is only 13.47 million tons, that is, only half of the total demand. The pulp and paper industry, construction, cottage industry and handloom, food, fuel, fodder and medicine annually consume about 13.4 million tons of bamboo amounting to Rs.2042 crores. Demand of bamboo for industrial use is met from state owned forests, while for non-industrial purpose it comes from private as well as state owned resources. Keeping abreast of versatility of bamboo uses and its potential to build up the rural economy, Government of India launched massive program viz. National Bamboo Mission for over all development of bamboo sector in the country and also to improve the Indian representation in global bamboo market. Bamboo has also been recommended for plantations for a greener, pollution free environment along with economic prosperity. Based on India's rich culture, bamboo utilization has triggered several programs in the country for economic and industrial development through the use of bamboo. Large targets for plantations across the country have been fixed. The National Bamboo Mission (India) envisages covering over 1.76-lakh hectare area through bamboo. This will need over 70 million field plantable saplings to raise bamboo plantations. The emphasis of the National Bamboo Mission is on an area based regionally differentiated strategy, for both forest and non-forest areas. A number of activities are proposed to be taken up for increasing production of bamboo through area specific species/varieties with high yield, plantation development and dissemination of technologies through a seamless blend of traditional wisdom and scientific knowledge, along with the convergence and synergy among stakeholders. Besides ensuring proper postharvest storage and treatment facilities, marketing and export National Bamboo Mission is committed to assure appropriate returns to growers/producers. Also, bamboo development is viewed as an instrument of poverty alleviation and employment generation for skilled and unskilled persons, especially unemployed youth particularly in the rural sector through ecorehabilitation purposes.

To support cause of the NBM for the development in the bamboo sector the situational as well as practical and legal constraints need to be removed. At present bamboo is classified as a 'tree' under the Indian Forest Act, 1927, and therefore, the various restrictions applicable to trees under the Act and its Rules apply to bamboo. The Supreme Court in *T.N. Godavarman Thirumulkpad vs. Union of India* (1977) ordered a complete ban on the felling of any trees. Further, the SC ordered a complete ban on the movement of cut trees and timber from any of the seven Northeastern states to any other state of the country either by rail, road or waterways. The cultivation, harvest and transport of bamboo are therefore constrained by the said judgment read with the Indian Forest Act. Nagaland state, however, has lifted restrictions on bamboo, but since the neighboring states have not lifted restrictions it is difficult to transport the produce beyond the state of Nagaland. While there is no restriction on the cultivation of bamboo on private lands, but transporting bamboo across the state boundaries or even within a state becomes problematic as forest and other officials at every check-point have to be satisfied that the bamboo being transported has been from private lands and not from forest lands. Therefore, the primary hindrance in respect of commercialization of bamboo is transport permits.

China successfully has been the main driving force in the global bamboo industry's development over the last 15 years, with a global share of almost 80%. The total world market for bamboo is worth USD 7 billion/ year with handicrafts taking up just over 40%. Oxfam Hong Kong predicts that the global market will continue to grow to an estimated value of USD 15–20 billion/year by 2017. India has opened its boundaries to the market economy and is promoting itself as being on the crossroads of trade between China and South-east Asia. Although policy reforms are continuing, doing business in India is still a challenge for the private sector. Many have recognized the great potential of bamboo as an industry, be it for handicrafts, shoots or industrial processing. Yet few countries have so far been able to develop their bamboo industries beyond their traditional handicraft markets to exploit the huge potential for rural economic growth and the resulting poverty reduction. While bamboo handicrafts exist around the world, the industrial bamboo subsector is too often notable only by its absence. Considerable challenges exist for those aspiring to replicate the success of regions such as Anji, Li'nan and Fujian in China. Such leading regions and their associated industries have set the benchmark in terms of cost and efficiency of production, exploiting competitive advantages from the development of dense industrial bamboo clusters. At the same time they have seemingly lowered technical barriers to entry by commercializing much of the technology and opening the markets to bamboo-based products. So why, with available technologies, developing markets and local bamboo resources, has it proved so difficult for others to replicate the success of China or Vietnam? Why have industrial bamboos clusters not emerged more widely and achieved the same large impacts on rural development and poverty reduction? The challenge for those outside China (countries and regions as well as private enterprises) has been how to make the transition from traditional, often small scale, processing industries to the efficiencies and scale needed to compete in the world market against China. China's leading bamboo regions have achieved remarkable efficiencies in using every part of the bamboo that leaves the forest – with raw material conversion rates often exceeding 95%, including branches and leaves as well as the main culms themselves. With every part of the bamboo being used somewhere in the industry, individual businesses are able to buy only the exact part of the bamboo they require for their particular product and achieve low unit costs of production despite some of the world's highest farm-gate prices for raw bamboo. The efficiency in material utilization has been made possible by the development of relatively complex and geographically concentrated supply chains. The strong competition for raw material among businesses means that every part of the bamboo is used for the products of greatest added value and businesses constantly strive to find ways to increase their efficiency and value addition to the bamboo. The Chinese industry has also been relentless in its innovations in processing and machinery. These have allowed it to achieve increasing labor productivity to offset the rising cost of workers. For those outside of China, the efficiencies of the Chinese industry have made it very difficult to compete with in export markets. Emerging industries elsewhere often struggle with raw material utilization rates for added value products of perhaps 15%–25%. The dramatically higher efficiencies in China mean that the sales prices of Chinese bamboo products are often lower than the cost of production of similar products elsewherewhere, despite big differences in the cost of raw materials and labor.

The financial viability of bamboo related commercial projects are another problem we need to look into. The financial institutions are wary of the facts regarding operational difficulties, logistics and supply chain problem, track record of stakeholders, lack of developed market for bamboo products and lack of work ethics. In such a situation banks want increased participation of all stakeholders and subsidy by the government. In such a situation development of suitable, small but profitable industrial investment opportunities in the bamboo and cane sector would be a necessary first step of the development of the sector, and thereby employment to the disgruntled youth. There is no National Policy on Bamboo. A National Policy or Policy on Bamboo for the NER will definitely boost the value added products made from bamboo. In NE State the state of Nagaland Government's initiative to create the Nagaland Bamboo Development Agency (NBDA) has made exemplary model in the country for institutionalizing bamboo for community. Also made its highest effort to make awareness of bamboo in the country by hosting the First "World Bamboo Day" Celebration at Kisama, Nagaland on 18th September 2010 which was graced by the Vice President of India H.E. Shri Hamid Ansari which was attended by delegates from all India and more than 15 countries.

To commoditize bamboo there is a great need to carry out institutional reform and restructuring. The examples of "Tea Board," "Coffee Board," "Coconut Development Board," "Spice Board," "Jute Development Board," "Silk Board," "Rubber Board," "Coir Board" under various ministries of Government of India are examples of commoditization. These bodies acts as "one-stop-shop" and looks after all the requirements from development, promotional, plantation to processing and marketing of the product including R&D activities and market promotion, etc.

The bamboo subsector in North East is undeveloped due to a lacking enabling environment to stimulate the private sector to invest in the states. Capacity of government staff to tackle reforms and necessary insights to promote and develop marketable products is limited. Consequently private sector and poor communities have few incentives to sustainable manage and gain from trading bamboo resources. Key intervention areas to support the development of the bamboo value chain are the enabling business climate, handicraft sector and piloting business models aiming at providing benefits for both the private sector and communities. Lessons from China, Vietnam and other South Asian Countries are needed to be learnt for India. India' Aggarbatti Industry (Incense Sticks) which is pegged at Rs.4000 crore market imports bulk of the required bamboo sticks from China and Vietnam whereas abundant bamboo resources in NE States are burnt by Jhumming or pulped for paper by the two giant Paper Mills in the region leaving less scope for the farmers for value addition and benefits. Bamboo is useful for ecological and environmental purposes as well. It is one of the best species for carbon sequestration; it conserves soil and water besides many other protective purposes Since, market for environment-friendly "green" products is growing, India must try to secure her due share in world bamboo market which is expected to grow from USD 10 billion to over USD 20 billion by 2015, if we could expand the bamboo economy steadily to Rs.26,000 Crore by 2015 from Rs.2000 Crore, as envisioned by the Building Materials and Technology Promotion Council (BMPTC), Government of India. To achieve this target the social and commercial aspects of bamboo and bamboo products need to be disseminated at the grass root level. Simultaneously, a program of awareness building is needed to influence the government (commerce, industry and various ministries), bankers, potential traders and investors on the potential of bamboo – economic, social and environmental. However, this awareness has to be built regarding benefits they can relate to- the value proposition, people and environmental friendliness of bamboo and connecting the poor.

KEYWORDS

- **Bamboo**
- **Climate Change**
- **Green Gold**
- **Mitigation Strategies**
- **National Bamboo Mission**
- **North East States**

REFERENCES

Planning Commission, (2003). Report of the National Mission on Bamboo Technology and Trade Development, Yojona Bhavan, New Delhi, Govt. of India.

Kamesh Salam (2002). Keynote address Expert consultation on sustainable utilization of bamboo resources subsequent to gregarious flowering in the northeast. Rain Forest Research Institute, Jorhat, 24th April 2002.

Brias, Victor. (2003). UNIDO Technical Report – Bamboo Plantation Feasibility Study for North East India, (Unpublished Report).

DPR of National Bamboo Mission by CBTC, (2004). Ministry of Agriculture and Co operation, Govt. of India.

Kamesh Salam (2009). Foreword. Proc. VIII World Bamboo Congress, Bankok, Thailand.

David, E. (2009). Sands, Bamboo and Climate Change: The Imperative, 9th World Bamboo Congress proceedings.

Dr. Walter Liese. (2009). Bamboo as Carbon-Sink Fact or Fiction? 9th WBC Proceedings.

Ajay Kakra1 & Nirvanjyoti Bhattacharjee (2009). Yes Bank, Marketing and Supply Chain Analysis of Bamboo Products from North Eastern India in Major Consumption Markets of India, 9th WBC Proceedings.

Olivier Renard, & Patrice Lamballe. Creating sustainable jobs and incomes to reduce poverty: lessons from bamboo supply chain development project in North West Vietnam. 9th WBC, Proceedings.

Adarsh Kumar. Macroproliferation Technology for Raising Large Scale Plantations of Sympodial Bamboos. 9th WBC Proceedings.

Arun Jyoti Nath & Ashesh Kumar Das. Carbon farming through village bamboos in rural landscape of northeast India as affected by traditional harvest regimes, 9th WBC Proceedings.

Haque, M. S. Planning, designing and implementing a Jati Bamboo (Bambusa tulda) plantation scheme through bank credit on small landholder's revenue wastelands in Assam, India for sustainable livelihood, 9th WBC Proceedings.

Rajasekharan, V. M. ITC Limited Incense Business, 9th WBC Proceedings.

Nigel Smith & Tim DeMestre. Establishing industrial bamboo enterprises through the value chain approach, 9th WBC Proceedings

Salam, K. CBTC Newsletters 2004–2007.

CHAPTER 15

CLIMATE CHANGE EFFECTS ON FRUIT QUALITY AND POST-HARVEST MANAGEMENT PRACTICES

M. S. LADANIYA

Principal Scientist, National Research Centre for Citrus, P.B. No. 464, Shankar Nagar P.O., Nagpur-440010, India; E-mail: msladaniya@gmail.com

CONTENTS

ABSTRACT

Climate change is taking place due to industrial emission of green house gases (GHG), use of fossil fuels, deforestation, and destruction of vegetation and extensive construction activities. This is destroying ecosystem, natural hydrological cycle and affecting ground water recharges. Climate change is causing rise in temperature, erratic rainfall, increased wind velocity and lower RH. These changes are resulting in scarring, sunburn and poor color development in apple, apricots, cherries and many other fruits. Climate change is likely to alter the balance between insect–pests and their natural enemies and this may increase blemishes on fruit and percentage of cull fruit. Any increase in temperature, relative humidity, rainfall and plant canopy growth would increase incidence of *Botrytis* bunch rot in grapes. Rains and cloudy weather just before harvest can increase postharvest spoilage in all the fruits. Wind and insect scarring affect fruit quality and fruit become unmarketable. Evaporative cooling of fruit on trees through sprinklers and Sunburn protectant spray can reduce sunburn in apple,that is, 0.2% sunburn in treated fruit as compared to 10.8% in control. Frost damages have been reported in Aonla and Ber due to unusual low temperatures in northwest India. After chilling temperatures at maturity, postharvest losses in avocado have increased. In mandarins grown under higher temperatures, color development is poor with less acidity while rains and hailstorms during maturity increased postharvest losses. Rains have increased losses in acid lime due to rotting and splitting. With increasing temperature due to climate change, cold chain has become unavoidable and it is must. Higher market standards of importing countries may also lead to wastage as produce that does not meet size, shape or appearance criteria has to be disposed of at throw away prices. Airfreighted fruit has more carbon footprint and may face trade–barrier and sea-freight protocols need to be developed. Green and ecofriendly postharvest technologies will help mitigate warming impact. Changing conditions demand orientation of postharvest strategy to improve infrastructure to reduce losses and increase availability of produce. Utilization of damaged fruit, strengthening of processing infrastructure, and cottage scale and home scale processing need to be strengthened. Several issues about postharvest research, policy and development need rethinking and action in the face of climatic change that are going to affect our lives in coming decades.

15.1   INTRODUCTION

Global warming and resultant climate change is a slow phenomenon affecting mankind over the past one and half centuries and its pace has increased with the pace of industrialization and population growth on the Earth. This definitely slow and steady change in weather conditions at regional levels affecting seasonal operations, crop phenology, yield and quality of the produce.

Climate refers to average atmospheric conditions including gases, barometric pressure and radiation. It is a weather condition viz. temperature, RH, precipitation, wind, and sunshine that occur for longer period (many years) over the larger areas/region. Example-Tropical climate (between tropics of Cancer and Capricorn), Arctic climate (Near Arctic pole). Weather refers to localized atmospheric conditions over smaller area and exists for shorter period (day-to-day) and it changes within several hours or days and forecasts are made daily or for few days.

Atmosphere is agaseous surrounding of the Earth surface with its composition (water vapor, particles and other gases), barometric pressure and light (radiation energy) that is changing with altitude. Most important gas with respect to climate change is carbon dioxide. However gases such as nitrous oxide (N_2O), other nitrogen oxides, methane, ozone and chlorofluorocarbons also contribute to greenhouse effect and global warming. All these gases are known as green house gases (GHG).

Natural Environment includes everything living and nonliving things surrounding mankind on Earth viz. weather (Temperature, RH, sunshine (light), precipitation, wind/air), atmospheric gases (Oxygen, nitrogen, GHG), soil (land and mountains), water (rivers, oceans, lakes and other water bodies), flora and fauna.

Climate change (it is also referred as Global warming) is occurring due to: (1) industrial emission of GHG; (2) use of fossil fuels; (3) extensive construction activities destroying ecosystem, natural hydrological cycle and affecting ground water recharge; and (4) deforestation. Natural factors like volcanoes, forest fires and dust storms contribute very little to global warming.

15.2   GENERAL EFFECTS OF CLIMATE CHANGE ON WEATHER PARAMETERS AND ATMOSPHERE

15.2.1   INCREASED AIR TEMPERATURE

This is due to higher GHG in the air since carbon absorbs more heat. Temperatures are increasing from March onwards throughout North and Central India and in first week of April, 42–45°C is recorded in most parts of this region in the country.

15.2.2   ERRATIC RAINFALL

Global warming has affected ocean wind current intensity and directions. Slowly and steadily annual rainfall is moving towards deficit with erratic monthly trend. Distribution of rainfall is disturbed and more rains are received in July–August and sometimes extending in to September–October. It disturbs entire crop cycle in the year. Higher rains/cloud bursts are causing floods in some parts and in other parts deficit rains are causing drought. There is a shift in rain pattern. Monsoon is arriving late in central and North India and 10–15% less rains received since last few years. There is a heavy downpour at some places while some places are dry. In central part of India very less or negligible rains are received in June and there is a deficit of 5% in July. Heavy rains are received in August causing floods in some parts. Precipitation trend during last 30 years in Vidarbha indicated that average rainfall is 1000 mm but actually average rainfall received during last 10 years is 772 mm with annual deficit of 256 mm (Huchche et al., 2010). There is a significant shift in distribution and total precipitation received during last 10 years. It has detrimental effect on citrus industry of Vidarbha. Ground water table has gone down substantially in this part of the country.

It was observed that monsoon rainfall during 1901 to 2003 was without any trend and mainly random in nature over a long period of time, particularly on the all India time scale (Table 15.1). But on the spatial scale, existence of trends was noticed. The monsoon rainfall in sub-Himalayan West Bengal and Sikkim and the Bihar Plains are having decreasing trends while Punjab, Konkan and Goa, West Madhya Pradesh and Telangana are having increasing trends (Guhathkarta and Rajeevan, 2006).

TABLE 15.1   Decadal Mean (% departure from normal), Frequency of Deficit and Excess Rainfall Years During Last Century in India

Decade	Decadal mean Per cent departure from normal	Frequency of Deficient year (in a decade)	Frequency of Excess rainfall year (in a decade)
1901–1910	−2.2	3	0
1911–1920	−2.3	4	3
1921–1930	−0.4	1	0
1931–1940	1.7	1	1
1941–1950	3.3	1	1
1951–1960	2.5	1	3
1961–1970	−0.1	2	1
1971–1980	−0.8	3	1
1981–1990	−0.3	2	2
1991–2000	0.6	0	1
2001–2003	−5.6	1	0

Source: Rainfall pattern (Indian Meteorological Department, Pune, 2006).

Monthly trend indicated that July rainfall is decreasing for most parts of central India while it is increasing for the northeastern parts of the country. However June and August rainfall is increasing for the central and southwestern parts of the country. During the southwest monsoon season, Jharkhand, Chhattisgarh, Kerala showed significant decreasing trend and Gangetic West Bengal, Western UP, Jammu and Kashmir, Konkan and Goa, Maharashtra, Andhra Pradesh and North Karnataka showed significant increasing trends. Contribution of July rainfall is decreasing in central and west peninsular India but August rainfall is increasing in all these areas.

15.2.3   MELTING OF GLACIERS AND ICE COVER AT HIGHER LATITUDES AND NEAR POLAR REGION

It will have devastating effects on temperate fruit culture in long run.

15.2.4   EMISSIONS ASSOCIATED WITH FOSSIL FUEL AND BIOMASS

Burning has acted to approximately double the global mean tropospheric ozone concentration, and further increases are expected over the 20-first century. Tropospheric ozone is known to damage plants, reducing plant primary productivity and crop yields, yet increasing atmospheric carbon dioxide concentrations are thought to stimulate plant primary productivity. Increased carbon dioxide and ozone levels can both lead to stomatal closure, which reduces the uptake of either gas, and in turn limits the damaging effect of ozone and the carbon dioxide fertilization of photosynthesis (Sitch et al., 2007).

Regional climate changes take place depending on geographic position on Earth and damage to environment and carbon emission in that part. Countries with tropical climate near equator are facing different types of problems than temperate climate countries away from equator.

The climate changes due to global effects in temperature and CO_2 rise will be further aggravated by deforestation at local levels. Vegetation of bushes, trees and other plants play important and positive role in preserving microclimate and protecting from extremes of temperature, wind velocity, rains and solar intensity. With these positive effects, vegetation and forests actually are blessings of nature to mankind. Horticultural and fruit production is protected with such vegetation around orchards or on open lands, waste lands and road sides.

15.3   BENEFICIAL EFFECTS OF CLIMATE CHANGE/GLOBAL WARMING

Increased CO_2 is likely to increase photosynthetic output thus increasing yields in some crops. There will be qualitative improvement also in fruit quality in temperate and subtropical regions with rising temperature due to increased sugar accumulation and early maturity.

15.3.1   EFFECTS OF CLIMATE CHANGE ON FRUITS

Maximum effect of rising temperatures and lower relative humidity will be on perishable horticultural produce. Storage life and shelf life of the produce will be affected most as the temperatures have shown rising trend right from February onwards in most parts of India. If temperature data of last 50 years is taken in to account, it is clear that average temperatures have risen by 2–3°C and this rise is erratic in nature. There is a sudden rise in temperatures in March and April rising up to 42–46°C in central India. Such temperatures were earlier experienced in the month of May. Maximum temperatures have increased up to 47.7°C and 48.2°C in May 2010 at Nagpur and Chandrapur, respectively. This trend is subtle and now in most parts of North and Central India temperatures rise above 41–42°C right from March onwards. These conditions are forcing orange growers to harvest the fruit in February–March itself as compared with harvesting season till April earlier (up to year 2000) in Central India. Fruit drop in 'Nagpur' mandarin may increase due to water scarcity and dry (stress) conditions.

Climate impact is most felt in temperate region because increasing minimum and maximum temperatures are causing devastating effects on flowering and fruiting. Changing degree-days would change heat summation and in turn it will affect fruit development and maturity duration that will have direct bearing on postharvest operations, packing, transport and distribution chain. Farmers in Himachal Pradesh are shifting to other crops. Apple cultivation is being replaced with Kiwi and pomegranate (Singh, 2010).

In arid region of Rajasthan (Bikaner), before year 2000, temperatures never dropped below 0°C in winter. In last 8–10 years, it is common feature that temperature drops below 0°C every year. During 2006–2009, temperatures remained below 0°C for 2–3 days consecutively in January. Region is experiencing wide variation in rainfall since last 20 years, that is, 98–472 mm rains annually. Floods have also occurred in recent past and their frequency increased.

There is a lot of heterogeneity in the opinion and estimates about effects of climate change on fruits. Escalating greenhouse gas levels may significantly boost production of fruits. But the effect may be a double-edged sword; the increase in yield appears to be linked to a decrease in the nutritional value of these crops, which needs to be examined.

15.3.2 GENERAL EFFECTS OF CLIMATE CHANGE ON FRUIT QUALITY

Fruits are grown in temperate, subtropical and tropical regions and the climate change will have different effects on fruit species of these regions. High temperatures mean more heat summation and early fruit maturity. High temperature means higher respiration, decline in acidity, lower sugars and more pH. This leads to poor quality. Warmer winters will lead to early fruit maturity in citrus and wetter summers may lead to more pre- and postharvest diseases.

Rainfed horticulture will be more affected because with rising temperatures conditions will be harsher—for example:
1. with deficit rainfall, drought conditions will occur adversely affecting fruit set and fruit quality;
2. there will be lower juice content, poor quality in terms of aroma and taste;
3. fruits will be of smaller size and lower fruit weight.

Studies have shown that the production and quality of fresh fruit crops can be directly and indirectly affected not only by high temperatures and exposure to elevated levels of carbon dioxide but also higher ozone. Temperature increase affects photosynthesis directly, causing alterations in sugars, organic acids, and flavonoids contents, firmness and antioxidant activity. High concentrations of atmospheric ozone can potentially cause reduction in the photosynthetic process, growth and biomass accumulation. Ozone-enriched atmospheres increased vitamin 'C' content and decreased emissions of volatile esters of strawberries. Tomatoes expose d to ozone concentrations ranging from 0.005–1.0 μmol/mol had a transient increase lycopene contents.

15.4 PRE-HARVEST FACTORS THAT AFFECT FRUIT QUALITY AND STORAGE ABILITY

Climate is an important preharvest factor that affects postharvest quality. Pre-harvest factors also have bearing on postharvest losses. Rains and cloudy weather just before harvest season can increase postharvest spoilage considerably. Wind and insect scarring affect fruit quality and fruit become unmarketable. Most important but much more difficult to quantify are influences that affect susceptibility to physiological and pathological breakdown in postharvest environment.

15.4.1 BIOTIC STRESSES

15.4.1.1 INSECT–PESTS

Climate change is likely to alter the balance between insect–pests and their natural enemies and their hosts. Incidence of insect–pests is going to increase with rise in temperature. Most damaging will be fruit fly (*Bactrocera* sp.) that can have preas well aspostharvest impact on fruit quality. In Australia, costs of apple, citrus and pear production are estimated to increase by 25, 38 and 95% with increase of 0.5°C, 1.0°C, and 2°C temperature, respectively due to changing pattern of fruit fly attack (Sutherst, 2000).

15.4.1.2 DISEASES

It is estimated that any increase in temperature, relative humidity, rainfall and plant canopy growth would increase incidence of *Botrytis* bunch rot in grapes. It is estimated that field and postharvest fungal diseases would depend on temperature, relative humidity and rainfall changes, which in turn would be affected by climate change.

15.4.2 ABIOTIC STRESSES

These stresses include water scarcity (drought), excess water (floods), excess temperature, excess evaporation and high CO_2 levels. Losses due to fruit drop will be more under high temperature and water scarcity conditions as water during fruit development is necessary. Floods have already started causing soil erosion, which is going to affect production and quality in long run.

15.5   IMPACT OF CLIMATE CHANGE ON SOME FRUITS

15.5.1   APPLE, APRICOTS AND CHERRIES

Higher temperatures during autumn are causing sunburn and poor development of color affecting marketability in apples, apricots and cherries. Pre-harvest stresses predispose the fruit to physiological disorders whichappear long after harvest. Heat absorbed by fruit is related to fruit surface temperature and depends on light intensity. In Washington state USA, during 2003 season, apple fruit temperature increased to more than 45 °C for at least 15 min in 39 days during fruit growth and maturity in June-September. Although air temperature varied from 26–37 °C, the fruit surface temperature varied between 38 °C and 48 °C. Sunburn (Necrosis and browning) was largest source of cullage and up to 25% fruit was affected and rendered unmarketable (Schrader et al., 2003). It was also observed that high temperatures in field increased incidence of 'Fuji stain' and 'lenticel marking' after harvest in many apple cvs. In cv. 'Jonagold,' bitterpit incidence (which is known to be caused due to calcium deficiency) increased and fruit had blotchy skin appearance with irregular color development due to high temperature and water stress just before harvest. Watercore in 'Honeycrisp' apples and splitting of 'Fuji' apples were also noticed due to water stress and high temperature before harvest. 'Granny Smith' apples are generally susceptible to Sunscald and it develops during cold storage. Losses due to Sunscald were higher during the years when air temperatures in field were higher than usual.

With the increase of hailstorms as a possible result of global warming, fruit crops are increasingly grown under hailnets near Bonn in Germany. This results in lesser fruit quality in terms of coloration, fruit mass, firmness, starch and taste, that is, lower sugar and acid, and vitamin content under hailnet due to altered microclimate and light deprivation. Under the translucent, 'white' hailnet, humidity was increased by 6%, air temperature reduced by 1.6 °C, soil temperature increased by 0.5 °C and light reduced by 11–15% resulting in lesser fruit quality (2.5% less sugar and less taste). Reflective mulches of synthetic material (trade name 'Extenday' and 'Daybright') were tried by Solomakhin and Blanke (2007) to improve fruit quality and light utilization under hailnet. The two reflective mulches increased light reflection by 2.5–6.3 fold. No differences in fruit ripening and firmness were observed, but fruit from trees under hailnet with reflective mulch contained up to 2.4% (from 13.3–15.7%) more sugar than those of the control (uncovered grass alleys).

15.5.2 BLUEBERRY

In general, high temperatures during blueberry harvest season hasten ripening. Heavy rains during periods of high temperature can further hasten ripening, cause splitting, and greatly reduce storage quality. Overripe blueberries are extremely susceptible to damage.

15.5.3 AONLA

In arid region of Bikaner, aonla fruit is generally harvested before severe winter but fruit of late varieties gets affected by chilling temperatures in January. Fruit becomes whitish in color with oozing of water and then it becomes dry and black (More and Bhargava, 2010). Frost affected fruit has no shelf life as the secondary infection starts immediately.

15.5.4 BER

Fruits are susceptible to chilling temperature and frost and losses are higher in years where continuous low temperature persists (More and Bhargava, 2010). Cultivars Sanaur-5, Syriya and Tikkadi showed tolerance to frost, Umran and Mundia were found to be susceptible while Sanaur-1, Kathaphal and Jogia were moderately affected. Frost damaged fruits are shriveled, brown and turn black afterwards.

15.5.5 AVOCADO

Changes in climate will affect avocado fruit quality and storage ability also. Avocado fruit shape is influenced by temperature and humidity. Cooler climate tend to make fruit rounded while high temperature with humidity causes fruit to be elongated (Arpaia et al., 2004). After mild freezes in California, avocado fruit exposed to chilling temperature had more decay, chilling injury and weight loss following storage. Increased temperatures during harvesting increased rotting by 0.2% for every 1 °C rise in temperature in field during harvesting season. Higher the temperature and more the delay in cooling, more is the decay during storage (McCarthy, 2009). Unseasonal rainfall during harvesting increased inoculum and storage decay (Smilanick and Margosan, 2002)

15.5.6 BANANA

Climate change has positive effect on banana and number of leaves produced by the plant are higher with highertemperature and that increased production (Singh et al., 2010). Higher temperature with reduced RH decreased incidence of leaf spot under subhumid climate.

15.5.7 CITRUS

In citrus grown under higher temperatures, color development is poor with less acidity. Cool temperatures also have adverse effects. During development of Satsuma mandarin (*Citrus unshiu* Marc.), lower temperaturecan limit the total soluble solids: titratable acidity ratio of fruit. This delays harvesting season and number of fruit that meet maturity criteria. Higher/ rising temperatures significantly affect size and composition of Satsuma mandarins through changes in vascular transport capacity, ability to attract carbohydrates (sink strength), duration of growth stages and partitioning of carbohydrates in fruit. Richardson et al. (2000) studied effect of increased temperature on Satsuma mandarin under controlled conditions in New Zealand. Raising maximum temperatures by 5 °C during early fruit development consistently increased final fruit volume by 50% and elevated sugars and TSS in juice sacs (TSS:TA ratio of 12.3 vs. 7.7 in control). These effects were due toboth an advance in fruit development and a greater capacity of fruit to import photosynthates. This can be considered as a positive effect of higher temperatures in the event of global warming on some citrus cvs. in New Zealand.

Kinnow mandarin, sweet oranges and other citrus grown in north-west parts of India (Punjab, Haryana and Rajasthan) are prone to chilling temperature. Most citrus is mature during December–January and increasing frost incidence in recent times caused freezing and hardening of some fruit. Hailstorms just before harvesting season in February–March has become regular feature in Vidarbha region and causing 50% or even total loss of 'Nagpur' mandarin crop in some parts.

The sour rot pathogen, *Geotrichum candidum* (now *Galactomyces citri-aurantii*), which grows at higher temperatures of 30–32 °C has been found to dominate the spoilage and consequently lead to heavy postharvest losses in 'Nagpur' mandarin during February–March as compared to October–November when temperatures are 20–25 °C. With rising temperature due to climate change this fungus may also affect fruit in October–November when spring blossom crop is harvested. Rains have increased the losses in acid lime during May–June due to rotting and splitting. Unseasonal rains during mandarin harvesting season increased decay losses (Ladaniya and Wanjari, 2003).

15.5.8   GRAPES

In grapes poor color development may occur if temperatures are high during fruit maturity. Increased maximum and minimum temperature lead to reduction in anthocyanin content. Increase in night temperature (minimum temperature) will severely affect pigment synthesis. Total flavonoid content was also adversely affected due to rise in temperature in cv. Cabernet Sauvignon. High temperatures caused reduction in titratable acidity and increase in pH (Anonymous, 2008).

Ashenfelter and Storchmann (2010) measured the effect of year-to-year changes in the weather on grape wine prices in the Mosel Valley of Germany in order to determine the effect that climate change is likely to have on the income of wine growers. The retail price for 1994–2008, wholesale price for 1993–2001 and auction prices for 1981–2008 were taken in to consideration. The empirical results of their study models indicated that the vineyards of the Mosel Valley will increase in value under a scenario of global warming, and perhaps by a considerable amount. Vineyard and grape prices increased more than proportionally with greater ripeness, and it is estimated that 3 °C increase in temperature would double the value of vineyards while1 °C increase would increase prices by more than 25%. This may be because German wines are classified and labeled according to the natural sugar content of the unfermented grape must (freshly pressed grape juice). In general, sweeter unfermented musts lead to higher alcohol volumes more aroma and thus higher quality. Increased temperatures are likely to increase sugar contents.

15.5.9   STRAWBERRY

There is a positive effect of higher atmospheric CO_2 level on strawberries. Elevated CO_2 conditions increased the content of aromatic compounds in strawberry fruit. The anthocyanin content of strawberries was decreased at high temperature in the growth chamber, that is, 35/20 °C (day/night).

15.5.10   POMEGRANATE

North-West parts of India experiences frost in winter and frequency of chilling temperatures below0 °C increased in recent past. Chilling/freezing temperatures in winter causes hardening of fruit and softening with dull color rendering fruit unmarketable.

15.5.11   RAMBUTAN

Rambutan production in Hawaii is affected due to different weather conditions in recent past. As a consequence production is very erratic due to inconsistent flowering and the occurrence of deformed or aborted fruit devoid of fleshy aril has increased.

15.5.12   PINEAPPLE

There has been a bumper crop of pineapple in Taiwan in 2007,which was attributed to climate change. Fruit developed faster, ripened earlier and were more sweet. Usually pineapples mature in May–June in southern coastal counties of Taiwan but fruits were ready in March in 2007 due to erratic rising temperatures/global warming.

15.6 IMPACT OF CLIMATE CHANGE ON POST-HARVEST MANAGEMENT PRACTICES

Production of fruits is likely to be affected as the weather changes will be sudden and more intense in its effects thus changing flowering, fruiting and yield patterns. Erratic and un-seasonal rainfall may affect cropping cycle of short duration crops. Lack of availability of produce is likely to affect processing industry and stable supply for processing will be difficult to maintain. Fruit industry is dependent on regular and timely rhythmic cycle of seasons every year. Slight changes in total precipitation, its duration, timing, and intensity is going to affect fruit production. Variations in temperatures, humidity and untimely rains have affected citrus, mango and cashew crops in the past.

In India, cold chain is absent in case of almost 95% of the produce that is handled and marketed. Water loss, shriveling and ultimate weight loss will increase financial loss with rising temperatures. Due to these factors, all stakeholders in the marketing chain will be affected including growers, wholesalers and retailers.

Respiration rate increases (two-three times) with doubling of temperature in citrus fruit (Table 15.2) and that exhausts stored substrates in fruit tissues mainly sugars, carbohydrates that contribute to quality of produce (Ladaniya, 2008). It is true for all fruits. Rate of deterioration is generally proportional to respiration rate. Most of the fruit in India is handled at ambient conditions and with rising ambient temperatures respiratory activity is going to increase with detrimental effects on harvested fruit.

TABLE 15.2 Respiratory Activity (CO_2 mg/kg/hr) of Citrus Fruits as Influenced by Temperatures

Fruit	5 °C	10°C	15°C	20°C	25°C	30–32°C
'Mosambi' orange	7–8	11–12	18.5–19	27–28	38–40	-
Kinnow mandarin	5–6	10–11	12–14	17–18	28–30	48–50
'Nagpur' mandarin	7–9	10–13	15–18	25–30	32–38	40–46
Grapefruit	-	7–10	10–18	13–26	19–34	-
Lemons	-	11	10–23	19–25	20–28	-
Limes	-	4–8	6–10	10–19	15–40	40–55

Volatile compounds that contribute to fruit flavor (aroma) get evaporated rapidly with higher temperatures and lower RH thus adversely affecting this vital quality attribute. With higher temperatures, more water is required for maintaining quality of the produce in distribution chain.

Post-harvest losses due to frosts are likely to be increased as fruits develop freezing injury. Untimely rains and hailstorms in February-March cause losses to mandarin growers every year since last 15–18 years in Central India. Rains during harvest season cause considerable losses due topostharvest diseases.

Main factors in quality deterioration of fresh fruits are temperature and RH with respect to time. Growth of microbes could be exponential with higher temperatures. Post harvest pathogens have their temperature optima.

With changing climate, food security challenges are going to aggravate and postharvest management strategy needs a retrospection and modifications. A country like India is facing twin challenges, that is,(1) Population explosion; and (2) Global warming and climate change. It is clear that our planning and policy-making have failed to tackle former issue completely. For later, time is running out. Earlier the corrective measures taken better it will be. It is the industrialized nations who are generating maximum GHG and they have to take initiative. India needs to take corrective measures on its own at least on its territory to meet challenges of global warming immediately without waiting for other countries. Climate change not only threatens production system and impairs our efforts for higher productivity but it also affects quality and postharvest storage ability.

At present, losses are 20–30% depending on season of production, commodity, chain of middlemen, distance to market, packaging and post-harvest management practices. Post-harvest losses are likely to increase further with extremes of climatic events like droughts, floods, storms and frosts.

The impact on postharvest responses, of preharvest exposure of fruit to direct sunlight, with associated high tissue temperatures, has been reviewed (Woolf and Ferguson, 2000). Fruit flesh temperatures well above 40 °C have been recorded in direct sunlight in a wide range of crops in both hot and temperate climates. These high temperatures, both in terms of diurnal fluctuations and long-term exposure, can result in differences in internal quality properties such as sugar contents, tissue firmness, and oil levels, as well as in mineral content differences. Fruit with different temperature histories will also respond differently to postharvest low temperatures and heat treatments used for insect disinfestation. For example, avocado fruit from exposed sites on a tree have less chilling injury, whereas more chilling damage is found in exposed tissues of citrus and persimmons. High temperatures in field affects heat shock proteins, membrane damage, and skin characteristics. Examples of fruit postharvest responses to high field temperatures include: (1) skin chilling injury of 'Hass' avocado fruit following four weeks storage;(2) skin pitting of 'Haywood' kiwifruit following 16 weeks storage;(3) internal chilling injury (flesh gelling/softening) of 'Fuyu' persimmons following six weeks storage; and (4) watercore in 'Cox's Orange Pippin' apples three days after harvest (no storage).

Sunburn affected and blemished (due to wind and insect scarring) citrus fruit are more prone to decay during storage and same is the case with many other fruits (Ladaniya and Shyam Singh, 1998, 1999).

Marketing problems are likely to arise due to climate change and erratic weather conditions. Farmers with contracts to supply a particular product to a supermarket chain are obliged to deliver specified quantities; with major penalties should they fail to do so. This may lead them to produce more than they expect to need, in order to be sure of meeting the contract.

In case of quality loss due to climate change, there is often no alternative market, particularly in countries where supermarkets account for a large share of retail sales. Supermarket standards also lead to waste as produce that does not meet size, shape or appearance criteria has to be disposed of. There may be few alternative markets and even if there are it may be too costly for the farmer to supply them on an *ad hoc* basis. Finally, supermarkets and other retailers may throw food away when it reaches its "sell by" date.

As climate change becomes a concern for consumers in wealthier nations, fruit growers in developing countries may be negatively impacted by the potential market access implications. With climate change and national strategies to mitigate it's effects increasingly becoming a foreign policy issue, and with large European supermarket chains such as Sainsbury's and Tesco, putting pressure on supply chains to account for reduced emissions, fruit growers have reason to be concerned. It may be possible that in the near future, carbon emissions could be used as an artificial trade barrier to fruit producers. In the developing world, fruit growers will have to start managing the perceptions of buyers and policy makers in developed countries like UK and other European nations by providing them with objective information on carbon emissions along the supply chain as well as keep them informed of measures to reduce these emissions. This demands sound research and carbon calculation mechanism in production and supply chain of developing countries.

Some supermarkets in Europe are alerting consumers about the "carbon footprint" that some imported and air-freighted fresh fruit products might be making on the Earth. In fact, such footprints are very low because most of the products are transported by highly efficient sea freight to Europe. Road freight across Europe can leave a greater "footprint" than does sea freight. Most of the fruit export from India will have to be by sea freight in future and therefore suitablepostharvest technologies and handling facilities will be needed. This is critical in case of many tropical fruits like mango, banana, papaya and sapota,which have short storage and shelf life.

15.7   STRATEGIES TO TACKLE CLIMATE CHANGE IMPACT ON POST-HARVEST MANAGEMENT

All steps to use fruit and reduce losses will eventually add to fruit availability, nutritional security and will help to meet the challenge of adverse effects of climate change.

15.7.1   FACILITIES FOR PACKINGHOUSES

Changing and shifting fruit cultivation patterndue to climate changewould require new facilities for packing and distribution system at new locations.

15.7.2   COLD STORAGE FACILITIES

Storage facilities will have to be strengthened in major consuming centers. Cold storage and controlled atmosphere storage can be used to facilitate long-term storage of fruit and extend availability till next season in case wheretwo crops are taken in a year viz. citrus, guava, pomegranate, etc. Temperate fruit can also be made available for longer period with better storage facilities.

15.7.3 TRANSPORT/LOGISTICS

With the given set-up of roads and transport vehicles in some areas of the country, losses would be more due to unseasonal rains, hailstorms, heavy rains, floods and high temperatures caused by climate change. Transport facility needs to be improved in terms of frequency of trucks and rail cars and also interior design and temperature management of the rail cars. Transport facility and roads are abysmally poor in remote villages and NEH region of the country.

15.7.4 HANDLING PRACTICES

Major portion of postharvest losses in the country are due to improper handling of fruits. Fruits are handled manually several times on the ground predisposing it to infection. Reduction of handling steps in a supply chain and introduction of mechanization in handling can minimize losses to a great extent as observed in Kinnow and many other fruits and vegetables.

15.7.5 PACKAGING

Unit handling of packaged fruit in boxes reduces infection, physical touch and softening, improves presentation and is convenient in handling.

15.7.6 MARKETING INFRASTRUCTURE

Auction sheds need to be provided to reduce weather impact on fruit. At many wholesale markets auction is conducted in open exposing fruit to intense heat and rains.

15.7.7 RETAILING

With increased temperatures of ambient air, shriveling and weight loss problem will be higher during February–June causing increased losses to retailers. Mostly fruits are sold under open non-air conditioned shops and under scorching Sun in India. Infrastructure development by public-private partnership or private companies is necessary for organized retailing at lower temperatures.

15.7.8 IMPORTS

In order to meet market demand, imports may have to be increased during the years when crop losses are higher. But imported fruit would be expensive and it may not be affordable for large portion of the population. Nevertheless, it is must for food security and such arrangements may well become strategically very important in the future. Such international transport/shipments need to be environmentally sustainable.

15.7.9 UTILIZATION OF DAMAGED FRUIT

Fruit affected by frost injury and hailstorm can be used in processing provided fruits are picked immediately after frost damage and sent for processing to avoid fungal infection. Total loss from frost and hailstorm injury to fruit can be alleviated to some extent by spray of fungicides like benomyl and carbendazim immediately to avoid fungal infection. Extent of damage by these vagaries of weather would determine the loss and utility of fruit.

15.7.10 STRENGTHENING PROCESSING INFRASTRUCTURE

Efficiency of the processes in product manufacture needs to be improved for less energy consumption. Outdated machinery needs to be modernized.

15.7.11 COTTAGE SCALE AND HOME SCALE PROCESSING

Large quantities of fruit do not reach market and processing units due to various reasons. Processing in villages at cottage scale would provide nutritional security to rural masses and avoid losses in the changing scenario of global warming.

15.8 ISSUES RELATED TO POLICY AND DEVELOPMENT

Government and its various agencies need to look in to policy and developmental issues with respect to fruit production and postharvest management in the face of global warming.

(1) Strengthening meteorological forecast facility for rains, frost, hailstorms, temperature (heat or cold wave), wind velocity and relative humidity during growing and harvesting seasons so that appropriate steps to reduce losses can be taken up at different levels.

(2) Build capacity by educating and training people to tackle climate change at all levels throughout the value chain. Capacity building in terms of fruit processing in the country to meet nutritional needs in the event of extreme conditions due to global warming.

(3) Policy support from Government agencies (Ministries of Agriculture, Education, Environment) to educate all stakeholders including farmers and general public about environment changes and how to preserve fruit quality and reduce losses.

(4) Promotion of underused and less known species of hardy fruits as alternative to main fruit species to meet nutritional demand.

(5) Providing enhanced and sustained market access for fruits.

(6) Disseminating successful case studies on market access by growers (reasons for success and failure).

(7) Contingent plans to save the crop and its marketing/distribution would be required for aberrant weather conditions with the help of simulation models, prediction and forecast systems at state agriculture/ horticulture department level.

(8) Increased credit facility and subsidy for postharvest infrastructure set-up are necessary.

(9) Large scale plantation of hardy fruit species on waste lands and salt affected lands to increase green cover and mitigate global warming through carbon sequestration and at the same time ensuring nutritional security.

KEYWORDS

- **Climate Change**
- **Erratic Rainfall**
- **Fruit Quality**
- **Global Warming**
- **Post-Harvest Management**
- **Weather Parameters**

REFERENCES

Anonymous, (2008). Network project on impact, adaptation and vulnerability of Indian agriculture to climate change. Annual Report for 2007–08, ICAR New Delhi.

Arpaia, M. L., Bower, J. P., Hofman, P. J., Rooyen, Z. V. & Woolf, A. B. (2004). Grower practices will influence post-harvest fruit quality. Second Int. fruit seminar. 29 Sept–1 Oct, 2004, Chile.

Ashenfelter, O. C., & Storchmann, K. (2010). Measuring the economic effect of Global warming on viticulture using auction, retail and wholesale prices in Germany. Working paper 16037, National Bureau of Economic Research, Cambridge, MA, USA, 1–23.

Guhathkarta, G., & Rajeevan, M. (2006). Trends in the rainfall pattern over India. Research Rep. No. 2/2006. National Climate Centre. India Meteorological Department, Pune. 1–23.

Huchche, A. D., Panigrahi, P. & Shivankar, V. J. (2010). Impact of climate change on citrus in India. In: Challenges of climate change-Indian Horticulture. Eds Singh, H. P., Singh, J. P. and Lal, S. S. Westville Publishing House, New Delhi. 65–75.

Ikeda, T., Yamazaki, K., Kumakura, H., & Hamamoto. H. (2007). Effect of high temp on pot grown strawberyy plant. VI International Strawberry Symposium. *Acta. Hort.* 842, 92–104.

Ladaniya, M. S. (2008). Citrus fruit- Biology, technology and evaluation. Published by Elsevier (Academic Press), San Diego, CA, USA. 558 p.

Ladaniya, M. S., & Shyam Singh (1998). Post-harvest technology of 'Nagpur' mandarin. Technical Bulletin No. 2. 144p.

Ladaniya, M. S., & Shyam Singh (1999). Post-harvest technology and processing of citrus fruits in India. National Research Centre for Citrus, 148 p.

Ladaniya, M. S., & Wanjari, V. (2003). Network project on marketing and post harvest loss assessment of fruits and vegetables in India. Final report submitted to ICAR. 135p.

McCarthy, A. (2009). Harvesting 'Hass' during high temperatures. Department of Agriculture and Food, Western Australia.

More, T. A., & Bhargava, R. (2010). Impact of climate change on productivity of fruit crops in arid regions of Rajasthan. In: Challenges of climate change-Indian Horticulture. Eds: Singh, H. P., Singh, J. P., & Lal, S. S. Westville Publishing House, New Delhi. 76–84.

Richardson, A. C., Marsh, K. B., & MacRae, E. A. (2000). Temperature effects on the composition of Satsuma mandarins in New Zealand. Proc. Intl. Soc. Citricul. IX Congress, Florida. Vol. I. 303–307.

Schrader, L., Sun, J., Felicetti, D., & Jedlow Zhang, J. (2003). Stress induced disorders: Effects on Apple fruit quality. Proceedings Washington tree fruit post-harvest Conference, 2–3 December, 2003, Wenatchee, Washington, USA. 1–7.

Singh, H. P. (2010). Impact of climate change on horticultural crops. In: Challenges of climate change-Indian Horticulture. Eds: Singh, H. P., Singh, J. P., & Lal, S. S. Westville Publishing House, New Delhi. 1–8.

Sitch, S., Cox, P. M., Collins, W. J. & Huntingford, C. (2007). Indirect radiative forcing of climate change through ozone effects on the land-carbon sink. *Nature.* 448, 791–794.

Smilanick, J. D., & Margosan, D. (2002). Avocado post-harvest disease management. California avocado research symposium, (October 6, 2002).

Solomakhin, A. A.,& Blanke, M. M. (2007). Overcoming adverse effects of hailnets on fruit quality and microclimate in an apple orchard. *J. Sci Food Agri.*, 87, 2625–2637.

Suthers, R. W., Collyer, B. S. & Yonov, T. (2000). The vulnerability of Australian horticulture to the Queensland fruit fly, *Bactrocera* (*Dacus*) *tryoni*, under climate change. *Australian J Agric Res.,* 51, 467–480.

Woolf, A. B., & Ferguson, I. B. (2000). Post-harvest responses to high fruit temperatures in the field. *Post Harvest Biol. Technol.*, 21, 7–20.

Zheng, X., & Eltahir, E. A. B. (1997). The response to deforestation and desertification in a model of West African monsoons. *Geophysical Res. Letters.* 24(2), 154–158.

CHAPTER 16

ECO-FRIENDLY POSTHARVEST TREATMENTS FOR FRUITS

RAM ASREY[1] and KALYAN BARMAN[2]

[1]Division of Post Harvest Technology, Indian Agricultural Research Institute, New Delhi 110 012, India; E-mail: ramu_211@yahoo.com

[2]Department of Horticulture (Fruit and Fruit Technology), Bihar Agricultural University, Sabour, Bhagalpur 813 210, Bihar, India

CONTENTS

ABSTRACT

Caring, protecting and conserving the environment for future generations of plants, human beings and other creatures has become a priority of today's world. The existing food laws have brought about several desirable changes in logistics of postharvest handling and value chain of fresh horticultural produce. Synthetic pre and post storage treatment agents/molecules have been replaced with ecofriendly products. In last couple of years, some environmental and consumer friendly post harvest treatment materials like nitric oxide (NO), salicylic acid, methyl jasmonates (MeJA), ethanol, polyamines, 1-Methylcyclopropene, ozone, edible waxes, essential oils, biocontrol agents and irradiation are gaining popularity across the globe. These ecofriendly agents are equally effective over traditional synthetic chemicals in respect of bringing down the physiological and biochemical changes in the harvested produce. Application of nitric oxide, salicylic acid, methyl jasmonates (MeJA), ethanol, polyamines, 1-Methylcyclopropene, ozone, edible waxes, essential oils, biocontrol agents and irradiation have been proved effective in minimizing storage disorders likechilling injury (CI), scald; fungal diseases like stem-end rot, blue mold rot, green mold rot, anthracnose; regulation of ripening and senescence, etc. Quality traits of fresh produce were also found retained for longer period in fruits by application of these compounds.

16.1 INTRODUCTION

The world has witnessed a very fast change in the ways of transportation and utilization of fruits and vegetables in last one and half decades. Moreover, developing and near to develop countries has made a significant progress on production, handling and utilization of fresh horticultural produce. With the increasing health and hygiene concerns among consumers; producers, traders and researchers are looking for green post harvest technologies, which can enhance the produce shelf-life and retain the nutritive value up to consumers end. Warnings about the dangers of the excessive use of chemicals for the human health and biodiversity have been well reported over the years. The World Health Organization (WHO) estimates the occurrence of some more than 27 million cases of excessive chemical uses causing slow poisoning globally each year (Anon, 2012).

In developing countries, sulfur and chlorine based formulations are often used for postharvest treatments of horticultural produce. Recently, some bio-based and other safe postharvest treatments have been formulated for loss reduction and shelf-life extension of fresh fruits.

16.2 NITRIC OXIDE (NO)

Nitric oxide is a gaseous free radical with relatively long half-life, lasting in biological systems up to 3–5 seconds. It is very reactive species and forms other oxides (NO_2, N_2O_3 and N_2O_4) in presence of atmospheric oxygen. Post harvest treatment of fruits with a low concentration of NO gas can extend post harvest shelf-life, but the application of NO by releasing from a gas cylinder is practically difficult. To solve this problem, the solid NO-donor compounds like diethylenetriamine (DETANO) or sodium nitroprusside (SNP) may be used in sachets or in treatment solution for liberating nitric oxide in storage chamber. The use of NO can be made for delaying fruit ripening and improvement in retention of texture during storage (Zhu et al., 2010). The better retention of cellular components such as pigments, phenolics and antioxidants can be achieved due to reduction in the degree of disintegration of cellular membranes with lesser electrolyte leakage. The chilling injury is reduced in the fruits kept in cold storage and symptoms like fresh browning and translucency are reduced (Singh et al., 2009). It is highly potent molecule for induction of resistance in produce against systemic infection (Manjunatha and Shetty, 2006). It also protect from microbial infections such as *Penicillium italicum, Rhizopus nigricans, Aspergillus niger* and *Monilinia fructicola* (Fan et al., 2008; Lazar et al., 2008). This resulted in post harvest shelf-life extension of treated fruits. Due to maintenance of cellular compartmentation, the enzymes do not come into contact with the substrate, thus preventing the rate of reaction of enzymes such as polyphenol oxidase. This new direction emanated from research on NO would allow more recent formulations for the use in enhancing fruit self-life in ecofriendly manner.

16.3 SALICYLIC ACID

Salicylic acid (SA) or ortho-hydroxybenzoic acid is a ubiquitous simple phenolic compound involved in the regulation of many processes in plant growth and development. Salicylic acid is considered as a plant hormone because of its role in regulating some aspects of disease resistance in plants (Raskin, 1992). More recently, the involvement of SA as a signal molecule in systemic acquired resistance associated with the production of pathogenesis-related proteins has been extensively shown (Beckers and Spoel, 2006). Moreover, dietary salicylates from fruit and vegetables are described as bioactive compounds with health care potential (Hooper and Cassidy, 2006), and considered as generally recognized as safe (GRAS). There are several reports on beneficial effects of SA treatment in fruits. During kiwifruit ripening, the pattern of decrease in endogenous SA levels was related to accelerated softening, while the application of acetylsalicylic acid (ASA, a derivative of SA) slowed down the softening rate of kiwifruit by inhibiting ethylene production and maintaining higher endogenous SA levels (Zhang et al., 2003). On the other hand, SA application either preharvest (Yao and Tian, 2005) or postharvest reduced fungal decay in sweet cherry through induction of the defense resistance system (Chan and Tian, 2006) and stimulation of antioxidant enzymes (Xu and Tian, 2008). In addition, in chilling injury sensitive fruit, pretreatment with SA reduced chilling injury symptoms in peaches (Wang et al., 2006) and pomegranates (Sayyari et al., 2009).

16.4 METHYL JASMONATE (MEJA)

Jasmonates are a class of endogenous plant growth regulators that have unique and potentially useful properties that affect plant growth and development in response to environmental stresses. Methyl Jasmonate (MeJA) was discovered in 1962 as a sweet-smelling compound in *Jasminium grandiflorum* flower extracts (Demole et al., 1962). The main effects of MeJA in post harvest management of fruits are- control postharvest disease and decay of fruits, alleviate chilling injury, regulation of fruit ripening and senescence, fruit quality maintenance, development of color and aroma volatiles, etc. Zhang et al. (2006) reported that MeJA can be effectively used to control gray mold rot in strawberry caused by *Botrytis cinerea* and it also enhances disease resistance in peach (Jin et al., 2009), loquat (Cao et al., 2008) and raspberries (Chanjirakul et al., 2006). Potential uses of MeJA in alleviating chilling injury of fruits have also been studied. In most of the experiments, loquat and pomegranate fruits treated with MeJA, exhibited higher levels of total phenolics, total antioxidant activity and significantly lower electrolyte leakage than the untreated control fruits, thus attributed to lower chilling injury symptoms. MeJA has also been implicated in delaying the onset of fruit ripening on the tree. Jasmonates have been found to have close interaction with plant hormone ethylene in this regard. Ziosi et al. (2009) reported jasmonate-induced ripening delay is associated with up-regulation of polyamine levels in peach fruit. Kondo (2009) studied the effect of MeJA on color development and aroma volatiles in apples. The expression of UDP-glucose flavonoid 3-O-glucozyltransferase (UF-GluT) anthocyanin biosynthetic gene was increased in the skin of fruits treated with jasmonates and these fruits also had much higher anthocyanin content than untreated controls. The impact of jasmonate application on volatile compound production was dependent on fruit ripening stage; jasmonates increased the volatiles in preclimacteric fruit, but decreased the volatiles in climacteric fruit.

16.5   ETHANOL

Ethanol, also called as ethyl alcohol is a volatile, flammable and colorless liquid. It is a small molecule produced either by chemical synthesis or by microbial fermentation. The production of two anaerobic metabolites acetaldehyde (AA) and ethanol in fruit, while still attached on the tree or during postharvest storage, leads to dramatic changes in fruit ripening. Ethanol is a volatile compound naturally produced by plant tissues under anaerobic conditions. It is also accumulated in a short period in anaerobically stored fruits without adversely affecting fruit quality. Ethanol can be applied by simply dipping fruit in an ethanol solution or as vapors. In some fruits (grapes) spray method can also be used. Ethanol can be used in controlling decay of fruits. Major target of ethanol is cell membrane. Polar moiety of ethanol anchor to the phospholipid head group and nonpolar part bounds to phospholipid acyl-chains. Lipid acyl-chains from the opposing monolayers are fully interpenetrated. This results to formation of voids in bilayer interior and permeability of the membrane to lipophilic compounds increases. Unbalanced cytoplasmic permeability and cytosol leakage ultimately results in disintegration of cell. It has been reported to effectively control postharvest table grape decay caused by *Botrytis cinerea, Alternaria alternata* and *Aspergillus niger* when applied after or before harvest. Wang et al. (2011) studied that ethanol vapor treatment (250 or 500 mL/L for 3 hr) alone or in combination with hot air treatment (48°C for 3 hr) in Chinese bayberries significantly lower decay incidence caused by *Verticicladiella abietina, Penicillium citrinum* or *Trichoderma viridae*compared to control. Ethanol acts as a precursor of natural aroma compounds. Ethanol is converted to acetaldehyde by enzyme alcohol dehydrogenase. Acetaldehyde is the precursor for the acetate esters. Thus Berger and Drawert (1984) reported that storage of 'Red Delicious' apples for 24 hr in an atmosphere containing ethanol vapors resulted in more than three-fold increase in the ethyl ester formation. The effect of ethylene on a range of climacteric fruit has been shown to enhance or inhibit ripening depending on the type of fruit. Another application is to enhance the anthocyanin content in fruit tissues as reported in bayberry fruit when treated with 1000 mL/L ethanol (Zhang et al., 2007). Spraying Cabernet Sauvignon grapes with 5% ethanol at veraison stage also enhances anthocyanin accumulation (Kereamy et al., 2002). Postharvest application of ethanol vapor in 'Amas' bananas and in persimmon removes the astringency (Esguerra et al., 1992; Khademi et al., 2010). Ethanol reacts with soluble tannins to form an insoluble gel, which is nonastringent. Ethanol vapor exposure of fruits also has insecticidal activity as reported by Dentener et al. (2000) in apples against *E. postvittana*. Application of ethanol vapors was also effective in reducing farnesene and conjugated trienes, involved in the development of scald in 'Granny Smith' apples (Ghahamani et al., 1999). Thus exogenous application of ethanol can be beneficially applied to many fruits for improving their aroma, controlling decay, delaying ripening and ethylene production and reduction of chilling injury symptoms.

16.6   POLYAMINES

Polyamines (PAs) are low molecular weight small aliphatic amines that are ubiquitous in living organisms and have been implicated in a wide range of biological processes, including plant growth, development and response to stress (Smith, 1985). In plants, they have been implicated in a wide range of biological processes, including growth, development and abiotic stress responses. The most common polyamines are Putrescine (PUT), Spermidine (SPD) and Spermine (SPM) found in every plant cell in titers ranging from approximately micromolar to millimolar, together with the enzymes regulating their metabolism and depend greatly on environmental conditions, especially stress.

Polyamines can be isolated commercially from plant as well as microbial sources. Among plant sources they can be isolated from leaves and stems of corn (*Zea mays* L.), cucumber (*Cucumis sativus* L.), oat (*Avena sativa* L.) and radish (*Raphanus sativus* L.). They can also be isolated from *Saccharomyces cerevisiae* and *Candida utilis*.

16.6.1 *POLYAMINES IN POSTHARVEST MANAGEMENT OF FRUITS*

Polyamines have been proposed to be a new category of plant growth regulator and are proposed to be involved in large spectrum of physiological and biological processes of fruits. Polyamines in their free forms have been reported as antisenescent agent, from both endogenous and exogenous applications (Valero et al., 1998). The main effects of PAs in post harvest management of fruits are- inhibit biosynthesis of ethylene, increase fruit firmness, reduce respiration rate, reduce chilling injury, retard color changes and reduce mechanical damage.

To investigate the role of putrescine in ethylene biosynthesis in plum, Khan et al. (2007) treated the fruits by dipping in aqueous solution containing different concentrations of putrescine (1 mM, 2 mM) and Tween-20 (0.01%) as a surfactant. Following PUT treatment fruits were stored at 0 ± 1 °C and $90 \pm 5\%$ RH for 6 weeks. Results found that PUT treatment reduced ethylene production during storage in 'Angelino' plum fruits. There was an inverse relationship between concentration of PUT and ethylene production rate. The reduction in ethylene production with PUT treatment attributed to competitive biosynthesis mechanism between ethylene and polyamine.

The effect of polyamine on maintaining fruit firmness can be attributed to their cross-linkage to the carboxyl group of the pectic substances in the cell wall, resulting in rigidification thus blocks the access of degrading enzymes and reducing rate of softening (Valero et al., 2002). Post harvest treatment of fruits with polyamine found effective on increasing fruit firmness in apple, plum mango, apricot, lemon, etc., however, the efficiency was generally greater for those molecules with higher number of available cations, that is $SPM^{+4} > SPD^{+3} > PUT^{+2}$.

Malik and Singh (2005) worked on the effect of prestorage application on improving fruit quality of mango ('Kensington Pride'). Fruits were
dipped for 6 min in aqueous solution containing different concentrations
of various polyamines including putrescine (1 mM), spermidine (0.5 mM)
and spermine (0.01 mM) with 0.01% Tween-20 as surfactant. Treated
fruits were stored at 13 ±1 °C and 85 ± 5% RH for 4 weeks. Following
storage, fruits were allowed to ripen at a temperature of 22 ± 1 °C. Fruit
firmness was significantly reduced in untreated fruits whereas PA treated
fruits exhibit a significantly higher firmness up to 28 days after storage and
spermine was found more effective.

It has been reported that under chilling condition, changes in cell membrane lipid from a liquid-crystalline to solid-gel state are induced in plant
tissue, which leads to increase in membrane permeability as wall as electrolyte leakage of ions (Gomez-Galindo et al., 2004). Polyamines when
applied exogenously they seem to induce cold acclimation, which would
lead to maintenance of membrane fluidity at low temperatures and could
be responsible for reducing electrolyte leakage and skin browning. These
polyamines mainly protect the membrane lipid from being conversion in
physical state, and as it has antioxidant property, it prevents lipid peroxidation and thus severity of chilling injury (Barman et al., 2011a). Mirdeghan
et al. (2007) studied the effect of prestorage application of polyamines by
pressure infiltration on reducing chilling injury in pomegranate fruit. Fruits
were treated with 1 mM putrescine, 1 mM spermidine or with distilled water, which serves as control. Treatments were performed by low pressure
infiltration method by applying pressure 0.05 bars for 4 min as well as immersion method with polyamine and Tween-20 (2 g/L) as surfactant. Following treatments they were stored at 2 °C and 90% RH. Before analysis of
the samples they were exposed to 20 °C temperature for 3 days. The results
found that control fruits reached the highest percentage of skin browning
without significance differences between them. However, the application
of putrescine of spermidine led to significant reduction in skin browning. It
also maintains the functional and sensory qualities during low temperature
storage of pomegranate fruits (Barman et al., 2011b).

16.7   1-METHYLCYCLOPROPENE

1-methylcyclopropene (1-MCP) is a synthetic cyclic olefin that inhibits
ethylene by blocking access to the ethylene-binding receptor (Sisler and

Serek, 1997). The application of ethylene action inhibitors lowers the endogenous levels of most maturation-associated genes. Successful application of 1-methylcyclopropene to permit extended storage requires prior assessment of the appropriate concentration range and storage conditions for each type of produce at particular maturity (Martinez-Romero et al., 2002).

16.7.1 EFFECTS OF 1-MCP ON FRUITS

Avocado treated with 1-MCP for 24 h showed significantly less weight loss and retained greener color than control fruit at the full-ripe stage. Dou et al. (2005) found that 1-MCP treated at the concentration of 50–500 µg/L effectively controls the blue mold rot and post harvest pitting in citrus. Treated fruit of pears with 1-MCP were more than 75 Newton firmer than control fruit after 6 days storage period. Lime (citrus) treated with 1-MCP at low concentrations (250 to 500 µL/L) effectively suppressed endogenous ethylene production and retarded yellowing at ambient temperature. 1-MCP bound at the ethylene receptors in pear fruit and inhibits cell wall-degrading enzymes such as PG secreted by pathogens and thus prevents pathogenesis. The 1-MCP treatment of 'd'Anjou' pear fruit effectively inhibited ethylene production after cold storage followed by ripening at 20 °C.

The use of 1-MCP suppresses the ethylene response pathway by permanently binding to a sufficient number of ethylene receptors. Combination of 1-MCP and controlled atmospheric storage did not affect the total soluble solids content and titratable acidity in the apple fruit. Vitamin C content in 1-MCP-treated fruits of guava was significantly higher than in nontreated fruits (Singh and Pal, 2008). Post harvest application of 1-MCP significantly delayed and suppressed the climacteric ethylene production of plum with reduction in the activities of ethylene biosynthesis enzymes and fruit softening enzymes in the skin as well as in pulp tissues. The onset of ethylene production and the rise in the respiration rate was delayed and suppressed in 1-MCP treated fruits of papaya (Manenoi et al., 2007).

Application of 1-MCP delayed the onset of the ethylene climacteric in immature green and mature green fruits of golden berry in a dose-dependent manner, and transiently decreased ethylene production in yellow and orange fruit (Gutierrez et al., 2008). Softening and respiration rates were decreased in response to 1-MCP treatment (1 μL/L for 6 h at 14 °C) of fresh-cut banana slices (after processing). 1-MCP has been reported to delay ripening of avocado, custard apple, mango and papaya. Apricot fruit treated with 1 μL/L of 1-MCP for 4 hours at 20 °C and subsequently stored at 0 or 20 °C, showed delayed onset of ethylene production and lower respiration rate. Banana ripening was delayed when exposed to 0.01–1 μL/L of 1-MCP for 24 hours and increasing concentrations of 1-MCP were generally more effective for longer periods of time.

16.8 OZONE

Ozone is the most effective natural bactericide of all the disinfecting agents. Activated oxygen is considered to be the best available technology and a much better alternative for water purification. Ozone is now being used in food processing or storage of perishables as an antimicrobial agent or as a food processing aid. Ozone is the safe and natural purification and disinfecting agent. It is strong and ideal, germicide, sanitizer, sterilizer and vermicide, antimicrobial, bactericide, fungicide and deodorizer, detoxifying agent.

16.8.1 EFFECTS ON FRUITS

Ozone oxidizes the metabolic products and neutralizes the odors generated during ripening of fruits in storage. This helps in preserving and enhancing shelf-life of fresh produce. Ozone enhances the taste of most perishables by oxidizing pesticides and by neutralizing ammonia and ethylene gases produced by ripening or decay. It also helps in retaining original flavor of the produce. The reduction of ethylene gas increases the shelf-life and reduces shrinkage. Ozone changes the complex molecular structure of chemicals back to its safe and original basic elements. Its use does not leave any toxic by-products or residues, does not affect healthy cells or alter its chemistry and is noncarcinogenic.

The shelf-life of perishable foods varies when they are subjected to O_3 treatment. Factors like age, crispness, quality, humidity, temperature, the condition during treatment, and the reduction of pathogens during ozonation determines the extension of shelf-life of fresh produce. Positive effect shows at low and constant levels between 0.05 ppm and 0.1 ppm and it allows workers to enter the storage area and carry out their work comfortably (Perez et al., 1999). Ozone should be constantly consumed and absorbed during the oxidation process. The effectiveness is influenced (lowered) due to the presence of steam or 100% humidity levels. The microorganisms have to be in a certain condition of swelling in order to be attacked. When the humidity level is below 50%, the efficiency of these microorganisms reduces as a bacterial medium (Castillo et al., 2003). Growers and processors can use ozonated water to wash fruits by replacing chlorine. Application of it can be done safely in process water for hydro-cooler systems, bin, dump and dip tanks, flumes, spray-wash systems, wastewater processing and storage areas, at an affordable cost. Making full use of ozonated water will increase the production of fresh produce (Mehta, 2003).

16.9   EDIBLE COATINGS AND FILMS

Edible coatings are thin layers of edible material applied to the product surface in addition to or as a replacement for natural protective waxy coatings and provide a barrier to moisture, oxygen and solute movement for the food. They are applied directly on the food surface by dipping, spraying or brushing to create a modified atmosphere. An ideal coating is defined as one that can extend storage life of fresh fruit without causing anaerobiosis and reduces decay without affecting the quality of the produce. Previously, edible coatings have been used to reduce water loss, but recent developments of formulated edible coatings with a wider range of permeability characteristics has extended their potential in using in handling the fresh produce. The effect of coatings on fruits depends greatly on temperature, alkalinity, thickness and type of coating, and the variety and condition of the produce.

Edible polymer film is defined as a thin layer of edible material formed on a product surface as a coating or placed (preformed) on or between food components. Several types of edible films have been applied successfully for preservation of fresh horticultural produce (Baldwin et al., 1995). Fruit based films provide enhanced nutrition for food products, while increasing their marketing value. Edible and biodegradable films must meet a number of special functional requirements, for example, moisture barrier, solute or gas barrier, water/lipid solubility, color and appearance, mechanical and rheological characteristics, nontoxicity, etc. These properties depend on the type of material used, its formation and application. The benefit of using selective films seems to be the reduction of water loss, which is one of the most important factors in the deterioration of highly perishable commodities. The films provide protection against moisture loss and maintain an attractive appearance of the produce. Specific coating materials used commercially for different fruit crops are enlisted in Table 16.1.

TABLE 16.1   Specific Coating Applications for Different Fruits

Coating Material	Fruits
Pro-long	Banana
Semperfresh	Banana
Semperfresh with organic acid	Banana
Ban-seel	Banana and plantains
Tal prolong, Semperfresh and applewax	Apple
Nutri-save	Golden delicious apple
Semperfresh	Granny smith apple
Brilloshine	Apple, avocado, melons and citrus fruits
Nu-coatflo, Brilloshine and Citrashine	Citrus fruits
Semperfresh	Guava
Palm oil	Guava
Vapor gard	Mango
Chitosan	Strawberry and raspberry
N,O-Carboxymethyl chitosan	Fruits

16.9.1   EFFECTS ON FRUIT ATTRIBUTES

16.9.1.1   APPLE WRAPS

Apple based wraps are made from apple puree with various concentrations of fatty acids, fatty alcohols, beeswax and vegetable oil and have a color of apple sauce. These wraps are excellent oxygen barriers, particularly at low to moderate relative humidity, but are not very good moisture barriers unless lipids are added. Wrapping apple based films formed around apple pieces significantly reduced moisture loss and browning in cut apples, increased the intensity of apple flavor, and maintained the texture during 12-days storage period at 5 °C (McHugh and Senesi, 2000).

16.9.1.2   NATURE SEAL (NS)

NS, a cellulose-based edible coating, has been used (in combination with antimicrobials, plasticizers, antioxidants, etc.) to coat fresh-cut apples. This coating significantly reduced weight loss of apples more than those treated with water solutions and was not objectionable in taste during several weeks of storage.

16.9.1.3   CHITOSAN COATINGS

Chitosan, a by-product from crustacean shell wastes, is a high molecular weight cationic polysaccharide, normally obtained by the alkaline deacetylation of chitin and refers to as a range of polymers that, unlike chitin are soluble in dilute organic acids. Chitosan-based coatings are effective in prolonging the shelf-life and improving quality of fruits by delaying ripening, reducing respiration rate, reducing desiccation, regulating gas exchange, decreasing transpiration losses, modifying the internal atmosphere, maintaining the quality of harvested fruits, retaining fruit firmness, freshness, titratable acidity, soluble carbohydrates and vitamin C. It is also highly effective in reducing decay (inhibited spore germination, germ tube elongation, and radial growth of *B. cinerea* and *Rhizopus* species in the culture) of strawberries, raspberries, etc. Chitosan coated berries, and peaches were better in quality when compared to controls. The application of chitosan coating delayed changes in contents of anthocyanin, flavonoid, total phenolics, and reduced weight loss and browning of litchi fruit, improved storability, and delayed the increase in polyphenol oxidase enzyme activity in litchi fruit. The coating partially inhibited PPO activity of longan fruit, which is associated with peel discoloration (Jiang and Li, 2001).

16.9.1.4   CORN-ZEIN COATINGS

Zein is a natural corn protein produced from corn gluten meal and is insoluble in water, but soluble in aqueous alcohol, glycols and glycol esters. It has good film- forming, binding and adhesive properties. Its water vapor permeability is about 800 times higher than a typical shrink-wrapping film.

16.9.1.5   WAX COATINGS

Wax, the first edible coating known, is the most effective coating to block moisture migration. There are a number of waxes used but the most effective one is paraffin wax, followed by beeswax. The resistance is related to their compositions. Paraffin wax consists of a mixture of long-chain saturated hydrocarbons while beeswax comprises a mixture of hydrophobic, long chain ester compounds, long chain hydrocarbons and fatty acids. The absence of polar groups in paraffin and presence in low levels in beeswax account for their resistance to moisture transport. An increase in ethanol, acetaldehyde, total soluble solids content and a decrease in total solids and titratable acidity was observed for waxed mandarins during storage relative to unwaxed fruits stored in film lined boxes. Storing waxed mandarins at room temperature can lead to anaerobic respiration with higher levels of ethanol and acetaldehyde. Oranges coated with commercial solvent type wax had less weight loss than those with comparable amounts of water wax or polyethylene coatings. There was an increase in storage life and a decrease in weight loss when mangoes were coated with a wax emulsion in water compared to mineral oil coated and control samples.

16.9.1.6   WHEY PROTEIN COATINGS

The mechanical properties of whey protein films adequately provide durability when used as coatings on food products or films separating layers of homogeneous foods. Adding glycerol and sorbitol reduces internal hydrogen bonding in films, thereby increasing film flexibility. Incorporating whey protein edible films in food product can results in reduction of food losses due to spoilage and thus extend the shelf-life.

16.9.1.7   CASEIN COATINGS

Research conducted in different countries of the world showed that casein-lipid coatings provide protection of fruits from moisture loss and oxidative browning. Calcium caseinate and whey protein solutions efficiently delays browning of apple by acting as oxygen barriers. Respiration rate of Red Delicious apples increased for both caesinate coated and uncoated apples. This indicated that caesin coated formulations cannot modify fruit respiration rate.

16.9.1.8   PRO-LONG

Pro-long is a mixture of sucrose fatty acid esters, sodium carboxymethyl cellulose and mono- and di-glycerides. The mode of action of Pro-long involves the creation of a selectively permeable barrier, which creates the internal atmospheres and preserve the fruit by reducing water loss and chilling injury characteristics, that might be used both in the storage of fruit and for the maintenance of quality during the marketing period. Treatment with 0.75% Pro-long significantly increased the storage life of mangoes, retarded ripening and reduced weight loss and chlorophyll loss, without adversely affecting the sensory quality of the limes. Apples treated with 1.25% sucrose ester formulation were stored in air at 3.5 °C for up to 5 months. When applied after storage, the coating reduced yellowing, loss of firmness and markedly increased internal carbon dioxide levels during 21-day simulated marketing period. Coating bananas with Pro-long reduces weight loss, oxygen uptake, ethylene release, chlorophyll loss and modifies their internal atmosphere by reducing the permeability of the fruit peel to gases (Banks, 2004).

16.9.1.9   SEMPERFRESH

Semperfresh, a food-grade coating used to retard moisture loss, ripening and spoilage of fruit is a mixture of sucrose esters with high proportion of short chain unsaturated fatty acid esters, sodium salts of CMC and mixed mono and diglycerides. Semperfresh is an improved formulation of earlier SPE (sucrose polyester). The major difference is improved dispersion due to incorporation of higher proportion of short chain USFA esters. These fruit coatings were found to be significantly effective in retention of reducing sugars, delaying loss of firmness, titratable acidity, pH, soluble solids, sugars, ascorbic acid and lycopene synthesis. Semperfresh significantly reduces water loss and internal carbon dioxide in several fruits and reduced color changes, retained acid, increased shelf-life and maintained the keeping quality of apples. The coating maintains titratable acidity, ascorbic acid, firmness, green color and decreased weight loss, total soluble solids and pH in mangoes when compared with the noncoated fruits.

16.10   ESSENTIAL OILS

A new approach to control the postharvest pathogens, while maintaining fruit quality as an alternative to chemical fungicides is by the application of essential oils to fruit. This approach eliminates the need for synthetic fungicides, thereby complying with consumer preferences, organic requirements and reducing environmental pollution. Essential oils (EOs) are volatile, natural, complex compounds characterized by a strong odor and are formed by aromatic plants in different parts (flowers, buds, seeds, leaves, twigs, bark, herbs, wood, fruits and roots) as secondary metabolites. These are registered food grade materials and have the potential to be applied as an alternative treatment to control post harvest decay of fruits. An important characteristic of EOs and their components is their hydrophobicity, which enables them to partition in the lipids of the bacterial cell membrane and mitochondria, disturbing the structures and rendering them more permeable. As a result, damage of membrane proteins and depletion of proton motive force takes place. Leakage of ions and other cell contents can then occur which after attaining extensive loss of cell contents or of critical molecules and ions lead to death of bacterial cells. A number of essential oil components have been identified as effective antibacterial agents, for example, eugenol, carvacrol, thymol, menthol which can be used in controlling postharvest diseases of several fruits like apple, peach, sweet cherry, strawberry, grape, citrus, mango, etc. (Regnier et al., 2008) (Table 16.2).

TABLE 16.2   Essential Oils Used for the Control of Postharvest Diseases of Fruit

Essential oil	Major component(s)	Bacteria	Fruit crops
Clove oil	Eugenol	*P. expansum, M. fructigena, B. cinerea, P. vagabunda*	Apple, grape
Mint oil	Menthol	*P. italicum, B. cinerea, R. stolonifer*	Orange, strawberry
Thyme oil	Thymol, carvacrol	*B. cinerea, R. stolonifer, M. fructicola*	Strawberry, grape, sweet cherry
Cinnamon oil	Cinnamaldehyde	Natural flora	Kiwifruit
Lemongrass oil	Citral	*C. gloeosporioides, P. expansum, B. cinerea, R. Stolonifer*	Mango, peach

16.11 BIOCONTROL AGENTS

Biocontrol strategy has made much progress during the last decade. At present, several biocontrol treatments have been approved for commercial applications, and now research is focused on improving the bioefficacy of the antagonist. One of the approaches has been the selection of combinations of antagonists, whichmay work more effectively (Barkai-Golan and Phillip, 2001). It is a very challenging work, as microorganisms have differential growth habits, requirements for nutrition and cultural conditions. Naturally occurring micro organisms, which are found to be adhered on the fruits and vegetables surface have been shown potential to protect the fresh produce against postharvest disease causing pathogens. During last decade several products *viz.* Serenade (*Bacillus subtilis* based), Messenger (*Erwinia amylovora* based), Biosave (*Pseudomonas syringae* strain 10 LP), Aspire (*Candida oleophila* strain 1–18), AQ-10 bio-fungicide (*Ampelomyces quisqualis*) have been isolatedand registered in the United States and Germany (Fravel, 2005; Zhao et al., 2007). Use of some safe bioactive compounds have been proved beneficial in bringing down the pathological activities in fruits during transportation, storage and minimizing the over all qualitative and quantitative losses (Table 16.3).

TABLE 16.3 Microbial Antagonists Used for the Control of Postharvest Diseases of Fruit

Antagonists	Disease (Pathogen)	Fruit crops
Bacillus subtilis	Brown rot (*Lasiodiplodia theobromae*), Gray mold (*Botrytis cinerea*), Green mold (*Penicillium digitatum*), Stem end rot (*Botryodiplodia theobromae*), Alternaria rot (*Alternaria alternata*)	Apricot, strawberry, citrus, avocado, cherry, Litchi
Bacillus licheniformis	Anthracnose (*Colletotrichum gloeosporioides*), stem end rot (*Dothiorella gregaria*)	Mango
Trichoderma viride	Green mold (*Penicillium digitatum*), Stem-end rot (*Botryodiplodia theobromae*), Gray mold (*Botrytis cinerea*)	Citrus, mango, strawberry
Pseudomonas syringae	Blue mold (*Penicillium expansum*), Green and blue mold (*Penicillium digitatum* and *P. italicum*), Gray mold (*Botrytis cinerea*), Brown rot (*Monilinia laxa*)	Apple, citrus, peach

Antagonists	Disease (Pathogen)	Fruit crops
Candida oleophila	Penicillium rots (*Penicillium digitatum and Penicillium italicum*), Crown rot (*Colletotrichum musae*), Anthracnose (*Colletotrichum gloeosporioides*)	Citrus, banana, papaya
Debaryomyceshansenii	Green and blue mold (*Penicillium digitatum* and *Penicillium italicum*), Blue mold (*Penicillium italicum*), Rhizopus rot (*Rhizopus stolonifer*), Sour rot (*Geotrichum candidum*)	Citrus, peach

16.12   IRRADIATION

Irradiation can be applied by exposing the fruit to ionizing radiations from radioisotopes (normally in the form of gamma-rays but X-rays can also be used) and from machines, which produce a high-energy electron beam. The FDA has approved two types of radiation sources for the treatment of foods: gamma rays produced by the natural decay of radioactive isotopes of cobalt-60 or cesium-137, X-rays with a maximum energy of 5 million electron volts (MeV), and electrons with a maximum energy of 10 MeV. The rays are directed onto the fruit being irradiated, but the fruit itself never comes into contact with sources of radiation. Irradiation may be referred to as a "cold pasteurization" process, as it does not significantly raise the temperature of the treated fruits. Since fruits contain 80–95% water and their intercellular spaces (about 20% of total volume) contain oxygen, the high-energy gamma rays generate copious amounts of free radicals from those of water and oxygen in the fruit immediately after radiation. The free radicals in turn bring about the breakage of the genetic material (DNA) of the insects and spoilage microorganisms, thus destroying them. But after a short period (2–3 days) the free radicals get scavenged off or converted into harmless molecules. Gamma radiation also delays ripening and senescence by decreasing the activity of the cell wall degrading enzyme pectin methyl esterase (PME) and the activity of ACC-oxidase involved in ethylene synthesis. Activities of some other enzymes likepolygalactosidase, cellulase and 1-aminocyclopropane-1-carboxylate oxidase are also negatively affected. Ripening of bananas is inhibited at irradiation doses of 0.25–0.35 kGy, and the irradiated fruit later can be ripened by treatment with ethylene. Similar results have been reported for mango, papaya, guava and several other tropical and subtropical fruits. Irradiation treatment can also be used to control several postharvest diseases of fruit crops (Table 16.4). But ionizing radiation at doses above 1 kGy can induce various types of physiological disorders in fruits like internal browning in avocados; skin discoloration and stem darkening in grapes; skin damage in bananas; internal cavities in lemon and lime, etc.

TABLE 16.4   Doses of Irradiation for Different Fruits for Postharvest Disease Control

Fruit crop	Minimum dose required (Gy)	Maximum dose tolerated (Gy)
Apple	150	100–150
Apricot, peach, nectarine	200	50–100
Avocado	—	15
Lemon	150–200	25
Orange	200	200
Strawberry	200	200
Grapes	—	25–50

For the control of postharvest diseases generally, a minimum dose of 1.75 kGy is required for effective inhibition of postharvest fungi. So to avoid such a high dose, combination treatment such as heat + irradiation is sometime used. Such combination has been shown to be effective for control of brown rot on stone fruits and anthracnose on mango and papaya. Irradiation at doses below 1 kGy is an effective insect disinfestations treatment against various species of fruit flies, mango seed weevil, navel orange worm, codling moth, scale insects and other insect species of quarantine significance in marketing fresh fruits. Most insects are sterilized at doses of 0.05–0.75 kGy; some adult moths survive 1 kGy, but their progeny are sterile. For mango irradiation at a dose between 0.25–0.75 kGy is used for quarantine treatment and shelf-life extension.

16.12.1   HOT WATER TREATMENT

Fruits may be dipped in hot water before marketing or storage to control various postharvest diseases and improving peel color of the fruit. In mangoes, the hot water treatment is recommended at 50–52°C for 5 min to reduce the fungal infection during ripening or storage. This treatment helps in attaining uniform ripening within 5–7 days. Fruits should not be handled immediately after heat treatment. Cool water showers or forced air should be provided to help return the fruit to their optimum temperature as soon as possible after completion of the treatment. Hot water treatment has been very useful in several other fruits (Table 16.5).

TABLE 16.5   Recommended Hot Water Treatments for Controlling Different Pathogens in Fruits and Vegetables

Commodity	Pathogens	Temp. (°C)	Time (min)	Possible injuries
Apple	*Gloeosporium sp.* *Penicillium expansum*	45	10	Reduced storage life
Grapefruit	*Phytophthora citrophthora*	48	3	
Green beans	*Pythium butleri* *Sclerotinia sclerotiorum*	52	0.5	
Lemon	*Penicillium digitatum* *Phytophthora sp.*	52	5–10	
Mango	*Collectotrichum gloeosporioides*	52	5	No stem rot control
Melon	*Fungi*	57–63	0.5	
Orange	*Diplodia sp.* *Phomopsis sp.* *Phytophthora sp.*	53	5	Poor degreening
Papaya	*Fungi*	48	20	
Peach	*Monolinia fructicola* *Rhizopus stolonifer*	52	2.5	Motile skin
Pepper (bell)	*Erwinia sp.*	53	1.5	Slight spotting

16.12.2 VAPOR HEAT TREATMENT (VHT)

This treatment proved very effective in controlling infection of fruit flies in fruits after harvest. For this treatment, the boxes containing fruits are stacked in a room, which are heated and humidified by injection of steam. The temperature and exposure time are adjusted to kill all stages of insects (egg, larva, pupa and adult), but fruit should not be damaged. A recommended treatment for citrus, mangoes, papaya and pineapple is 43°C in saturated air for 8 h and then holding the temperature for further 6 h. VHT is mandatory for export of mangoes

16.13 CONCLUSION

In past, there were several chemicals in use for post harvest treatment of fruits, but due to advancement in technology, health consciousness of consumers and environmental concern has phased out these chemicals with the adoptions of improved ones. In last couple of years, some environmental and consumer friendly post harvest treatment materials like nitric oxide (NO), salicylic acid, methyl jasmonates (MeJA), ethanol, polyamines, 1-Methylcyclopropane, ozone, edible waxes, essential oils, biocontrol agents irradiation and thermal treatment are gaining popularity across the globe.

KEYWORDS

- 1-Methylcyclopropene
- Biocontrol Agents
- Ecofriendly treatments
- Edible Coatings
- Essential Oils
- Fruit Attributes
- Polyamines
- Postharvest Quality

REFERENCES

Anon. (2012). UN food agency, updated pesticide standards. Press release from U.S. Embassy, Tokyo, Japan.

Baldwin, E. A., Nisperos-Carriedo, M. O. & Baker, R. A. (1995). Use of edible coatings to preserve quality of lightly (and slightly) processed products. *Crit. Rev. Food Sci. Nutr.*, 35(6), 509–552.

Banks, N. H. (2004). Some effects of TAL Pro-long coating on ripening bananas. *J. Expt.Bot.*, 35(150), 127–137.

Barkai-Golan, R., & Phillip, D. J. (1991). Postharvest fruit treatment of fresh fruits and vegetables for decay control. *Plant Diseases*, 75, 1085–1089.

Barman, K., Asrey, R., & Pal, R. K. (2011a). Putrescine and carnauba wax pretreatments alleviate chilling injury, enhance shelf life and preserve pomegranate fruit quality during cold storage. *Sci. Hort.*, 130, 795–800.

Barman, K., Asrey, R., Pal, R. K., Kaur, C., & Jha, S. K. (2011b). Influence of putrescine and carnauba wax on functional and sensory quality of pomegranate (*Punica granatum* L.) fruits during storage. *J. Food Sci. Technol.*, DOI: 10.1007/s13197–011–0483–0.

Beckers, G. J. M., & Spoel, S. H. (2006). Fine-tuning plant defense signaling: salicylate versus jasmonate. *Plant Biol.*, 8, 1–10.

Berger, R. G. & Drawert, F. (1984). Changes in the composition of volatiles by postharvest application of alcohols to Red Delicious apples. J. Sci. Food Agric., 35, 1318–1325.

Cao, S. F., Zheng, Y. H., Yang, Z. F., Tang, S. S., Jin, P. & Wang, K. T. (2008). Effect of methyl jasmonate on inhibition of *Colletotrichum acutatum* infection in loquat fruit and the possible mechanisms. *PostharvestBiol. Technol.* 49, 301–307.

Castillo, A., Mckenzie, K. S., Lucia, L. M. & Acuff, G. R. (2003). Ozone treatment for reduction of *Escherichia coli* and *Salmonella* serotype *Typhimurium* on beef carcass surfaces. *J. Food Prot.*, 66(5), 775–779.

Chan, Z., & Tian, S. (2006). Induction of H_2O_2–metabolizing enzymes and total protein synthesis by antagonist and salicylic acid in harvested sweet cherry fruit. *Postharvest Biol. Technol.*, 39, 314–320.

Chanjirakul, K., Wang, S. Y., Wang, C. Y., & Siriphanich, J. (2006). Effect of natural volatile compounds on antioxidant capacity and antioxidant enzymes in raspberries. *Postharvest Biol. Technol.*, 40, 106–115.

Demole, E., Lederer, E., & Mercier, D. E. (1962). Isolement et détermination de la structure du jasmonate de méthyle, constituant odorant charactéristique de l'essence de jasmin. *Helvetica Chimica Acta,* 45, 675–685.

Dentener, P. R., Lewthwaite, S. E., Bennett, K. V., Maindonald, J. H. & Connolly, P. G. (2000). Effect of temperature and treatment conditions on the mortality of Epiphyas postvittana (Lepidoptera: Tortricidae) exposed to ethanol. *J. Econ. Entomol.* 93, 519–525.

Dou, H., Jones, S., & Ritenour, M. (2005). Influence of 1–MCP application and concentration on postharvest peel disorders and incidence of decay in citrus fruit. *J. Hort. Sci. Biotech.*, 80(6), 786–792.

Esguerra, E. B., Kawada, K., Kitagawa, H., & Subhadrabandhu, S. (1992). Removal of astringency in 'Amas' banana (Musa AA group) with postharvest ethanol treatment. *Acta Hort.*, 321, 811–820.

Fan, B., Shen, L., Liu, K., Zhao, D. & Yu, M. (2008). Interaction between nitric oxide and hydrogen peroxide in post harvest tomato resistance response to *Rhizopus nigricans. J. Sci. Food Agric.*, 88 1238–1244.

Fravel, D. R. (2005). Commercialization and implementation of biocontrol. *Annu. Rev. Plant Biol.*, 43, 337–359.

Ghahamani, F., Scott, K. J., Buckle, A. & Paton, J. E. (1999). A comparison of the effects of ethanol and higher alcohols for the control of superficial scald in apples. *J. Horti. Sci. Biotechnol.*, 74, 87–93.

Gomez-Galindo, F., Herpppich, W., Gekas, V., & Sjoholm, I. (2004). Factors affecting quality and postharvest properties of vegetables: Integration of water relations and metabolism. *Crit. Rev. Food Sci. Nutir.*, 44, 139–154.

Gutierrez, S. M., Trinchero, D. G., Cerri, M. A., Vilella, F., & Sozzi, O. G. (2008). Different responses of golden berry fruit treated at four maturity stages with the ethylene antagonist 1-methylcyclopropene. *Postharvest Biol. Technol.,* 48(2), 199–205.

Hooper, L., & Cassidy, A. (2006). A review of the health care potential of bioactive compounds. *J. Sci. Food Agric.*, 86, 1805–1813.

Jiang, Y., & Li, Y. (2001). Effects of chitosan on postharvest life and quality of long an fruit. *Food Chem.*, 73, 139–143.

Jin, P., Zheng, Y., Tang, S., Rui, H., & Wang, C. Y. (2009). Enhancing disease resistance in peach fruit with methyl jasmonate. *J. Sci. Food Agric.,* 89, 802–808.

Kereamy, E. A., Chervin, C., Souquet, J. M., Moutounet, M., Monje, C. M., Nepveu, F., Mondies, H., Ford, M. C., Heeswijck, V. R., & Roustan, P. J. (2002). Ethanol triggers grape gene expression leading to anthocyanin accumulation during berry ripening. *Plant Sci.*, 163, 449–454.

Khademi, O., Mostofi, Y., Zamani, Z., & Fatahi, R. (2010). The effect of deastringency treatments on increasing the marketability of persimmon fruit. Proceedings of 6th International Postharvest Symposium. *Acta Hort.*, 877, 687–691.

Khan, A. S., Singh, Z., & Abbasi, N. A. (2007). Pre-storage putrescine application suppresses ethylene biosynthesis and retards fruit softening during low temperature storage in 'Angelino' plum. *Postharvest Biol. Technol.*, 46, 36–46.

Kondo, S. (2009). Fruit ripening and characteristics regulated by physiologically active substances. *Fresh Produce* 3 (Special Issue 1), 7–11.

Lazar, E. E., Wills, R. B. H., Ho, B. T., Harris, A. M., & Spohr, L. J. (2008). Antifungal effect of gaseous nitric oxide on mycelium growth, sporulation and spore germination of the postharvest horticulture pathogens, *Aspergillus niger, Monilinia fructicola* and *Penicillium italicum. Lett. Appl. Microb.*, 46, 688–692.

Malik, A. U., & Singh, Z. (2005). Pre-storage application of polyamines improves shelf-life and fruit quality of mango. *J. Hort. Sci. Biotech.*, 80(3), 363–369.

Manenoi, A., Bayogan, E. R. V., Thumdee, S. & Paull, R. E. (2007). Utility of 1-methylcyclopropene as a papaya postharvest treatment. *Postharvest Biol. Technol.,* 44(1), 55–62.

Manjunatha, G., & Shetty, H. S. (2006). Involvement of nitric oxide signaling in induction of systemic resistance in pearl millet by endophytic cell wall protein of Penicillium sp. *J. Mycol. Plant Pathol.*, 36, 392–404.

Martinez-Romero, D., Serrano, M., Carbonell, L., Burgos, F., & Valero, D. (2002). Effects of postharvest putrescine treatment on extending shelf life and reducing mechanical damage in apricot. *J. Food sci.*, 67(5), 1706–1711.

McHugh, T. H., & Senesi, E. (2000). Apple wraps: A novel method to improve the quality and extend the shelf life of fresh-cut apples. *J. Food Sci.*, 65(3), 480–485.

Mehta, K. (2003). Use of ozone for fresh and processed foods. *Proc. Food Indus.*, 6(8), 16–19.

Mirdehghan, S. H., Rahemi, M., Castillo, S., Martínez-Romero, D., Serrano, M. & Valero, D. (2007). Pre-storage application of polyamines by pressure or immersion improves shelf-life of pomegranate stored at chilling temperature by increasing endogenous polyamine levels. *Postharvest Biol. Technol.*, 44(1), 26–33.

Perez, A. G., Sanz, C., Rios, J., & Olias, J. M. (1999). Effects of ozone treatment on postharvest strawberry quality. *J Agri. Food Chem.*, 47, 1652–1656.

Raskin, I. (1992). Salicylate, a new plant hormone. *Plant Physiol.*, 99, 799–803.

Regnier, T., duPlooy, W., Combrinck, S., & Botha, B. (2008). Fungitoxicity of *Lippia scaberrima* essential oil and selected terpenoid components on two mango postharvestspoilage pathogens. *Postharvest Biol. Technol.* 48 254–258.

Sayyari, M., Babalar, M., Kalantari, S., Serrano, M., & Valero, D. (2009). Effect of salicylic acid treatment on reducing chilling injury in stored pomegranates. *Postharvest Biol. Technol.*, 53, 152–154.

Singh, S. P., & Pal, R. K. (2008). Response of climacteric-type guava (*Psidium guajava* L.) to postharvest treatment with 1-MCP. *Postharvest Biol. Technol.,* 47(3), 307–314.

Singh, S. P., Zora, S., & Swinny, E. E. (2009). Postharvest nitric oxide fumigation delays fruit ripening and alleviates chilling injury during cold storage of Japanese plum (*Prunus saliciana*). *Postharvest Biol. Technol.,* 53, 101–108.

Sisler, E. C., & Serek, M. (1997). Inhibitors of ethylene responses in plants at the receptor level: recent developments. *Physiol. Plant.,* 100, 577–582.

Smith, T. A. (1985). Polyamines. *Annu Rev. Plant Physiol.,* 36, 117–143.

Valero, D., Martinez, D., Riquelme, F., & Serrano, M. (1998). Polyamine response to external mechanical bruizing in two mandarin cultivars. *HortSci.*, 33(7), 1220–1223.

Valero, D., Martínez-Romero, D., & Serrano, M. (2002). The role of polyamines in the improvement of the shelf life of fruit. *Trends in Food Sci. Technol.*, 13, 228–234.

Wang, K., Jin, P., Tang, S., Shang, H., Rui, H., Di, H., Cai, Y., & Zheng, Y. (2011). Improved control of postharvest decay in Chinese bayberries by a combination treatment of ethanol vapor with hot air. *Food Cont.*, 22, 82–87.

Wang, L., Chen, S., Kong, W., Li, S., & Archbold, D. D. (2006). Salicylic acid pretreatment alleviates chilling injury and affects the antioxidant system and heat shock proteins of peaches during cold storage. *Postharvest Biol. Technol.*, 41, 244–251.

Xu, X., & Tian, S. (2008). Salicylic acid alleviated pathogen-induced oxidative stress in harvested sweet cherry fruit. *Postharvest Biol. Technol.*, 49, 379–385.

Yao, H., & Tian, S. (2005). Effects of preharvest application of salicylic acid or methyl jasmonate on inducing disease resistance of sweet cherry fruit in storage. *Postharvest Biol. Technol.*, 35, 253–262.

Zhang, F. S., Wang, X. Q., Ma, S. J., Cao, S. F., Ji, N., Wang, X. X., & Zheng, Y. H. (2006). Effects of methyl jasmonate on postharvest decay in strawberry fruit and the possible mechanisms involved. *Acta Hort.,* 712, 693–698.

Zhang, W. S., Li, X., Wang, X. X., Wang, G. Y., Zheng, J. T., & Abeysinghe, D. C. (2007). Ethanol vapor treatment alleviates postharvest decay and maintains fruit quality in Chinese bayberry. *Postharvest Biol Technol.*, 46, 195–198.

Zhang, Y., Chen, K., Zhang, S., & Ferguson, I. (2003). The role of salicylic acid in postharvest ripening of kiwifruit. *Postharvest Biol. Technol.*, 28, 67–74.

Zhao, Y., Shao, X. F., Tu, K., & Chen, J. K. (2007). Inhibitory effect of *Bacillus subtilis* B10 on the diseases of postharvest strawberry. *J. Fruit Sci.*, 24, 339–343.

Zhu, L., Zhou, J., & Zhu, S. (2010). Effects of a combination of nitric oxide treatment and intermittent warming on prevention of chilling injury of peach fruit during storage. *Postharvest Biol. Technol.*, 121, 165–170.

Ziosi, V., Bregoli, A. M., Fregola, F., Costa, G. & Torrigiani, P. (2009). Jasmonate-induced ripening delay is associated with up-regulation of polyamine levels in peach fruit. *J. Plant Physiol.,* 166, 938–946.

CHAPTER 17

EFFECT OF CLIMATE CHANGE ON POSTHARVEST QUALITY OF FRUITS

MOHAMMED WASIM SIDDIQUI[1,3], V. B. PATEL[2], and
M. S. AHMAD[1]

[1]Department of Food Science and Technology; Agricultural University,
Sabour, Bhagalpur, Bihar, India.
[3]E-mail: wasim_serene@yahoo.com

[2]Department of Horticulture (Fruit and Fruit Technology), Bihar Agricultural
University, Sabour, Bhagalpur (813210) Bihar, India.

CONTENTS

17.1   INTRODUCTION

Climate on Earth has changed many times during the existence of our planet, ranging from the ice ages to periods of warmness. During the last several decades, increases in average air temperatures have been reported and associated effects on climate have been debated worldwide in a variety of forums. Due to its importance around the globe, agriculture was one of the first sectors to be studied in terms of potential impacts of climate change (Adams et al., 1990). Environmental factors include climate (temperature, wind, and rainfall), air quality, and positional effects both within a planting and within the tree. Elements such as wind, heavy precipitation, and frost may result in direct loss of the fruit from the postharvest chain due to fruit scarring; increased incidence of plant pathogens associated with high rainfall, especially during flowering (i.e., anthracnose); and loss of fruit related to freeze damage. Temperature during fruit growth and maturation may also influence fruit quality by either hastening or delaying horticultural maturity. From preindustrial levels of 280 ppm, carbon dioxide (CO_2) has increased steadily to 384 ppm in 2009, and average temperature has increased by 0.76 °C over the same period. Projections to the end of this century suggest that atmospheric (CO_2) will top 700 ppm or more, whereas global temperature will increase by 1.8–4.0 °C, depending on the greenhouse emission scenario (IPCC, 2007).

Temperature increase and the effects of green house gases (GHG) are among the most important issues associated with climate change. High temperatures and exposure to elevated levels of carbon dioxide and ozone can affect the production and quality of fresh fruit crops, both directly and/or indirectly. Temperature increase affects photosynthesis directly, causing alterations in sugars, organic acids, and flavonoids contents, firmness, and antioxidant activity. Taste of fruit is highly dependent on the balance between organic acids and soluble sugars, which are predominantly represented by organic acids (citric, malic and tartaric acids) and sugars (sucrose, fructose, and glucose), respectively (Medlicott and Thompson, 1985). Another quality trait for fruits is their shelf life, which can vary with pre as well as postharvest conditions, the best known of which is temperature. However, this attribute can be influenced by conditions during fruit growth that affect the supply of minerals to the fruit. Special attention must also be paid to the influence of environmental factors on ingoing fruit fluxes as well as on the balance between mineral ions in fruits (Ferguson et al., 1999). Moreover, shelf life can be discussed in terms of dry matter content, which is directly affected by carbohydrate and water fluxes at the fruit level during its growth. Nowadays, climate changes, their causes and consequences, gained importance in many other areas of interest for sustainable life on Earth. The subject is, however, controversial.

17.2 PHYSIOLOGICAL BACKGROUND OF CLIMATE CHANGE

Crops sense and respond directly to rising (CO_2) through photosynthesis and stomatal conductance, and this is the basis for the CO_2 fertilization effect on crop yield (Long et al., 2006). These responses are highly dependent on temperature (Polley, 2002). It has been suggested that higher temperatures reduce net carbon gain by increasing plant respiration more than photosynthesis (DaMatta et al., 2009). Therefore, understanding how crop species will respond to these environmental changes is crucial for maximizing the potential benefits of elevated CO_2, for which agronomic practice needs to adapt as both temperature and CO_2 rise (Challinor and Wheeler, 2008b).

17.3 IMPACTS OF CLIMATE CHANGE ON POSTHARVEST QUALITY OF FRUITS

17.3.1 TEMPERATURE

All horticultural crops are sensitive to temperature, and most have specific temperature requirements for the development of high yields and quality. Decisions on the location of production and crop and cultivar selection, are influenced by temperature, access to and timing of markets, suitable soils and availability and reliability of irrigation water. The temperature response of crop growth and yield must be considered to predict the (CO_2) effects (Polley, 2002; Porter and Semenov, 2005). The threshold developmental responses of crops to temperature are often well defined, changing direction over a narrow temperature (Porter and Semenov, 2005). High temperatures reduce the net carbon gain in crops, particularly in C_3, by increasing photorespiration; by reducing photorespiration, (CO_2) enrichment is expected to increase photo-synthesis more at high than at low temperatures, and thus at least partially offsetting the temperature effects of supra-optimal temperatures on yield (Polley, 2002).

Higher temperatures can increase the capacity of air to absorb water vapor and, consequently, generate a higher demand for water. Water stress is of great concern in fruit production, because trees are not irrigated in many production areas around the world. It is well documented that water stress not only reduces crop productivity but also tends to accelerate fruit ripening (Henson, 2008). Exposure to elevated temperatures can cause morphological, anatomical, physiological, and, ultimately, biochemical changes in plant tissues and, as a consequence, can affect growth and development of different plant organs. These events can cause drastic reductions in commercial yield. However, by understanding plant tissues physiological responses to high temperatures, mechanisms of heat tolerances and possible strategies to improve yield, it is possible to predict reactions that will take place in the different steps of fruit and vegetable crops production, harvest and postharvest (Kays, 1997).

Photosynthetic activity is proportional to temperature variations. High temperatures can increase the rate of biochemical reactions catalyzed by different enzymes. However, above a certain temperature threshold, many enzymes lose their function, potentially changing plant tissue tolerance to heat stresses (Bieto and Talon, 1996). Pome and stone fruits require a specific amount of winter chilling to develop fruitful buds and satisfactorily break dormancy in the spring. Increasing minimum temperatures under climate change may induce insufficient chilling accumulation resulting in uneven or delayed bud break. Color development in apples occurs through the production of anthocyanin. Anthocyanin production is reduced by high temperatures.

Simulations have shown that temperature influences processes involved in fruit growth at the sink level, that is, fruit demand and growth rate. The contribution of temperature to fruit demand can be associated with the daily variation in degree-days used to compute fruit demand in the model of mango growth in dry mass (Léchaudel et al., 2005a). It has been suggested in other species, including Satsuma mandarin (Marsh et al., 1999) and apples (Austin et al., 1999), that temperature may affect the rate of cell division. Other preharvest factors such as resource limitation during cell division due, for example, to carbon competition, can be a source of variation of the initial fruit dry mass (Léchaudel and Joas, 2007).

Apple fruits exposed to direct sunlight had a higher sugar content compared to those fruits grown on shaded sides (Brooks and Fisher, 1926). Grapes also had higher sugar content and lower levels of tartaric acid when grown under high temperatures (Kliewer and Lider, 1970). Coombe (1987) observed that a 10 °C increase in growth temperature caused a 50% reduction in tartaric acid content. Kliewer and Lider (1970) and Lakso and Kliewer (1975) verified that malic acid synthesis was more sensitive to high temperature exposure during growth than was the synthesis of tartaric acid. 'Hass' avocados grown under high ambient temperatures (45 ± 2 °C), had higher moisture con- tent at harvest than fruit grown under lower temperatures (30 ± 2 °C) (Woolf et al., 1999). They also noted that higher temperature influenced oil composition, where the concentration of certain specific fatty acids increased (e.g., palmitic acid by 30%) whereas others did not (e.g., oleic acid). Avocados with higher dry matter content take longer to ripen which could pose a serious problem for growers planning to market their fruits immediately after harvest (Woolf and Ferguson, 2000; Woolf et al., 1999). Mineral accumulation was also reported to be affected by high temperatures and/or direct sunlight. 'Hass' avocado fruits exposed to direct sunlight showed higher calcium (100%), magnesium (51%) and potassium (60%) contents when compared to fruits grown under shaded conditions (Woolf et al., 1999).

Wang and Zheng (2001) observed that 'Kent' strawberries grown in warmer nights (18–22 °C) and warmer days (25 °C) had a higher antioxidant activity than berries grown under cooler (12 °C) days. The investigators also observed that high temperature conditions significantly increased the levels of flavonoids and, consequently, antioxidant capacity. Galletta and Bringhurst (1990) verified that higher day and night temperatures had a direct influence in strawberry fruit color. Berries grown under those conditions were redder and darker. McKeon et al. (2006) also addressed the effects of climate changes in functional components. They verified that higher temperatures tended to reduce vitamin content in fruit and vegetablecrops. Frequent exposure of apple fruit to high temperatures, such as 40 °C, can result in sunburn, development of watercore and loss of texture (Ferguson et al., 1999). Moreover, exposure to high temperatures on the tree, notably close to or at harvest, may induce tolerance to low-temperatures in postharvest storage (Woolf et al., 1999). In general, higher temperatures increases pest and disease activity, alter their development rate, including that of host crops, and increase survivability of some organisms, especially in warmer winters.

17.3.2   CARBON DIOXIDE

The Earth's atmosphere consists of nitrogen (78.1%) and oxygen (20.9%), with argon (0.93%) and carbon dioxide (0.031%) comprising next most abundant gases (Lide, 2009). Nitrogen and oxygen are not considered to play a significant role in global warming because both gases are virtually transparent to terrestrial radiation. The greenhouse effect is primarily a combination of the effects of water vapor, CO_2 and minute amounts of other gases (methane, nitrous oxide, and ozone) that absorb the radiation leaving the Earth's surface (IPCC, 2007). Because of climate change, the CO_2 concentrations in the atmosphere have increased approximately 35% from preindustrial times to 2005 (IPCC, 2007).

There is growing evidence suggesting that many crops (notably C_3 crops) may respond positively to increased atmospheric (CO_2) in the absence of other stressful conditions (Long et al., 2004). On the other hand, the beneficial impact of elevated (CO_2) can be equalized by other effects of climate change, such as elevated temperatures, higher tropospheric ozone concentrations and altered patterns of precipitation (Easterling et al., 2007). For photosynthesis to occur, CO_2 must diffuse from the atmosphere towards the chloroplasts. The main gate of entry of CO_2 into the leaf is via the stomatal pore. Changes in CO_2 concentration in the atmosphere can alter plant tissues in terms of growth and physiological behavior. Several studies concluded that increased atmospheric CO_2 alters net photosynthesis, biomass production, sugars and organic acids contents, stomatal conductance, firmness, seed yield, light, water, and nutrient use efficiency and plant water potential.

Jablonski et al. (2002) evaluated 79 crop and native species at elevated CO_2. They found, on average, increases in the numbers of flowers (19%) and fruits (18%), in biomass per seed (4%), in total seed biomass (25%), and in total plant biomass (31%). Bindi et al. (2001) studied the effects of high atmospheric CO_2 during growth on the quality of wines. These authors observed that elevated atmospheric CO_2 levels had a significant effect on fruit dry weight, with increases ranging from 40 to 45% in the 550 mmol CO_2/mol treatment and from 45% to 50% in the 700 mmol CO_2/

mol treatment. Tartaric acid and total sugars contents increased around 8% and 14%, respectively, by rising CO_2 levels up to a maximum increase in the middle of the ripening season. However, as the grapes reached the maturity stage, the CO_2 effect on both quality parameters almost completely disappeared. In sour orange (*Citrus aurantium*) plants grown for 17 years in elevated (CO_2) (at 300 ppm CO_2 above ambient), Kimball et al. (2007) showed that instead of acclimation, the plants continued to respond by increasing fruit production by 70%. During the last years of the experiment, increased wood growth resulted in extra (70%) biomass accumulation.

Changing carbon availability to mango fruit influenced both the dry mass and the water mass of its three main compartments: skin, pulp and stone (Léchaudel et al., 2002). Since dry matter accumulation is affected by the availability of assimilate supply, changes in its structural component, including cell walls, and its nonstructural one, consisting of soluble sugars, acids, minerals and starch, have been investigated according to their sensitivities to leaf-to-fruit ratio treatments. The influence of dilution during fruit growth on quality traits such as flesh taste or shelf life has been considered by expressing concentrations of the main biochemical and mineral compounds per unit of fresh mass (Léchaudel et al., 2005).

A recent study by Sun et al. (2012) demonstrated that elevated CO_2 could alleviate the negative effect of high temperature on fruit yield of strawberry (Fragaria × ananassa Duch. cv. Toyonoka) at different levels of nitrogen. Their results confirmed that elevated CO_2 and high temperature caused a further 12% and 35% decrease in fruit yield at low and high nitrogen, respectively. The fewer inflorescences and smaller umbel size during flower induction caused the reduction of fruit yield at elevated CO_2 and high temperature. Interestingly, nitrogen application has no beneficial effect on fruit yield, and this may be because of decreased sucrose export to the shoot apical meristem at floral transition. Moreover, elevated CO_2 increased the levels of dry matter-content, fructose, glucose, total sugar and sweetness index per dry matter, but decreased fruit nitrogen content, total antioxidant capacity and all antioxidant compounds per dry matter in strawberry fruit. They concluded that elevated CO_2 improved the production of strawberry (including yield and quality) at low temperature, but decreased it at high temperature.

Wang and Bunce (2004) reported the effect of elevated carbon dioxide on fruit quality and aroma volatile composition in field-grown strawberries (Fragaria × ananassa Duch). Elevating the ambient CO_2 concentration (ambient+300, and ambient+600 μmolmol−1 CO_2) resulted in high fruit dry matter, fructose, glucose and total sugar contents and low citric and malic acid contents. High CO_2 growing conditions significantly enhanced the fruit content of ethyl hexanoate, ethyl butanoate, methyl hexanoate, methyl butanoate, hexyl acetate, hexyl hexanoate, furanol, linalool and methyl octanoate. Thus, the total amounts of these compounds were higher in berries grown in CO_2-enriched conditions than those grown in ambient conditions. The highest CO_2 enrichment (600 μmolmol−1) condition yielded fruit with the highest levels of these aroma compounds.

The positive effects of elevated CO_2 in the fruit-storing environment have been established. One of the most notable effects of high CO_2 levels in postharvest handling is to inhibit ethylene binding and action competitively hence delaying ripening in climacteric fruits (Kanellis at al., 2009). High CO_2 will directly inhibit succinate dehydrogenase, thus impairing the functioning of the tricarboxylic acid cycle and aerobic respiration. There are numerous physiological disorders that can be attributed to high CO_2 stress, including black heart of potatoes (Davis, 1926), brown heart or core in apples and pears (Meheriuk et al., 1994), surface bronzing in apples (Meheriuk et al., 1994) and brown stain of lettuce (Kader and Saltveit, 2003). High CO_2 can also modulate chilling stress, ethylene induced disorders and susceptibility to pathogenic attack (Kader and Saltveit, 2003).

17.3.3 RAINFALL

The majority of horticultural industries have a dependence on irrigation, and very little rain grown production occurs. Rainfall has positive and negative effects on horticultural production. When 'normal in-season' rainfall events occur, irrigation storages (dams and aquifers) are replenished, and the amount of irrigation required to grow crops is reduced. 'Out of season' rainfall events, especially if high in intensity, often have devastating consequences for product quality and production. All horticultural regions will continue normal production for extended periods as droughts develop. This situation has occurred in many horticultural regions for a number of years without significant rainfall, which produces storage replenishing runoff. This is in direct contrast with much of the broad-acre and grazing industries, which depend on preplanting and in-crop rainfall to produce economic yields.

Erratic rainfall distribution due climate change effect has deterring effect on fruit quality especially at growth and development stages. Onset of unusual rains during flowering of mango cause negative effects and the incidence of several pests. The occurrence of anthracnose and powdery mildew pathogens has been reported. The anthracnose pathogens has been reported to infect fruits quiescently and become visible during postharvest conditions resulting poor appearance of fruits. Sudden rains after a dry spell result the fruit cracking in several fruits (litchi, pomegranate, apple, etc.) affecting postharvest quality, storability, and acceptability. Changing rainfall amounts and patterns may modify this temperature effect for each organism. Pathogens, which require free moisture or high humidity to reproduce and establish (e.g., black spot in apple, brown rot in stone fruit and a range of leaf and soil-borne pathogens of fruit and vegetables), are less prevalent in those regions where spring and summer rainfall is likely to reduce and dry conditions prevail for longer periods between rainfall events, and vice versa for those regions where rainfall increases. As the number of extreme rainfall events increases, soil conditions will favor the establishment and reproduction of soil borne pathogens such as Phytophthora cinnamomi in avocado.

The main effect of water stress on fruit growth according to the quantity of water shortage and the period when stress occurred was to alter the final mango size. The effect of water shortage by the partial root zone drying technique on other quality traits was not significant, as shown by the total soluble solids-to-acidity ratio or firmness measurements. Changing water availability by reducing irrigation (40% of the daily evapotranspiration) enhanced the increase of pulp dry matter content with fruit development (Diczbalis et al., 1995). Late water stress accentuated the decline in fruit Ca concentration (Simmons et al., 1995) and strongly affected fructose concentration on a fresh mass basis (Léchaudel et al., 2005). The increased levels of fructose indicated that this soluble sugar may have contribute to osmotic adjustment of mango, as has been demonstrated for other fruits under water stress condition (Mills et al., 1997).

17.3.4   OZONE

Ozone in the troposphere is the result of a series of photochemical reactions involving carbon monoxide (CO), methane (CH_4) and other hydrocarbons in the presence of nitrogen species ($NO + NO_2$) (Schlesinger, 1991). It forms during periods of high temperature and solar irradiation, normally during summer seasons (Mauzerall and Wang, 2001). It is also formed, naturally during other seasons, reaching the peak of natural production in the spring (Singh et al., 1978). However, higher concentrations of atmospheric ozone are found during summer due to increase in nitrogen species and emission of volatile organic compounds (Mauzerall and Wang, 2001). Ozone enters plant tissues through the stomates, causing direct cellular damage, especially in the palisade cells (Mauzerall and Wang, 2001). The damage is probably due to changes in membrane permeability and may or may not result in visible injury, reduced growth and, ultimately, reduced yield (Krupa and Manning, 1988).

Strawberries cv. Camarosa stored for three days under refrigerated storage (2 °C) in a ozone-enriched atmosphere (0.35 µL/L) showed a 3-fold increase in vitamin C content when compared to berries stored at the same temperature under normal atmosphere as well as a 40% reduction in emissions of volatile esters in ozonized fruits (Perez et al., 1999).

The quality of persimmon (*Diospyros kaki* L. F.) fruits (cv. Fuyu) harvested at two different harvest dates was evaluated after ozone exposure. Fruits were exposed to 0.15 µmol/mol (vol/vol) of ozone for 30 days at 15 °C and 90% relative humidity (RH). Ozone exposure was capable to maintain firmness of second harvested fruits, which were naturally softer that first harvested fruits, over commercial limits even after 30 days at 15 °C plus shelf life. Ozone-treated fruit showed the highest values of weight loss and maximum electrolyte leakage. However, ozone exposure had no significant effect on color, ethanol, soluble solids, and pH (Salvador et al., 2006).

17.5 CONCLUSIONS

Despite the significant uncertainty regarding the scale, type, and interactions of climate change impacts, mitigation and adoption strategies are needed if we are to avoid the most serious consequences of global warming. It is not known if horticulture is a net emitter or sequester of GHG. Many factors will need to be understood to be able to determine this for the large range of commodities, regions, and farming systems utilized by growers. Temperature variation can directly affect crop photosynthesis, and a rise in global temperatures can be expected to have significant impact on postharvest quality by altering important quality parameters such as synthesis of sugars, organic acids, antioxidant compounds, and firmness. Rising levels of carbon dioxide also contribute to global warming, by entrapping heat in the atmosphere. Increased levels of ozone in the atmosphere can lead to detrimental effects on postharvest quality of fruit and vegetable crops. Elevated levels of ozone can induce visual injury and physiological disorders in different species, as well as significant changes in dry matter, reducing sugars, citric and malic acid, among other important quality parameters. Erratic and unusual rainfall pattern is also affecting the fruit production and quality to cause enough economic loss.

KEYWORDS

- **Adverse effects**
- **Climate Change**
- **Fruits**
- **Physiological Background**
- **Postharvest Quality**

REFERENCES

Adams, R. M., Rosenzweig, C., Peart, R. M., Ritchie, J. T., McCarl, B. A. & Glyer, J. D. (1990). Global climate change and US agriculture. *Nature*, 345, 219–224.

Challinor, A. J., & Wheeler, T. R. (2008). Crop yield reduction in the tropics under climate change: Processes and uncertainties. *Agric. Forest Meteorol.,* 148, 343–356.

Damatta, F. M., Grandis, A., Arenque, B. C., & Buckeridge, M. S. (2009). Impacts of climate changes on crop physiology and food quality. *Food Res. Int., In Press.* doi: 10.1016/j. foodres.2009.11.001

Ferguson, I., Volz, R., & Woolf, A. (1999). Pre harvest factors affecting physiological disorders of fruit. *Postharvest Biol. Technol.,* 15, 255–262.

IPCC. (2007). Climate change. In: Solomon, S., Qin, D., Manning, M., Chen, Z., Marquis, M., Averyt, K. B., Tignor, M., Miller, H. L. (Eds.), the physical science basis. Contribution of working group I to the fourth assessment report of the intergovernmental panel on climate change (996 p.). Cambridge, United Kingdom: Cambridge University Press.

Long, S. P., Ainsworth, E. A., Leakey, A. D. B., & Ort, D. R. (2006). Food for thought: Lower-than expected crop yield stimulation with rising CO_2 conditions. *Science*, 312, 1918–1921.

Medlicott, A. P. & Thompson, A. K. (1985). Analysis of sugars and organic acids in ripening mango fruits (*Mangifera indica* L. var Keitt) by high performance liquid chromatography. *J. Sci. Food Agric.*, 36, 561–566.

Polley, H. W. (2002). Implications of atmospheric and climate change for crop yield. *Crop Science*, 42, 131–140.

Porter, J. R., & Semenov, M. A. (2005). Crop responses to climatic variation. Philosophical Transactions of the Royal Society B. *Biol. Sci.*, 360, 2021–2035.

Henson, R. (2008). The rough guide to climate changes (2nd ed.). London: Penguim Books (p. 384).

Kays, S. J. (1997). Postharvest physiology of perishable plant products. Athens: AVI. 532 p.

Bieto, J. A., & Talon, M. (1996). Fisiologia y bioquimica vegetal. Madrid: Inter americana, McGraw-Hill. 581 p.

Léchaudel, M., & Joas, J. (2007). An overview of pre harvest factors influencing mango fruit growth, quality and postharvest behaviour. *Braz. J. Plant Physiol.*, 19(4), 287–298.

Marsh, K. B., Richardson, A. C., & Macrae, E. A. (1999). Early and mid-season temperature effects on the growth and composition of satsuma mandarins. *J. Hort. Sci. Biotechnol.* 74, 443–451.

Austin, P. T., Hall, A. J., Gandar, P. W., Warrington, I. J., Fulton, T. A. & Halligan, E. A. (1999). Compartment model of the effect of early-season temperatures on potential size and growth of 'Delicious' apple fruits. *Ann. Bot.* 83, 129–143.

Brooks, C., & Fisher, D. F. (1926). Some high temperature effects in apples: Contrasts in the two sides of an apple. *J. Agric. Res.*, 23, 1–16.

Kliewer, M. W. & Lider, L. A. (1970). Effects of day temperature and light intensity on growth and composition of Vitis vinifera L. fruits. *J. Am. Soc. Hortic. Sci.*, 95, 766–769.

Coombe, B. G. (1987). Influence of temperature on composition and quality of grapes. *Acta Hortic.* 206, 23–35.

Lakso, A. N., & Kliewer, W. M. (1975). The influences of temperature on malic acid metabolism in grape berries I. Enzyme responses. *Plant Physiol.*, 56, 370–372.

Woolf, A. B., Ferguson, I. B., Requejo-Tapia, L. C., Boyd, L., Laing, W. A., & White, A. (1999). 'Impact of sun exposure on harvest quality of 'Hass' avocado fruit. *Revista Chaingo Serie Horticultura*, 5, 352–358.

Woolf, A. B., & Ferguson, I. B. (2000). Postharvest responses to high fruit temperatures in the field. *Postharvest Biol. Technol.*, 21, 7–20.

Wang, S. Y., & Zheng, W. (2001). Effect of plant growth temperature on antioxidant capacity in strawberry. *J. Agric. Food Chem.*, 49, 4977–4982.

Galletta, G. J., & Bringhurst, R. S. (1990). Strawberry management. In Galletta, G. J., Bringhurst, R. S. (Eds.). Small fruit crop management (pp. 83–156). Prentice Hall: Englewood Cliffs.

McKeon, A. W., Warland, J., & McDonald, M. R. (2006). Long-term climate and weather patterns in reaction to crop yield: A mini review. *Canadian J. Bot.*, 84, 1031–1036.

Ferguson, I., Volz, R., & Woolf, A. (1999). Pre-harvest factors affecting physiological disorders of fruit. *Postharvest Biol. Technol.*, 15, 255–262.

Lide, D. R. (2009). Handbook of chemistry and physics (90th ed.). Boca Raton: CRC Press. 2804.

Easterling, W. E., Aggarwal, P. K., Batima, P., Brander, L. M., Erda, L., & Howden, S. M. (2007). Food, fiber and forest products. In Parry, M. L., Canziani, O. F., Palutikof, J. P., van der Linden, P. J., Hanson, C. E. (Eds.). Climate change 2007: Impacts, adaptation and vulnerability. Contribution of Working Group II to the fourth assessment report of the intergovernmental panel on climate change (pp. 273–313). Cambridge: Cambridge University Press.

Jablonski, L. M., Xianzhong, W. & Curtis, P. S. (2002). Plant reproduction under elevated CO2 conditions: A meta-analysis of reports on 79, crop and wild species. *New Phytologist*, 156, 9–26.

Bindi, M., Fibbi, L. & Miglietta, F. (2001). Free air CO_2 Enrichment (FACE) of grapevine (Vitis vinifera L.): II. Growth and quality of grape and wine in response to elevated CO_2 concentrations. *Euro. J. Agron.*, 14, 145–155.

Kimball, B. A., Idso, S. B., Johnson, S., & Rillig, M. T. (2007). Seventeen years of carbon dioxide enrichment of sour orange trees: Final results. *Global Change Biol.*, 13, 2171–2183.

Léchaudel, M., Génard, M., Lescourret, F., Urban, L., & Jannoyer, M. (2002). Leaf-to-fruit ratio affects water and dry matter content of mango fruit. *J. Hort. Sci. Biotechnol.*, 77, 773–777.

Léchaudel, M., Génard, M., Lescourret, F., Urban, L., & Jannoyer, M. (2005). Modeling effects of weather and source sink relationships on mango fruit growth. *Tree Physiol.*, 25, 583–597.

Wang, S. Y. & Bunce, J. A. (2004). Elevated carbon dioxide affects fruit flavor in field-grown strawberries (*Fragaria* × *ananassa* Duch). *J. Sci. Food Agric.*, 84, 1464–1468. doi: 10.1002/jsfa.1824.

Diczbalis, Y., Hofman, P., Landrigan, M., Kulkarni, V., & Smith, L. (1995). Mango irrigation management for fruit yield, maturity and quality. In: Proceedings of Mango 2000 marketing seminar and production workshop. *Brisbane Australia*, 85–90.

Simmons, S. L., Hofman, P. J., & Hetherington, S. E. (1995). The effects of water stress on mango fruit quality. In: Proceedings of Mango 2000 marketing seminar and production workshop. *Brisbane Australia*, 191–197.

Mills, T. M., Behboudian, M. H., & Clothier, B. E. (1997). The diurnal and seasonal water relations, and composition, of 'Braeburn' apple fruit under reduced plant water status. Plant Sci. 126, 145–154.

Schlesinger, W. H. (1991). Biogeochemistry: An analysis of global change. New York: Academic Press. 443 p.

Mauzerall, D. L. & Wang, X. (2001). Protecting agricultural crops from the effects of tropospheric ozone exposure: Reconciling science and standard setting in the United States, Europe, and Asia. *Ann. Rev. Ener. Environ*, 26, 237–268.

Krupa, S. V., & Manning, W. J. (1988). Atmospheric ozone: Formation and effects on vegetation. *Environ. Pollu.*, 50, 101–137.

Perez, A. G., Sanz, C., Rios, J. J., Olias, R. & Olias, J. M. (1999). Effects of ozone treatment on postharvest strawberry quality. *J. Agric. Food Chem.*, 47(4), 1652–1656.

Salvador, A., Abad, I., Arnal, L., & Martinez Javegam, J. M. (2006). Effect of ozone on postharvest quality of persimmon. *J. Food Sci.*, 71(6), 443–446.

CHAPTER 18

IMPACT OF CLIMATE CHANGE ON FOOD SAFETY

H. R. NAIK[1] and S. SHERAZ MAHDI[2]

[1]Division of Food Technology, Sher-e-Kashmir University of Agricultural Sciences and Technology of Kashmir Shalimar Campus - 191121 (J &K), India; E-mail: haroonnaik@gmail.com

[2]Deparment of Agronomy, Bihar Agricultural University, Sabour - 810 213, Bhagalpur (Bihar), India; E-mail: sheerazsyed@rediffmail.com

CONTENTS

ABSTRACT

Climate change and variability in environment may have impact on occurrence of food safety hazards at different stages of food chain. There are multiple pathways through which climate related factors mayimpact food safety including: change in temperature and precipitation patterns, increased frequency and intensity of extreme weather events, ocean warming and acidification and changes in contaminants, green house effects. Climate change may also affect socioeconomic aspects related to food systems such as agriculture, animal production, global trade, and post harvest quality and human behavior which ultimately influence food safety. Temperature increase and the effects of greenhouse gases are among the most important issues associated with climate change. A rise in temperature will increase the risk of food poisoning and food spoilage unless the cold-chain is extended and improved. The little data that is available suggests that currently the cold-chain accounts for approximately 1% of CO_2 production in the world; however this is likely to increase if global temperature increases significantly. Using the most energy efficient refrigeration technologies would be possible to substantially extend and improve the cold-chain without any increase in CO_2, and possibly even a decrease. Studies have shown that the production and quality of fresh fruit and vegetable crops can be directly and indirectly affected by high temperatures and exposure to elevated levels of carbon dioxide and ozone. Temperature increase affects photosynthesis directly, causing alterations in sugars, organic acids, and flavonoids contents, firmness and antioxidant activity. Carbon dioxide accumulation in the atmosphere has direct effects on postharvest quality causing tuber malformation, occurrence of common scab, and changes in reducing sugars contents on potatoes. High concentrations of atmospheric ozone can potentially cause reduction in the photosynthetic process, growth and biomass accumulation. Ozone-enriched atmospheres increased vitamin C content and decreased emissions of volatile esters on strawberries. Tomatoes exposed to ozone concentrations ranging from 0.005 to 1.0 micro mol/mol had a transient increase in b-carotene, lutein and lycopene contents. Production of harmful algal blooms affecting fish production and poisoning thereof is not a debatable subject now

18.1 INTRODUCTION

Assuring food safety is a complex task. Food safety hazards can arise at any stage of the food chain from primary production through to consumption. Foods are governed by food laws and regulations, which are collectively known as the *food control system.* The ultimate goal of this system is to ensure that food presented to consumers is safe and honestly presented. It is in the interest of all stakeholders to optimize the efficiency of the system in order to make the best possible public health impact with limited resources available. Major principles that underlie strategies for improving the efficiency and effectiveness of food control are:

- that efforts are focused on issues that pose the greatest risk;
- that the responsibility for producing safe food rest unambiguously with the food businesses who are best placed to design and implement controls at the most appropriate point within the food production systems to *prevent*or minimize food safety risks;
- that government establishes food safety requirements, facilitate industry's compliance with these and then ensure that the requirements are met through a range of regulatory and nonregulatory measures.

The Fourth Assessment Report of the Intergovernmental Panel on Climate Change (IPCC 2007) dispelled many uncertainties about climate change. Warming of the climate system is now unequivocal and according to IPCC the increase in global temperatures observed since the mid-twentieth century is predominantly due to human activities such as fuel burning and land use changes. Projections for the twenty-first century show that global warming will accelerate with predictions of the average increase in global temperature ranging from 1.8°C to 4 °C. Other effects of climate change include trends towards stronger storm systems, increased frequency of heavy precipitation events and extended dry periods. The contraction of the ice sheets will lead to rising sea-levels. These changes have implications for food production, food security and food safety.

18.1.1 EFFECT ON FOOD CROPS AND ANIMALS

Crop production is extremely susceptible to climate change. It has been estimated that climate changes are likely to reduce yields and/or damage crops in the twenty-first century (IPCC, 2007). While the impact of biotic (microbial population of fungi, bacteria, viruses of the macroenvironment, soil, air and water) and abiotic factors (nutrient deficiencies, air pollutants and temperature/moisture extremes) on crop production and food security are more obvious, it is important to note that these factors may also have significant impact on the safety of food crops. Further concern is the impact of climate change on the prevalence of environmental contaminants and chemical residues in the food chain. Climate change may affect zoonoses (diseases and infections which are naturally transmitted between vertebrate animals and man) in a number of ways. It may increase, the transmission cycle of many vectors;

* the range and prevalence of vectors and animal reservoirs;
* in some regions it may result in the establishment of new diseases.

18.1.2 EFFECT ON FISHERIES

Climate change has implications for food safety. From a microbiological perspective, climate change exacerbates eutrophication (nutrient loading) causing phytoplankton growth, increased frequencies of harmful algal blooms, particularly of toxic species. Accumulation of these toxins by filter feeders (bivalve molluscs) and the subsequent consumption of these products have serious implications for humans. Furthermore, an increase in water temperatures promotes the growth of organisms such as *Vibrio vulnificus* leading to an increased risk from handling or consuming fish grown in these waters (Paz et al., 2007). Climate change (in particular temperature increase) facilitates methylation of mercury and subsequent uptake by fish.

18.1.3 EFFECT ON FOOD HANDLING, PROCESSING, TRADING

Climate change impacts not only on primary production but also on food manufacturing and trade. Emerging hazards in primary production could influence the design of the safety management systems required to effectively control those hazards and ensure the safety of the final product. Furthermore, increasing average temperatures could increase hygiene risks associated with storage and distribution of food commodities. It is important, therefore, that the food industry be vigilant to the need to modify hygiene programs. Reduced availability and quality of water in food handling and processing operations will also give rise to new challenges to hygiene management.

Using the classic epidemiologic triad (host, agent, and environment), it is clear that climate, whichimpacts all three sectors of the triad, can have a dramatic effect on infectious disease. This is well documented and even predictable for some food and waterborne diseases of the developing world (e.g., bacillary dysentery, cholera) and perhaps less so for the developed world, where stringent public health measures (sewage disposal, clean water and hygiene) moderate the risk of diarrheal disease. Evidence of the impact of climate change on the transmission of food and waterborne diseases comes from a number of sources, for example, the seasonality of foodborne and diarrheal disease, changes in disease patterns that occur as a consequence of temperature, and associations between increased incidence of food and waterborne illness and severe weather events (Hall et al., 2002; Rose et al., 2001).

18.1.4   SOURCES AND MODES OF TRANSMISSION

The microflora of a food consists of the microorganisms associated with the raw material, those acquired during handling and processing and those surviving preservation technique and storage. Bacteria, viruses, and parasitic protozoa (BVP) that are pathogenic to humans and frequently contaminate the food supply can be subcategorized based on their ultimate source. They are predominantly associated with fecal matter (animals/human), on the skin, nose and throat of healthy individuals and in nature. Based on the above categorization, the general scenarios by which foods become contaminated with pathogens include:
- contact with human/animal sewage/feces;
- contact with infected food handlers;
- environmental contamination (from air, water, food contact materials, etc.);
- contact with raw foods, etc.

Such contamination can arise along any part of the farm-to-fork continuum and may arise from any number of sources. Climate constrains the range of infectious diseases, while weather, which is impacted by cli-

mate, affects the timing and intensity of outbreaks (Epstein, 2001). Therefore, the two early manifestations of climate change, particularly global
warming, would be expansion in the geographic range and seasonality of
disease, and the emergence of outbreaks occurring as a consequence of
extreme weather events (Epstein, 2001).

18.1.5 CLIMATIC INFLUENCES (E.G., TEMPERATURE, HUMIDITY) ON THE PREVALENCE OF SOME DISEASES

- Increases in disease notifications, particularly salmonellosis
  (D'Souza et al., 2004) and to a lesser extent campylobacteriosis
  (Kovats et al., 2005) are frequently preceded by weeks of elevated
  ambient temperature.
- Higher temperature and humidity in the week before infection has
  been correlated with decreased hospitalization rates for children
  diagnosed with rotavirus. This is particularly interesting because
  survival of the virus is favored at lower temperature and humidity
  (D'Souza et al., 2008). Rotavirus is considered a significant cause of
  food borne illness (FAO, 2008a).
- El Nino-associated rises in cholera have been documented for both
  Peru and Bangladesh, as have been increases in diarrheal disease in
  Peruvians (reviewed by Hall et al., 2002).
- Cholera is perhaps the best model for understanding the potential
  for climate-induced changes in the transmission of foodborne disease. *Vibrio cholerae* is the causative agent of this disease, which
  produces substantial morbidity and mortality, particularly in the developing world.

Extreme weather conditions (e.g., flooding, drought, hurricanes, etc.)
can impact on the transmission of disease. For example, periods of excessive precipitation and periods of drought influence both the availability
and quality of water and have been linked to the transmission of water
and food borne disease. Furthermore, extreme weather events can result in
forced evacuation of refugees into close quarters. This frequently results
in extreme stress, malnutrition, and limited access to medical care, all of
which contribute to increased susceptibility and severity of disease.

18.2 EFFECT OF CLIMATE CHANGE ON ZOONOTIC DISEASE

Zoonotic diseases are transmitted from animals to people in a number of ways. Some diseases are acquired by people through direct contact with infected animals or animal products and wastes. Other zoonoses are transmitted by vectors; while others are transmitted through the consumption of contaminated food or water (Table 18.1). The proliferation of zoonoses and other animal diseases may result in an increased use of veterinary drugs that could lead to increased and possibly unacceptable levels of veterinary drugs in foods (FAO, 2008b).

TABLE 18.1 Examples of Some Zoonotic Agents That Are Expected to be Affected by Climate Change and Their Mode of Transmission

Virus	Host	Mode of transmission to humans
Rift Valley fever Virus	Multiple species of livestock and wildlife	Blood or organs of infected animals (handling of animal tissue), unpasteurized or uncooked milk of infected animals, mosquito, hematophagous flies
Nipah virus	Bats, and pigs	Directly from bats to humans through food in the consumption of date palm sap (Luby et al., 2006). Infected pigs present a serious risk to farmers and abattoir workers
Hendra virus	Bats, and horse	Secretions from infected horses
Rotavirus	Humans	Faecal-oral route, spread through contaminated water and also by infected food-handlers who do not wash their hands properly.
Hepatitis E virus	Wild and domestic Animals	Faecal-oral.pig manure is a possible source through contamination of irrigation water and shellfish in coastal waters

Transmission of Bacteria

Bacterium	Host	Mode of transmission
Salmonella	Poultry and pigs	Faecal/oral
Campylobacter	Poultry	Faecal/oral
E. coli O157	Cattle and other ruminants	Faecal/oral

Anaerobic spore forming Bacteria	Birds, mammals and livestock	Ingestion of spores through environmental routes, water, soil and feeds. This has been associated with outbreaks of anthrax in livestock and wild animals, blackleg (*Clostridium chauvoei*) in cattle and botulism in wild birds after droughts. The meat and milk from cattle that have botulism should not be used for human consumption.
Yersinia	Birds and rodents with regional differences in the species of animal, infected. Pigs are a major livestock reservoir	Handling pigs at slaughter is a risk to humans
Listeria monocytogenes	Livestock	In the northern hemisphere, listeriosis has a distinct seasonal occurrence in livestock probably associated with feeding of silage
Leptospirosis	All farm animal species	Leptospirae shed in urine to contaminate pasture, drinking water and feed

Transmission of Protozoas

Protozoan	Host	Mode of transmission
Toxoplasma gondii	Cats, pigs, sheep	Cat feces are a major source of infection. Handling and consuming raw meat from infected sheep and pigs pose a zoonotic risk
Cyptosporidium and *Giardia*	Cattle, sheep	Faecal-oral transmission. (Oo)cysts are highly infectious and with high loadings, livestock feces pose a risk to animal handlers

Transmission of parasites

Parasite	Host	Mode of transmission
Tapeworm (*Cysticercusbovis*)	Cattle	Faecal-oral
Liver fluke (*Fasciola hepatica*)	Sheep, cattle	Eggs are excreted in feces, and life cycle involves lymnaeid snail hosts. Human cases generally associated with the ingestion of marsh plants such as watercress

18.2.1 CLIMATE CHANGE EFFECTS ON BVP

- Increase in the susceptibility of animals to disease
- Increase in the range or abundance of vectors/animal reservoirs
- prolonging the transmission cycles of vectors

18.2.2 CLIMATE CHANGE AND ITS INFLUENCE ON MOULD AND MYCOTOXIN CONTAMINATION

Mycotoxins are a group of highly toxic chemical substances that are produced by toxigenic molds that commonly grow on a number of crops. These toxins can be produced before harvest in the standing crop and many can increase, even dramatically, after harvest if the postharvest conditions are favorable for further fungal growth. Human dietary exposure to mycotoxins can be directly through consumption of contaminated crops. Mycotoxins can also reach the human food supply through livestock that have consumed contaminated feed. The problem of mycotoxin contamination of foods and the resulting public health impact is not new – it is likely that mycotoxins have plagued mankind since the beginning of organized crop production (FAO, 2001). At high doses mycotoxins produce acute symptoms and deaths but, arguably, lower doses that produce no clinical symptoms are more significant to public health due to the greater extent of this level of exposure. Particular mycotoxins maypossess carcinogenic, immunosuppressive, neurotoxic, estrogenic or teratogenic activity, some more than one of these. Table 18.2 lists mycotoxins that are of world-wide importance meaning that they have been demonstrated to have significant impact on public health and animal productivity in a variety of countries. There are several other mycotoxins that are considered to be of regional significance (FAO, 2001).

TABLE 18.2 Moulds and Mycotoxins of World-Wide Importance

Mould Species	Mycotoxins Produced
Aspergillus parasiticus	Aflatoxins B1, B2, G1, G2
Aspergillus flavus	Aflatoxins B1, B2
Fusarium sporotrichioides	T-2 toxin
Fusarium graminearum	Deoxynivalenol (or nivalenol) Zearalenone
Fusarium moniliforme (F. verticillioides)	Fumonisin B1
Penicillium verrucosum	Ochratoxin A
Aspergillums' ochraceus	Ochratoxin A

Although the impact of climate change on fungal colonization has not been yet specifically and thoroughly addressed, temperature, humidity and precipitation are known to have an effect on toxigenic molds and on their interaction with the plant hosts. In general we know that fungi have temperature ranges within which they perform better and therefore increasing average temperatures could lead to changes in the range of latitudes at which certain fungi are able to compete. Since 2003, frequent hot and dry summers in Italy have resulted in increased occurrence of *A. flavus*, the most xerophilic of the *Aspergillus* genus, with consequent unexpected and serious outbreak of aflatoxin contamination, uncommon in Europe. Also in United States serious outbreaks of *A. Flavus* have been reported for similar reasons. Generally moist, humid conditions favormold growth – moist conditions following periods of heavy precipitation or floods would be expected to favormold growth. Generally speaking, conditions adverse to the plant (drought stress, stress induced by pest attack, poor nutrient status, etc.) encourages the fungal partner to develop more than under favorable plant conditions with the expectation of greater production of mycotoxins.

18.2.3   INFLUENCE CLIMATE CHANGE ON POST-HARVEST CONDITIONS

It is common for commodities to contain mycotoxigenic fungi at harvest. Up to the point of harvest, the status of the plant will play a major role in determining the degree of mycotoxin contamination. Thereafter, fungal development and mycotoxin production will be controlled by postharvest handling techniques and practice. In the simplest terms this will consist of some kind of cleaning, whichmay be conducted concomitantly with harvest, drying and storage where stability is maintained by restricting water availability to a level well below that required for fungal growth. Climate change could impinge on this part of the food chain especially in regions where capital investment on such production infrastructure is lacking.

18.2.4   EFFECT ON POST HARVEST QUALITY OF FRUITS AND VEGETABLES

Climatic change results in Green house effect on fruits/ vegetables. Higher temperature, Co_2 level, Ozone depletion has direct or indirect effect on

Quality of horticultural produce at different stages from production to consumer (Moretti et al., 2010).

18.2.4.1   PHOTOSYNTHESIS

Due to Global Warmingalterations in Sugars, Acids, flavonoids and firmness in crops has been reported. For example:
- Increased CO_2 level has caused tuber malformation, common scab, changed reducing Sugar content in Potatoes.
- Increased ozone level resulted in:
- decreases photosynthesis process, Decrease growth and biomass accumulation.
- increased Vitamin C accumulation, Decreased emission of volatile esters in Strawberries.
- increased β carotene, lutein and lycopene in tomato.

18.3   EFFECT OF TEMPERATURE ON FRUITS/VEGETABLES

- Photosynthesis, respiration, aquos relations in membrane stability affected.
- Normal Physical process temperature (0–40 °C). Best reported photosynthesis respiration ratio >10 at 15 °C.
- Leaf to air vapor pressure difference (D) change effects photosynthesis.
- Enzyme catalyzed Biochemical reactions affects photosynthesis.
- Harvest Index affected. Crops mature earlier.
- Rapid cooling to remove field heat to increase shelf life 2–3 fold for each 10 °C increase in temperature. This reduces Respiration rate and enzyme activity, which slows ripening, senescence, maintains firmness, minimizes water loss and pathogens but higher pulp temperature Produce shall need more energy for cooling thus raising product prices.

18.3.1   EFFECT OF HIGH TEMPERATURE

a)   On Quality
- Flavor is affected, for example, higher sugar content and lower organic acids of apple groups.
- Firmness two times more in crops grown under high temperature due to change in cell wall composition, cell number and cell turgor properties.
- Increase in moisture content and certain fatty acids like Palmitic, oleic, etc.
- Higher concentration of minerals like Ca, Mg, Cu and K due to less water movement through crop.
- Increase in antioxidant activity (flavonoids) in berries and but decrease in vitamin content.

b)   Physiological disorders
- Tomatoes growing at >30 °C temperature affects their color develop (yellowish white color rather than red), cause softening, increased respiration rate and ethylene production.
- Fruits exposed to 40 °C have induced metabolic disorders, fungal and bacteria invasion.
- Greater than 40 °C results in Sun burn of apples, water core and loss of texture and decreased tolerance to low temperature of apple on storage.
- Due to delay in winter, yields dropped significantly in crops like cauliflower, soybean.

18.3.2   EFFECT OF CO_2 EXPOSURE

Composition of air (78% N, 21°% O, 0.93% Ar and 0.03% CO_2) gets changed due to climate change resulting in Green house combination effect of water vapor, CO_2 and minute amounts of other gases like (CH_4, Nitrous oxide and Ozone) which absorb radiations leaving earth's surface. CO_2 or other gases absorb earth's infrared radiations thus trapping heat resulting in global warming. Their effect on crops is:
- Alter plant tissues in tenure of growth and physiological behavior.
- Photosynthesis, biomass product, sugars and organic acid content, Stomatal-conductance, firmness, yield, light, water, nutrient use efficiency and plant water potential.

- Decreased tuber formation in potatoes (63%,) resulting poor processing quality and lower tuber greening (12%). About 550 micro mol CO_2 |mol concentration of CO_2 decreased glucose, fructose, and RS concentrations reducing tuber quality due to decreased browning and acryl amide formation in French fries.
- About 34% common scab reported.
- Increased CO_2 concentration however, stimulates grapevine produce with affecting quality of grapes.

18.3.3 EFFECT OF HIGH OZONE FORMATION

Forms during periods of high temperature and solar irradiation, normally during summer season. Concentration maximum in late afternoon and minimum in early morning hours. Increased ozone may also be by movement of local winds or downdrafts from the Stratosphere. The physiological effects are:

- The Stomata conductance and ambient concentration is most important for O_3 uptake by plants. Ozone enters plant through Stomata's causing direct cellular damage. Damage due to charges in membrane permeability.
- May or may not result in visible injury, reduced growth and ultimately reduced yield.

18.3.4 VISIBLE INJURY

a) Change in pigmentation called bronzing.
b) Leaf chlorosis, for example, yellowing green vegetable leaves affecting pH and quality likeappearance, color, flavor compounds.
c) Premature senesce.
  - Increased or decreased in yield of different crops due to impaired conductance
  - Change in Carbon transport of root, tubers, bulbs results in less accumulation of starch and sugars.
  - Decreased biomass produce directly impacts size, appearance, etc.

18.3.5 EFFECT OF OZONE ON QUALITY

- Crops like mushroom, seedless cucumber and broccoli stored at 3–10 °C and exposed to O_3 showed min. response to storage temperatures.
- O_3 removes ethylene from environment so useful in closed rooms, for example, 0.4HL/L O_3 removes 1.5 – 2.0 HL/L Ethylene from Apple store room.
- O_3 increased Vitamin C and decreased volatiles esters emission in strawberries
- O_3 has germicidal effect and overall no bad effect observed.

18.4 IMPACTS OF HARMFUL ALGAL BLOOMS

During recent decades, there has been an apparent increase in the occurrence of Harmful Algal Blooms (HABs) in many marine and coastal regions (Fig. 18.1; Hallegraeff, 1993). Toxin-producing HAB species are particularly dangerous to humans. A number of human illnesses are caused by ingesting seafood (primarily shellfish) contaminated with natural toxins produced by HAB organisms; these include amnesic shellfish poisoning (ASP), diarrheic shellfish poisoning (DSP), neurotoxic shellfish poisoning (NSP), azaspiracid shellfish poisoning (AZP), paralytic shellfish poisoning (PSP), and ciguatera fish poisoning. These toxins may cause respiratory and digestive problems, memory loss, seizures, lesions and skin irritation, or even fatalities in fish, birds, and mammals (including humans) (Anderson et al., 2002; Sellner et al., 2003). Some of these toxins can be acutely lethal and are some of the most powerful natural substances known; additionally, no antidote exists to any HAB toxin (Glibert et al., 2005). Because these toxins are tasteless, odorless, and heat and acid stable, normal screening and food preparation procedures will not prevent intoxication if the fish or shellfish is contaminated (Baden et al., 1995; Fleming et al., 2006).

In addition to human health effects, HABs also have detrimental economic impacts due to closure of commercial fisheries, public health costs and other related environmental and sociocultural impacts (Trainer and Suddleson, 2005; NOAA – CSCOR, 2008) (Table 18.3).

TABLE 18.3   Examples of HAB-Forming Algae and Their Effects on Food Safety (Faust and Gulledge, 2002; Sellner et al., 2003).

Poisoning	Functional group	Species
Diarrheic shellfish poisoning (DSP)	Dinoflagellates	*Prorocentrum* spp.
		Dinophysis spp.
		Protoperidinium spp.
Paralytic shellfish poisoning (PSP)	Dinoflagellate	*Alexandrium* spp.
Neurotoxic shellfish poisoning (NSP)	Dinoflagellate	*Gymnodinium* spp.
Amnesic shellfish poisoning (ASP)	Diatoms	*Pseudo-nitzschia* spp.
Ciguatera fish poisoning	Dinoflagellate	*Gambierdiscus* spp.

Harmful Algal Bloom is because of:

a) Acidification of Waters

It is possible that ocean acidification may cause changes in HAB dynamics through changes in phytoplankton community composition, however, there are insufficient data to draw any conclusions about the impacts that increasing CO_2 might have on the growth and composition of HAB-causing marine phytoplankton

b) Impact of Sea-level Rise, Increased Precipitation and Flash Floods on Harmful Algal Communities

Sea level rise, increased precipitation and flash floods are most likely to affect harmful algal communities through increased nutrient release to coastal and marine waters. Two key nutrients required for phytoplankton growth, nitrogen (N) and phosphorus (P), are found in fertilizers and animal and human waste; however silicon (Si), which is only required for diatom growth, is not added to the environment through human activity. Increased concentrations of N and P without a corresponding increase in Si may cause changes in phytoplankton community composition, favoring dinoflagellates, which have no biological requirement for Si, at the expense of diatoms (Smayda, 1990). Such a shift in the phytoplankton community towards dinoflagellate dominance may result in increased numbers of HAB species in regions prone to increased anthropogenic nutrients. Flash flooding and sudden storm events may release 'pulses' of nutrients into coastal waters. As the sea reclaims low-lying land, areas that are currently intensively farmed or urbanized may be drowned, causing the addition of nutrients, particularly N and P, to coastal systems. Additionally, as sea levels rise, wetland habitats are lost. Wetlands act as natural filters for anthropogenic nutrients and are therefore important in regulating nutrient loads to coastal waters. Wetlands and mangrove habitats also provide a natural form of protection from storm surges and flooding (Nicholls et al., 2007). Without these habitats, coastal waters may be more prone to increased levels of nutrients and nutrient imbalance.

18.5 ENVIRONMENTAL CONTAMINANTS AND CHEMICAL RESIDUES IN THE FOOD CHAIN

There are many pathways through which global climate change and variability mayimpact environmental contamination and chemical hazards in foods. Contamination of agricultural and pastureland soil with dioxins have been associated with climate change related extreme events, particularly with the increased frequency of inland floods. Soil contamination can be attributed to remobilization of contaminated river sediments, which are subsequently deposited on the flooded areas. In other cases, contamination of the river water bodies, and subsequently of the flooded soils, may have resulted from mobilization in upstream-contaminated terrestrial areas such as industrial sites, landfills, sewage treatment plants, etc. Results showed very high levels of polychlorinated dibenzo-p-dioxins and dibenzofurans (PCDD/Fs) present in soil in periodically flooded pastureland riverside of the dikes, and grazing on the floodplains revealed a significant transfer of PCDD/Fs into milk (Umlauf et al., 2005). While the uptake of contaminated soils during grazing is an important factor considering the transfer into the food chain, barn feeding of properly harvested greens from the same floodplains is less critical (Umlauf et al., 2005). Sources of chemical contamination of flood water included oil spills from refineries and storage tanks, pesticides, metals and hazardous waste. Several chemicals, such as hexavalent chromium, manganese, p-cresol, toluene, phenol, 2, 4-D (an herbicide), nickel, aluminum, copper, vanadium, zinc, andBenzidine were detected in flood water. Trace levels of some organic acids, phenols, trace cresols, metals, sulfur chemicals, and minerals associated with seawater were also detected (EPA, 2005). Concentrations of most contaminants were within acceptable short-term levels, except for lead and volatile organic compounds in some areas (Pardue et al., 2005).

18.5.1   CONTAMINATION OF WATERS

Higher water temperatures, increased precipitation intensity, and longer periods of low flows exacerbate many forms of water pollution, including sediments, nutrients, dissolved organic carbon, pathogens, pesticides and salts (Kundzewicz, et al., 2007). In regions where intense rainfall is expected to increase, pollutants (pesticides, fertilizers, organic matter, heavy metals, etc.) will be increasingly washed from soils to water bodies (Boorman, 2003). Higher runoff is expected to mobilizefertilizers and pesticides to water bodies in regions where their application time and low vegetation growth coincide with an increase in runoff (Soil and Water Conservation Society, 2003). Because of compaction, heavy rainfall after drought can result in more severe runoff and increased risk of certain types of contamination. Alternating periods of floods and drought can therefore aggravate the problem. Increasing ocean temperatures may indirectly influence human exposure to environmental contaminants in some foods (e.g., fish and mammal fats). Ocean warming facilitates methylation of mercury and subsequent uptake of methyl mercury in fish and mammals has been found to increase by 3–5% for each 1°C rise in water temperature. Temperature increases in the North Atlantic are projected to increase rates of mercury methylation in fish and marine mammals, thus increasing human dietary exposure (Booth and Zeller, 2005).

Sea level rise related to climate change is expected to lead to saltwater intrusion into aquifers/water tables in coastal areas. This will extend areas of salinization of groundwater and estuaries, resulting in a decrease in freshwater availability for humans, agriculture and ecosystems in coastal areas. One-quarter of the global population lives in coastal regions; these are water-scarce <10% of the global renewable water supply (WHO, 2005) and are undergoing rapid population growth.

18.5.2   THE EFFECT OF CLIMATIC CHANGE ON THE COLD-CHAIN

The food manufacturing industry utilizes chilling and freezing processes as a means of preserving foods. Refrigeration of these foods is continued during transportation, retail distribution and home storage to maintain the foods at the desired temperatures. These are important steps in maintaining the safety, quality and shelf life of foods for the consumer, and the processes from primary cooling through to domestic storage make up the 'food cold-chain.' If climatic change results in a substantial rise in average ambient temperatures this will impose higher heat loads on all systems in the cold-chain. In systems that have capacity to cope with these higher loads this will just require the refrigeration plants to run for longer periods and use more energy (James and James, 2010). In addition to the generation of CO_2 the refrigerants currently used in cold-chain have considerable global warming potential (GWP). Use of alternative refrigerants and alternative refrigeration cycles with a reduced GWP is need of our.

About 20% of the global-warming impact of refrigeration plants is due to refrigerant leakage. The dominant types of refrigerant used in the food industry in the last 60 years have belonged to a group of chemicals known as halogenated hydrocarbons, for example, chlorofluorocarbons (CFCs) and the hydrochlorofluorocarbons (HCFCs). Scientific evidence clearly shows that emissions of CFCs have been damaging the ozone layer and contributing significantly to global warming. The little data that is available suggests that currently the cold-chain accounts for approximately 1% of CO_2 production in the world. However this is likely to increase if global temperatures increase significantly. Until recently the major concern in the refrigeration industry regarding climate change has been the impact of refrigerants on the ozone layer and the replacement of current refrigerants with "greener" alternatives. Energy efficiency is increasingly of concern to the food industry mainly due to substantially increased energy costs and pressure from retailers to operate zero carbon production systems. Reducing energy in the cold-chain has a big part to plays since worldwide it is estimated that 40% of all food requires refrigeration and 15% of the electricity consumed worldwide is used for refrigeration. Simple solutions such as the maintenance of food refrigeration systems will reduce energy consumption. Repairing door seals and door curtains, ensuring that doors can be closed and cleaning condensers produce significant reductions in energy consumption. In large cold storage sites it has been shown that energy can be substantially reduced if door protection is improved, pedestrian doors fitted, liquid pressure amplification pumps fitted, defrosts optimized, suction liquid heat exchangers fitted and other minor issues corrected.

New/alternative refrigeration systems/cycles, such as Trigeneration, Air Cycle, Sorption-Adsorption Systems, Thermoelectric, Stirling Cycle, Thermoacoustic and Magnetic refrigeration, have the potential to save energy in the future if applied to food refrigeration.

18.6 ADDRESSING FOOD SAFETY IMPLICATIONS OF CLIMATE CHANGE

a) Understanding Foodborne Disease including Zoonosis
b) Predictive Models
c) Foodborne Disease: Surveillance/Animal Disease Surveillance
d) Foodborne Pathogens: Monitoring and Surveillance
e) Improved Coordination among Public Health, Veterinary Health, Environmental Health and Food Safety Services, that is, one health concept.
f) Prevention of Mycotoxin Contamination
g) Agricultural Policy and Public Information Review for Mycotoxins
h) Good Horticultural, Agricultural, Animal husbandry, Aquaculture and Veterinary Practices
i) Replacement of hydroflourocarbon refrigerants in refrigeration.
j) Data Exchange: Good data exchange mechanisms are required at both national and international level. These should cover the distribution of animal and plant diseases, pests, ecological conditions including climate, and associated usage of pesticides, veterinary drugs and chemotherapeutants will be needed to enable risk assessment, prevention, monitoring and control.
k) Development of Tools for Rapid Detection or Removal of Contaminants

18.7 SUMMARY AND CONCLUSIONS

Assuring food safety is a complex issue as it involves considerations from preproduction through to final home preparation of the food product. Recommendations on food safety management emphasize the need for broad input and coordination even though this remains a challenge in many countries. Recognizing, understanding and preparing for the impacts of climate change further highlight the need to promote interdisciplinary approaches to addressing challenges affecting food safety given the interrelationships among environmental impacts, animal and plant health impacts and food hygiene. These interrelationships are further complicated by the broader public health implications of climate change as well as the food security implications. To address the challenges of climate change all the above-mentioned practices need to be followed in letter and spirit and also new crop varieties and technologies need to be generated through research.

KEYWORDS

- **Agricultural Crops**
- **Chemical Residues**
- **Climatic Influences**
- **Food Handling**
- **Food Safety**
- **Processing**
- **Quality**

REFERENCES

Anderson, D. M., Glibert, P. M., & Burkholder, J. M. (2002). Harmful algal blooms and eutrophication. Nutrient sources, composition, and consequences. Estuaries 25, 704–726.

Baden, D., Fleming, L. E., & Bean, J. A. (1995). Marine toxins. In deWolff, F. A. (ed.) Handbook of Clinical Neurology: Intoxications of the Nervous System Part II Natural Toxins and Drugs. Elsevier Press, Amsterdam, Netherlands 141–175.

Boorman, D. B. (2003). LOIS in-stream water quality modeling. Part 2. Results and scenarios. Sci. Total Environ, 314–316, 397–409.

CAST (2003). *Mycotoxins: Risks in Plant, Animal and Human Systems, Task Force Report,* ISSN 0194 N. 139. CDC (2008). Congressional Testimony. Select Committee on Energy independence and Global Warming United States House of Representatives Climate Change and Public Health. Statement of Howard Frumkin, MD.

D'Souza, R. M., Becker, N. G., Hall, G., & Moodie, K. B. A. (2004). Does ambient temperature affect foodborne disease?" Epidemiol. 15, 86–92.

D'Souza, R. M., Hall, G., & Becker, N. G. (2008). Climatic factors associated with hospitalizations for rotavirus diarrhoea in children under 5, years of age. Epidemiol. Infect. 136, 56–64.

Environmental Protection Agency (2005). Environmental Assessment Summary for Areas of Jefferson, Orleans, St. Bernard, and Plaquemines Parishes Flooded as a Result of Hurricane. Katrina. http://www.epa.gov/katrina/testresults/katrina_env_assessment_summary.htm.

Epstein, P. R. (2001). Climate change and emerging infectious diseases. Microbes Infect. 3, 747–754.

FAO (2001). Manual on the Application of the HACCP System in Mycotoxin Prevention and Control. FAO Food and Nutrition Paper.

FAO (2008a). Viruses in Food: Scientific Advice to support risk management activities. Microbiological Risk Assessment Series No. 7.

FAO/IOC/WHO (2005) Report of ad hoc Expert Consultation on Biotoxins in Bivalve Mollusco. September 2004, Rome.

Faust, M. A., & Gulledge, R. A. (2002) Identifying Harmful Marine Dino flagellates. The Smithsonian National Museum of Natural History, Washington D. C. http://botany. si.edu/references/dinoflag/index.htm

Fleming, L. E., Broad, K., Clement, A., Dewailly, E., Elmir, S., Knap, A., Pomponi, S. A., Smith, S., Gabriele, H. S., & Walsh, P. (2006). Oceans and human health: Emerging public health risks in the marine environment. Marine Pollution Bulletin 53:545–560

Glibert, P. M., Anderson, D. M., Gentien, P., Granéli, E., & Sellner, K. G. (2005) The global, complex phenomena of Harmful Algal Blooms. Oceanography 18:136–147

Hall, G. V., D'Souza, R. M., & Kirk, M. D. (2002). Food borne disease in the new millennium: out of the frying pan and into the fire? Med. J. Aust. 177:614–618.

IPCC (Intergovernmental Panel on Climate Change) (2007). "Summary for Policymakers. " In *Climate Change 2007: Impacts, Adaptation and vulnerability. Contribution of Working Group II to the Fourth Assessment Report of the Intergovernmental Panel on Climate Change*, Parry, M. L., Canziani, O. F., Palutikot, J. P., van der Linden, P. J., & Hanson, C. E., eds. Cambridge, UK: Cambridge University Press.

James, S. J., & James, C. (2010). The food cold-chain and climate change. Food Research International. In press.

Kundzewicz, Z. W., Mata, L. J., Arnell, N. W., Döll, P., Kabat, P., Jiménez, B., Miller, K. A., Oki, T., Sen, Z., & Shiklomanov, I. A. (2007). Freshwater resources and their management. *Climate Change 2007: Impacts, Adaptation and Vulnerability. Contribution of Working Group II to the Fourth Assessment Report of the Intergovernmental Panel on Climate Change*, M.L. 47.

Kovats, R. S., Edwards, S. J., Charron, D., Cowden, J., D'Souza, R. M., Ebi, K. L., Gauci, C., Gerner- Smidt, P., Hajit, S., Hales, S., Hernandez Pezzi, G., Kriz, B., Kutsar, K., McKeown, P., Mellou, K., Meene, B., O'Brien, S., van Pelt, W., & Schmid, H. (2005) Climate variability and campylobacter infection: an international study. Int. J. Bio. Meteorol. 49, 207–214.

Luby, S. P. et al. (2006) Food borne transmission of Nipah virus, Bangladesh. *Emerging Infectious Diseases*, 12(12), 1888–1894.

Moretti, C. L., Mattos, L. M., Calbo, A. G, & Sargent, S. A. (2010) Food Research International 43, 1824–1832.

Nicholls, R. J., Wong, P. P., Burkett, V. R., Codignotto, J. O., Hay, J. E., McLean, R. F., Ragoonaden, S., Woodroffe, C. D. (2007). Coastal systems and low-lying areas. In: Climate Change 2007: Impacts, Adaptation and Vulnerability Contribution of Working Group II to the Fourth Assessment Report of the Intergovernmental Panel on Climate Change. Cambridge University Press, Cambridge, UK, 315–356.

NOAA–CSCOR (2008). Economic Impacts of Harmful Algal Blooms (HABs) fact sheet.

http://www.cop.noaa.gov/stressors/extremeevents/hab/current/HAB_Econ.html

Pardue, J. H., Moe, W. M., McInnis, D., Thibodeaux, L. J., Valsaraj, K. T., Maciasz, E., van Heerden, I., Korevec, N., & Yuan, Q. Z. (2005). Chemical and microbiological parameters in New Orleans floodwater following Hurricane Katrina. Environ. Sci. Technol. 39, 8591–8599.

Paz, S., Bisharat, N., Paz, E., Kidar, O., & Cohen, D. (2007). Climate change and the emergence of *Vibrio vulnificus* disease in Israel. Environ. Res. 103, 390–396.

Rose, J. B., Epstein, P. R., Lipp, E. K., Sherman, B. H., Bernard, S. M., & Patz, J. A. (2001). Climate variability and change in the United States: potential impacts on water and food borne diseases caused by microbiologic agent. Environ. Health Perspectives 109, 211–221.

Sellner, K. G., Doucette, G. J., & Kirkpatrick, G. J. (2003). Harmful algal blooms causes, impacts and detection. Journal of Industrial Microbiology and Biotechnology 30, 383–406.

Smayda, T. J. (1990) Novel and nuisance phytoplankton blooms in the Sea: Evidence for a global epidemic. In: Toxic Marine Phytoplankton, 4th International Conference. Elsevier, Amsterdam, 29–40.

Soil and Water Conservation Society (2003). Soil erosion and runoff from cropland Report from the USA, Soil and Water Conservation Society, 63. 49.

Trainer, V. L., & Suddleson, M. (2005). Monitoring approaches for early warning of domoic acid events in Washington State. Oceanography 18, 228–237

Umlauf, G., Bidoglio, G., Christoph, E., Kampheus, J., Krüger, F., Landmann, D., Schulz, A. J., Schwartz, R., Severin, K., Stachel, B., & Dorit, S. (2005). The situation of PCDD/Fs and Dioxin-like PCBs after the flooding of river Elbe and Mulde in 2002. Acta Hydrochim. Hydrobiol. 33(5), 543−554.

WHO (2005). Ecosystems and human well-being: health synthesis. A report of the Millennium Ecosystem Assessment, World Health Organization, Geneva, 54, pp.

WHO (2005a). Ensuring Food Safety in the Aftermath of Natural Disasters World Health Organization, Geneva.

CHAPTER 19

CLIMATE CHANGE, FOOD SECURITY, AND LIVELIHOOD OPPORTUNITIES IN MOUNTAIN AGRICULTURE

AMIT KUMAR[1], NIRMAL SHARMA[2], M. S. AHMAD[3], and MOHAMMED WASIM SIDDIQUI[3]

[1]Division of Fruit Science, Sher-e-Kashmir University of Agricultural Science and Technology - Kashmir, Shalimar, India;
E-mail: khokherak@rediffmail.com

[2]Division of Fruit Science, Sher-e-Kashmir University of Agricultural Science and Technology - Jammu, Chatha, India.

[3]Department of Food Science and Technology, Bihar Agricultural University, Sabour, Bhagalpur, Bihar, India.

CONTENTS

19.1 INTRODUCTION

The advancement of any country depends upon agriculture that is the backbone of most of the many countries of the world. Correspondingly, agriculture is the main sector having potential to eliminate poverty because proportion of people living less than $1.25/day had dropped (FAO, 2012). Consequently, to exterminate starvation and paucity, focus must be toward agriculture sector especially in climate change that is most concerned issue of the era. World population is projected to be 8 Billion in 2025 therefore, understanding the climate change effects on agriculture needs extra consideration to feed billion of population. It was estimated that around 60% of the world population live in Asia and population is rising with double swift, that is,1% per year (UN, 2012).

Climate change is perceived to be the greatest threat to the food security and mankind in twenty-first century. Over the past few decades, increase in average air temperature on earth and its associated effects on climate and crops have became a concern worldwide, particularly after the 4th Assessment Report of Intergovernmental Panel on Climate Change (IPCC, 2007). Since preindustrial era to the year 2009, the carbon dioxide (CO_2) concentration in the atmosphere has increased from 280 ppm to 384 ppm coupled with an increase in mean temperature of 0.76°C. Further, according to the studies carried out by IPCC, it is predicted that by the end of this century increase in average air temperature in Asian countries could range between 1.8 to 6.0°C and CO_2 concentration may reach up to 700 ppm or more (IPCC, 2007). South Asian countries are predicted to have least increase in temperature, in the range of 1.8 to 5.0°C except for the Himalayas (IPCC, 2007). On the other hand, changes in rainfall pattern have also been predicted in the range of −5 to 20% during the winter season and −40 to 15% during the summer season (IPCC, 2007). As a consequence of rising atmospheric temperature, there will be frequent occurrence of drought, flood and heat waves.

The major greenhouse gasses (GHGs) responsible for climate change are carbon dioxide (CO_2), methane (CH_4) and nitrous oxide (N_2O). Collectively, these three gasses are responsible for 99% of global warming. Moreover, perfluorocarbons (PFCs), hydroflourocarbons (HFCs), sulfur hexafluoride (SF_6) and ozone (O_3) also make a small contribution to global warming. Among the different GHGs, carbon dioxide is the most important GHG and is alone responsible for 77% of global warming. The global warming potential of different GHGs is measured in terms of carbon dioxide equivalence (CO_{2eq}), which is the warming effect exerted by one molecule of CO_2 over a given period of time, usually 20 or 100 years. Methane and nitrous oxide have a global warming potential over a 100 year period are about 23 CO_{2eq} and 298 CO_{2eq}, respectively.

Horticulture depends on natural calamities for example favorable climate leads to good crop yield and ensures food safety. Since horticulture is the most affecting sector by climate change therefore, adaptation approaches need to be considered for survival of horticultural crops without affecting the nutritional status. This manuscript reviews current knowledge about the relationships between climate change, water, food and livelihood security.

19.2   AGRICULTURE, CLIMATE AND FOOD SECURITY

Agriculture is important for food security in two ways: it produces the food, people eat and (perhaps even more important) it provides the primary source of livelihood for 36 percent of the world's total workforce. In the heavily populated countries of Asia and the Pacific, this share ranges from 40 to 50%, and in sub-Saharan Africa, two-thirds of the working population still make their living from agriculture (ILO, 2007). If agricultural production in the low-income developing countries of Asia and Africa is adversely affected by climate change, the livelihoods of large numbers of the rural poor will be put at risk and their vulnerability to food insecurity increased. Agriculture, forestry and fisheries are all sensitive to climate. Their production processes are therefore likely to be affected by climate change. In general, impacts are expected to be positive in temperate regions and negative in tropical ones, but there is still uncertainly about how projected changes will play out at the local level, and potential impacts may be altered by the adoption of risk management measures and adaptation strategies that strengthen preparedness and resilience.

The food security implications of changes in agricultural production patterns and performance are of two kinds:

- Impacts on the production of food will affect food supply at the global and local levels. Globally, higher yields in temperate regions could offset lower yields in tropical regions. However, in many low-income countries with limited financial capacity to trade and high dependence on their own production to cover food requirements, it may not be possible to offset declines in local supply without increasing reliance on food aid.

- Impacts on all forms of agricultural production will affect livelihoods and access to food. Producer groups that are less able to deal with climate change, such as the rural poor in developing countries, risk having their safety and welfare compromised.

Other food system processes, such as food processing, distribution, acquisition, preparation and consumption, are as important for food security as food and agricultural production. Technological advances and the development of long-distance marketing chains that move produce and packaged foods throughout the world at high speed and relatively low cost have made overall food system performance far less dependent on climate than it was 200 years ago. However, as the frequency and intensity of severe weather increase, there is a growing risk of storm damage to transport and distribution infrastructure, with consequent disruption of food supply chains. The rising cost of energy and the need to reduce fossil fuel usage along the food chain have led to a new calculus – "food miles,"which should be kept as low as possible to reduce emissions. These factors could result in more local responsibility for food security, which needs to be considered in the formulation of adaptation strategies for people who are currently vulnerable or who could become so within the foreseeable future.

19.3 POTENTIAL IMPACTS OF CLIMATE CHANGE ON FOOD AVAILABILITY

Productionof food and other agricultural commodities may keep pace with aggregate demand, but there are likely to be significant changes in local cropping patterns and farming practices. There has been a lot of research on the impacts that climate change might have on agricultural production, particularly cultivated crops. Some 50% of total crop production comes from forest and mountain ecosystems, including all tree crops, while crops cultivated on open, arable flat land account for only 13% of annual global crop production. Production from both rainfed and irrigated agriculture in dry land ecosystems accounts for approximately 25%, and rice produced in coastal ecosystems for about 12% (Millennium Ecosystem Assessment, 2005).

19.4 POTENTIAL IMPACTS OF CLIMATE CHANGE ON FOOD ACCESS

19.4.1 ALLOCATION

Food is allocated through markets and nonmarket distribution mechanisms. Political and social power relationships are key factors influencing allocation decisions in times of scarcity. If agricultural production declines and households find alternative livelihood activities, social processes and reciprocal relations in which locally produced food is given to other family members in exchange for their support may change or disappear altogether. Public and charitable food distribution schemes reallocate food to the most needy, but are subject to public perceptions about who needs help, and social values about what kind of help it is incumbent on more wealthy segments of society to provide. If climate change creates other more urgent claims on public resources, support for food distribution schemes may decline, with consequent increases in the incidence of food insecurity, hunger and famine related deaths.

19.4.2 AFFORDABILITY

Climate impacts on income-earning opportunities can affect the ability to buy food, and a change in climate or climate extremes may affect the availability of certain food products, whichmay influence their price. High prices may make certain foods unaffordable and can have an impact on individual's nutrition and health.

19.4.3 PREFERENCE

Food preferences determine the kinds of food households will attempt to obtain. Changing climatic conditions may affect both the physical and the economic availability of certain preferred food items, which might make it impossible to meet some preferences. Changes in availability and relative prices for major food items may result in people either changing their food basket, or spending a greater percentage of their income on food when prices of preferred food items increase.

19.5   POTENTIAL IMPACTS OF CLIMATE CHANGE ON FOOD UTILIZATION

19.5.1   NUTRITIONAL VALUE

Food insecurity is usually associated with malnutrition, because the diets of people who are unable to satisfy all of their food needs usually contain a high proportion of staple foods and lack the variety needed to satisfy nutritional requirements. Declines in the availability of wild foods, and limits on small-scale horticultural production due to scarcity of water or labor resulting from climate change could affect nutritional status adversely. In general, however, the main impact of climate change on nutrition is likely to be felt indirectly, through its effects on income and capacity to purchase a diversity of foods.

19.6   POTENTIAL IMPACTS OF CLIMATE CHANGE ON FOOD SYSTEM STABILITY

19.6.1   STABILITY OF SUPPLY

Many crops have annual cycles and yields fluctuate with climate variability, particularly rainfall and temperature. Maintaining the continuity of food supply when production is seasonal, is therefore challenging. Droughts and floods are a particular threat to food stability and could bring about both chronic and transitory food insecurity. Both are expected to become more frequent, more intense and less predictable as a consequence of climate change. In rural areas that depend on rainfed agriculture for an important part of their local food supply, changes in the amount and timing of rainfall within the season and an increase in weather variability are likely to aggravate the precariousness of local food systems.

19.6.2   STABILITY OF ACCESS

As already noted, the affordability of food is determined by the relationship between household income and the cost of a typical food basket. Global food markets mayexhibit greater price volatility, jeopardizing the stability of returns to farmers and the access to purchased food of both farming and nonfarming poor people.

19.6.3   FOOD EMERGENCIES

Increasing instability of supply, attributable to the consequences of climate change, will most likely lead to increases in the frequency and magnitude of food emergencies with which the global food system is ill-equipped to cope. An increase in human conflict, caused in part by migration and resource competition attributable to changing climatic conditions, would also be destabilizing for food systems at all levels. Climate change might exacerbate conflict in numerous ways, although links between climate change and conflict should be presented with care. Increasing incidence of drought may force people to migrate from one area to another, giving rise to conflict over access to resources in the receiving area. Resource scarcity can also trigger conflict and could be driven by global environmental change.

19.6.4   LIVELIHOOD VULNERABILITY

Agriculture is often at the heart of the livelihood strategies of these marginal groups; agricultural employment, whether farming their own land or working on that of others, is key to their survival. In many areas, the challenges of rural livelihoods drive urban migration. As the number of poor and vulnerable people living in urban slums grows, the availability of nonfarm employment opportunities and the access of urban dwellers to adequate food from the market will become increasingly important drivers of food security.

Livelihood groups that warrant special attention in the context of climate change include:

- Low-income groups in drought and flood-prone areas with poor food distribution infrastructure and limited access to emergency response.
- Low to middle income groups in flood prone areas that may lose homes, stored food, personal possessions and means of obtaining their livelihood, particularly when water rises very quickly and with great force, as in sea surges or flash floods.

- Farmers whose land becomes submerged or damaged by sea-level rise or saltwater intrusions.
- Producers of crops that may not be sustainable under changing temperature and rainfall regimes.
- Producers of crops at risk from high winds.
- Poor livestock keepers in dry lands where changes in rainfall patterns will affect forage availability and quality.
- Managers of forest ecosystems that provide forest products and environmental services.
- Fishers whose infrastructure for fishing activities, such as port and landing facilities, storage facilities, fish ponds and processing areas, becomes submerged or damaged by sea-level rise, flooding or extreme weather events.
- Fishing communities that depend heavily on coral reefs for food and protection from natural disasters.
- Fishers/aqua-farmers who suffer diminishing catches from shifts in fish distribution and the productivity of aquatic ecosystems, caused by changes in ocean currents or increased discharge of freshwater into oceans.

19.7   CLIMATE CHANGE IN HIMALAYAS

Out of about 34 million people inhabiting the mountainous region of Himalaya, the major communities belong to hill, mountain and highland farming. They sustain on largely subsistence farming which they practice on marginal rainfed and some irrigated farmlands occupying 15.8% of the total area of the Himalayas. The rest of the Himalaya includes rangelands, pastures, wasteland and the forests, which account for nearly 69% of the Himalayan area. Another 15.2% is under permanent snow cover and Rocky Mountains and serves as perennial source of clean water to the hill people as well as to rest of the nation. Agriculture is the primary sector of the economy, contributing 45% to the total regional income of the inhabitants. The majority of the farming households in the hilly areas have landholdings of less than 0.5 ha and these are continuously becoming smaller. In Indian Himalaya livelihood through agriculture has been threatened by declining per capita available cropland coupled with poor productivity, poor production management, labor shortages, poor post production management, poor marketing and networks (lack of market development) and lack of entrepreneurship. Number of reports has indicated that mountainous regions are noticeably impacted by climate change. The most widely reported impact is the rapid reduction in glaciers (which has profound future implications for downstream water resources), crop shift, decline in productivity and nonperformance of existing crop varieties. The impacts of climate change are superimposed on a variety of other environmental and social stresses, many already recognized as severe (Ives and Messerli, 1989).

The Himalayan region is the source of ten of the largest rivers in Asia (Table 19.1). The basins of these rivers are inhabited by 1.3 billion people and contain seven megacities. Natural resources in these basins provide the basis for a substantial part of the region's total GDP and important environmental services, which are also of importance beyond the region (Penland and Kulp, 2005; Macintosh, 2005;Nicholls, 1995; Sanlaville and Prieur, 2005; She, 2004; Woodroffe et al., 2006). Continuing climate change is predicted to lead to major change in the strength and timing of the Asian monsoon, inner Asian high-pressure systems, and winter westerlies – the main systems affecting the climate of the Himalayan region. The impacts on river flows, groundwater recharge, natural hazards, and the ecosystem, as well as on people and their livelihoods, could be dramatic, although not the same in terms of rate, intensity or direction in all parts of the region. Given the current state of knowledge about climate change, determining the diversity of impacts is a challenge for researchers, and risk assessment is needed to guide future action.

TABLE 19.1 Principal Rivers of the Himalayan Region – Basin Statistics

River	River basin					
River	Annual mean discharge m³/sec	% of glacier melt in river flow	Basin area (km²)	Population density (per/km²)	Population ×1000	Water availability (m³/person/year)
Amu Darya	1376[a]	Not available	534,739	39	20,855	2081
Brahmaputra	21,261[a]	~ 12	651,335	182	118,543	5656
Ganges	12,037[a]	~ 9	1,016,124	401	407,466	932
Indus	5533[a]	Up to 50	1,081,718	165	178,483	978

Irrawaddy	8024[a]	Not available	413,710	79	32,683	7742
Mekong	9001[a]	~ 7	805,604	71	57,198	4963
Salween	1494[a]	~ 9	271,914	22	5982	7876
Tarim	1262[a]	Up to 50	1,152,448	7	8067	4933
Yangtze	28,811[a]	~ 18	1,722,193	214	368,549	2465
Yellow	1438[a]	~ 2	944,970	156	147,415	308
Total			1,345,241			

[a]The data were collected by the Global Runoff Data Centre (GRDC) from the following most downstream stations of the river basins:Chatly (Amu Darya), Bahadurabad (Brahmaputra), Farakka (Ganges), Pakse (Mekong), Datong (Yangtze), Huayuankou (Yellow).

[b]Estimation of the melt water contribution is difficult and varies in an upstream and downstream situation; approximates are given here.

Source: Chalise and Khanal (2001); Tarar (1982); Chen et al. (2005).

19.8   HIMALAYAN CLIMATE AND WATER

The Himalayas display great climatic variability. The mountains act as a barrier to atmospheric circulation for both the summer monsoon and the winter westerlies. The summer monsoon dominates the climate, lasting eight months (March-October) in the eastern Himalayas, four months (June-September) in the central Himalayas, and two months (July–August) in the western Himalayas (Chalise and Khanal, 2001). The east-west variation is based on the dominance of different weather systems, which in turn cause the monsoon to weaken from east to west. The monsoon penetrates northwards along the Brahmaputra River into the south-east Tibetan Plateau, but rarely as far as the Karakoram (Hofer and Messerli, 2006; Rees and Collins, 2006). The highest annual rainfall in the region occurs in Cherrapunji in India, amounting to more than 12,000 mm. The monsoon rainfall is mainly of an aerographic nature, resulting in distinct variations in rainfall with elevation between the southern slopes of the Himalayas and the rain shadow areas on the Tibetan Plateau (Mei'e et al., 1985). On the meso-scale, the impacts of climate are mainly due to local topographic characteristics (Chalise and Khanal, 2001) with dry inner valleys receiving much less rainfall than the adjacent mountain slopes as a result of the lee effect. This suggests that the currently measured rainfall, which is mainly based on measurements of rainfall in the valley bottoms, is not representative for the area, and the use of these data results in significant underestimates.

A substantial portion of the annual precipitation falls as snow, particularly at high altitudes (above 3000 amsl) feeding the Himalayan glaciers. The high Himalayan and inner Asian ranges have the most highly glaciated areas outside the polar regions (Dyurgerov and Meier, 2005; Owen et al., 2002). Glaciated areas in the greater Himalayan region cover an area of more than 112,000 km^2. The Himalayan range alone (a subregion) has a total area of approximately 33,000 km^2 of glaciers or 17% of the mountain area (as compared to 2.2% in the Swiss Alps) with a total ice volume of 3,420 km^3 which provides important short and long-term water storage facilities. These figures are very tentative, however, and need to be followed up with more research. Glaciers undergo winter accumulation and summer ablation in the west, but predominantly synchronous summer accumulation and summer melt in the east. The main melting occurs in high summer; however, when this coincides with the monsoon, it may not be as critical for water supply as when the melting occurs in the shoulder seasons: spring and autumn. When the monsoon is weak, delayed, or fails, melt water from snow and ice may limit or avert catastrophic drought (Table 19.2).

TABLE 19.2 Glaciated Areas in the Greater Himalayan Region (Dyurgerov and Meier, 2005)

Mountain range	Area (km^2)
Tien Shan	15,417
Pamir	12,260
Qilian Shan	1,930
Kunlun Shan	12,260
Karakoram	16,600
Qiantang Plateau	3,360
Tanggulla	2,210
Gandishi	620
Nianqingtangla	7,540
Hengduan	1,620
Himalayas	33,050
Hindu Kush	3,200
Hinduradsh	2,700
Total	112,767

Water from both permanent snow and ice and seasonal snow is released by melting, some to be temporarily stored in high altitude wetlands and lakes, but most flowing directly downstream in the large river systems, giving a distinct seasonal rhythm to annual stream flow regimes in these rivers. The contribution of snow and glacial melt to the major rivers in the region ranges from 2 to 50% of the average flow. In the 'shoulder seasons,' before and after precipitation from the summer monsoon, snow and ice melt contribute about 70% of the flow of the main Ganges, Indus, Tarim, and Kabul rivers (Barnett et al., 2005; Kattelmann, 1987; Singh and Bengtsson, 2004). The rivers of Nepal contribute about 40% of the average annual flow in the Ganges Basin, which alone is home to 500 million people, about 10% of the total human population of the region. Even more importantly, they contribute about 70% of the flow in the dry season (Alford, 1992). In western China, glacial melt provides the principal water source in the dry season for 25% of the population (Xu, 2008). The Indus Irrigation Scheme in Pakistan depends 50 percent or more on runoff originating from snow melt and glacial melt from the eastern Hindu Kush, Karakoram, and western Himalayas (Winiger et al., 2005) (Table 19.3).

TABLE 19.3   Glaciated Areas in the Himalayan Range (Qin, 2002)

Drainage basin	Number of glaciers	Total area (km^2)	Total ice reserves (km^3)
Ganges River	6694	16,677	1971
Brahmaputra River	4366	6579	600
Indus River	5057	8926	850
Total	16,117	32,182	3421

19.9   OBSERVED AND PROJECTED EFFECTS OF CLIMATE CHANGE

Climate change is currently taking place at an unprecedented rate and is projected to compound the pressures on natural resources and the environment associated with rapid urbanization, industrialization, and economic development. It will potentially have profound and widespread effects on the availability of and access to, water resources. By the 2050s, access to freshwater in Asia, particularly in large basins, is projected to decrease.

19.9.1   RISING TEMPERATURES

IPCC's Fourth Assessment Report (IPCC 2007a; 2007b) concludes that there is a more than 90 percent probability that the observed warming since the 1950s is due to the emission of greenhouse gasses from human activity. Temperature projections for the twenty-first century suggest a significant acceleration of warming over that observed in the twentieth century (Ruosteenoja et al., 2003). Warming is least rapid, similar to the global mean warming, in South-east Asia, stronger over South Asia and Eastern Asia, and greatest in the continental interior of Asia (Central, Western, and Northern Asia). Warming will be significant in arid regions of Asia and the Himalayan highlands, including the Tibetan Plateau (Gao et al., 2003; Yao et al., 2008). Based on regional climate models, it is predicted that the temperatures in the Indian subcontinent will rise between 3.5 and 5.5°C by 2100, and on the Tibetan Plateau by 2.5°C by 2050 and 5°C by 2100 (Rupa Kumar et al., 2006). However, because of the extreme topography and complex reactions to the greenhouse effect, even high-resolution climatic models cannot give reliable projections of climate change in the Himalayas.

Various studies suggest that warming in the Himalayas has been much greater than the global average of 0.7 °C over the last 100 years (Dutoit and Ziervogel, 2004; IPCC, 2007a). For example, warming in Nepal was 0.6 °C per decade between 1977 and 2000 (Shrestha et al., 1999). Warming in Nepal and on the Tibetan Plateau has been progressively greater with elevation (Tables 4a and 4b; Fig. 19.1) and suggests that progressively higher warming with higher altitude is a phenomenon prevalent over the whole of the greater Himalayan region (New et al., 2002). The major effect on agriculture in Himalayan region is that of crop shift and under performance of crop varieties. Apple growing area is shrinking because chilling requirement is not being met in some areas where apple was once successfully grown. Similar is the case with horticultural crops. Increasing temperature has also disturbed the annual cycles of agricultural crops in hilly areas. In many areas, a greater proportion of total precipitation appears to be falling as rain than before. As a result, snowmelt begins earlier

and winter is shorter; this affects river regimes, natural hazards, water supplies, and people's livelihoods and infrastructure. The extent and health
of high altitude wetlands, green water flows from terrestrial ecosystems,
reservoirs, and water flow and sediment transport along rivers and in lakes
are also affected.

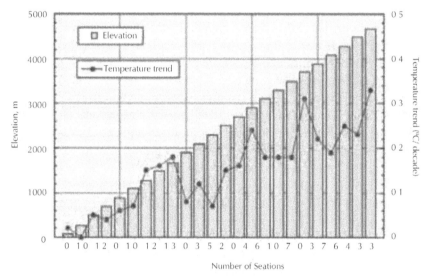

FIGURE 19.1   Dependence of warming on elevation on the Tibetan Plateau (New et al.,
2002).

19.9.2   PRECIPITATION TRENDS

Long-term paleo-climatic studies (e.g., ice core studies on the Tibetan Plateau) show that both wet and dry periods have occurred in the last millennium (Tan et al., 2008, Yao et al., 2008). During the last few decades, inter
seasonal, inter annual, and spatial variability in rainfall trends have been
observed across Asia. In the Himalayan region, both increasing and decreasing trends have been detected. Increasing trends are found on the Tibetan
Plateau in the north-east region (Zhao et al., 2004) and eastern and central
parts (Xu et al., 2007), while the western Tibetan region exhibits a decreasing trend; northern Pakistan also has an increasing trend (Farooq and Khan,
2004). Nepal showed no long-term trend in precipitation between 1948 and
1994 (Shrestha, 2004; Shrestha et al., 2000). A decrease in monsoon precipitation by up to 20% is projected by the end of the century in most parts of

Pakistan and in south-eastern Afghanistan. A similar reduction in precipitation is projected for the southern and eastern Tibetan Plateau and for the central Himalayan range. Increases in the range of 20 to 30% are projected for the western Himalayan Kunlun Shan range and Tien Shan range (Rupa Kumar et al., 2006) (Table 19.4a).

TABLE 19.4A Regional Mean Maximum Temperature Trends in Nepal from 1977–2000 (°C per year)

Region	Seasonal				Annual
	Winter (Dec–Feb)	Pre-monsoon (Mar–May)	Monsoon (Jun–Sep)	Post-monsoon (Oct–Nov)	(Jan–Dec)
Trans-Himalayas	0.12	0.01	0.11	0.1	0.09
Himalayas	0.09	0.05	0.06	0.08	0.06
Middle Mountains	0.06	0.05	0.06	0.09	0.08
Siwaliks	0.02	0.01	0.02	0.08	0.04
Terai	0.01	0	0.01	0.07	0.04
All Nepal	0.06	0.03	0.051	0.08	0.06

Source: updated from Shrestha et al. (1999).

There is a major need for more research on Himalayan precipitation processes, as most studies have excluded the Himalayan region due to the region's extreme, complex topography and lack of adequate rain-gage data (Shrestha et al., 2000) (Table 19.4b).

TABLE 19.4B Average annual increase in temperature at different altitudes on the Tibetan Plateau and surrounding areas 1961–1990 (°C per decade)

Altitude (m)	No. of stations	Spring	Summer	Autumn	Winter	Annual average change
<500	34	−0.18	−0.07	0.08	0.16	0.00
500–1500	37	−0.11	−0.02	0.16	0.42	0.11
1500–2500	26	−0.17	0.03	0.15	0.46	0.12
2500–3500	38	−0.01	0.02	0.19	0.63	0.19
>3500	30	0.12	0.14	0.28	0.46	0.25

19.9.3   GLACIAL RETREAT

Himalayan glaciers are receding faster today than the world average (Dyurgerov and Meier, 2005) (Fig. 19.2). In the last half of the twentieth century, 82% of the glaciers in western China have retreated (Liu et al., 2006). On the Tibetan Plateau, the glacial area has decreased by 4.5% over the last 20 years and by 7% over the last 40 years indicating an increased retreat rate (Ren et al., 2003). Glacier retreat in the Himalayas results from "Precipitation decrease in combination with temperature increase. The glacier shrinkage will speed up if the climatic warming and drying continues" (Ren et al., 2003).

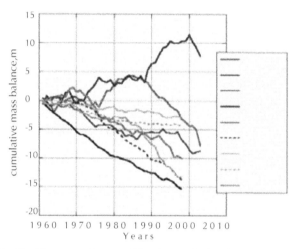

FIGURE 19.2   Rapid retreat of greater Himalayan glaciers in comparison to the global average (Dyurgerov and Meier, 2005).

The IPCC Fourth Assessment Report (IPCC, 2007a, 2007b) states that there is a high measure of confidence that in the coming decades many glaciers in the region will retreat, while smaller glaciers may disappear altogether. Various attempts to model changes in the ice cover and discharge of glacial melt have been made by assuming different climate change scenarios. One concludes that with a 2°C increase by 2050, 35% of the present glaciers will disappear and runoff will increase, peaking between 2030 and 2050 (Qin, 2002).

Retreat in glaciers can destabilize surrounding slopes and may give rise to catastrophic landslides (Ballantyne and Benn, 1994; Dadson and Church, 2005) which can dam streams and sometimes lead to outbreak floods. Excessive melt waters, often in combination with liquid precipitation, may trigger flash floods or debris flows. In the Karakoram, there is growing evidence that catastrophic rockslides have a substantial influence on glaciers and may have triggered glacial surges (Hewitt, 2005).

19.10   CONSEQUENCES FOR LIVELIHOODS AND THE ENVIRONMENT

19.10.1   PASTURES AND AGRICULTURE

The location and area of natural vegetation zones on the Tibetan Plateau will change substantially under projected climate change scenarios. Areas of temperate grassland and cold temperate Coniferous forest could expand, while temperate and cold deserts may shrink. The vertical distribution of vegetation zones could move to higher altitude. Climate change may also result in a shift of the boundary of the farming-pastoral transition region to the south in North-east China, whichmay increase grassland areas and provide more favorable conditions for livestock production. However, the transition area of the farming-pastoral region is also an area of potential desertification, and if protection measures are not taken in new transition areas, desertification may occur (Li and Zhou, 2001; Qiu et al., 2001). More frequent and prolonged droughts as a consequence of climate change together with other anthropogenic factors may also result in desertification.

There is significant uncertainty about the effects of global warming on the vegetation and animal productivity of large dryland ecosystems. Although high altitude drylands might enjoy increases in net primary productivity (NPP), locally, the greatest confidence is in predicting implications for vegetation production, with lesser confidence in implications for vegetation composition, animal production, and adaptation options (Campbell and Stafford Smith, 2000). Climate change has been reported to impact on grassland productivity, ecosystems, and the distribution and composition of plant communities (Wilkes, 2008). Some rangelands might suffer from degradation due to the warmer and drier climate (Dirnbock et al., 2003). Degraded rangeland already accounts for over 40 percent of dryland on the Tibetan Plateau (Gao et al., 2003); and it is expanding at a rate of 3 to 5 percent each year (Ma and Wang, 1999). Increases in evapo-ration, reduction in snow cover, and fluctuations in precipitation are key factors contributing to the degradation of dryland ecosystems.

The possibility of alterations in the overall albedo, water balance and surface energy balance in high-altitude grasslands and the increasing degradation and desertification of arid areas is causing concern. Signs of the effects of climate change on grasslands have been documented in the north-east Tibetan Plateau where *Kobresia* sedge and alpine turf commu-nities are changing to semiarid alpine steppes, known as 'black bleaching' (Ma and Wang, 1999; Miller, 2000). Upward movement of the tree line and encroachment of woody vegetation on alpine meadows are reported widely. In the eastern Himalayas, the tree line is rising at a rate of 5 to 10 m per decade (Baker and Moseley, 2007). As temperature rises, species shift their ranges to follow their principal habitats and climatic optima. Increas-ing temperatures and water stress are expected to lead to a 30% decrease in crop yields in Central and South Asia by the mid-twenty-first century (UNDP, 2008). At high altitudes and latitudes, crop yields should increase because of reductions in frost and cold damage. It will be possible to grow rice and wheat at higher latitudes than is currently the case in China.

Irrigated lowland agriculture found in all of the large basins receiving their runoff from the Hindu Kush-Himalayan system, is projected to suffer negatively from lack of dry season water. Considering that the reported or projected glacial melt water component amounts to, for example, 20 to 40% in rivers in Western China (Tao et al., 2005), 50% or more in the Indus (Tarar, 1982) and 30% in the major rivers in Nepal during the premonsoon season (Sharma, 1993), the implications of dry season wa-ter stress are likely to be massive. In addition, an increase in agricultural water demand by 6 to 10% or more is projected for every 1 °C rise in temperature (IPCC, 2007a). As a result, the net cereal production in South Asian countries is projected to decline by at least between 4 to 10% by the end of this century, under the most conservative climate production projections (IPCC, 2007a).

19.10.2   ECOSYSTEMS

Mountain ecosystems contain a series of climatically very different zones within short distances and elevations. They display a range of microhabitats with great biodiversity (Korner, 2004). Mountain ecosystems are sensitive to global warming and show signs of fragmentation and degradation (Korner, 2004; Xu and Wilkes, 2004). Species in high-elevation ecosystems are projected to shift to higher altitudes, although alpine plant species with restricted habitat availability above the tree line are projected to experience severe fragmentation, habitat loss or even extinction if they cannot move to higher elevations (Dirnbock et al., 2003). Climate warming may increase suitable habitats for the water hyacinth *(Eichhornia crassipes)*, a noxious weed able to survive winter temperatures.

The impacts of climate change on forest ecosystems include shifts in the latitude of forest boundaries and the upward movement of tree lines to higher elevations; changes in species composition and in vegetation types; and an increase in net primary productivity (NPP) (Ramakrishna et al., 2003). In the eastern Himalayas, forest vegetation will expand significantly; forest productivity will increase from 1 to 10 percent and it is expected that forest fires and pests such as the North American pinewood nematode *(Bursaphelenchus xylophilus)* will increase as dryness and warmth increase (Rebetez and Dobbertin, 2004).

19.10.3   HUMAN HEALTH

The impact of climate change on health conditions can be broken into three main categories: (1) direct impacts of for example, drought, heat waves, and flash floods, (2) indirect effects due to climate-induced economic dislocation, decline, conflict, crop failure and associated malnutrition and hunger, and (3) indirect effects due to the spread and aggravated intensity of infectious diseases due to changing environmental conditions (WHO, 2005). The latter effect includes the expansion of vector-borne diseases such as malaria and dengue and water-related diseases such as diarrhea. Regions such as the Hindu Kush-Himalayas, located at the fringe of the current geographic distribution of these and many other diseases are particularity susceptible to the negative effect of rising temperatures. It is projected that the spread of malaria, Bartonellosis, tick-borne diseases and infectious diseases linked to the rate of pathogen replication will all be enhanced. Malaria mosquitoes have recently been observed at high altitudes in the region (Eriksson et al., 2008).

Endemic morbidity and mortality due todiarrheal disease associated with floods and droughts are expected to rise in East, South and South-east Asia due to projected changes in the hydrological cycle (IPCC, 2007a). This will be in addition to an already very high global burden of climate change attributablediarrhea. Empirical studies project that the population at risk of dengue fever will be larger in India and China. In these countries, a high increase in mortality due to heat stress is also projected. However, there are also expected to be positive climate change induced effects on the health status of certain populations in the Himalayan region. High altitude areas will open up to new types of agricultural production and new livelihood opportunities, people will find their homes and villages more comfortabledue to less cold conditions and the risks associated with cold and respiratory diseases will be reduced as the use of fuel wood for heating is reduced.

19.10.4 MOUNTAIN INFRASTRUCTURE

Valuable infrastructure, such as hydropower plants, roads, bridges and communication systems will be increasingly at risk from climate change. Entire hydropower generation systems established on many rivers will be in jeopardy if landslides and flash floods increase, and hydropower generation will be affected if there is a decrease in the already low flows during the dry season. Engineers will have to consider how to respond to these challenges (OECD, 2003).

A specific hazard related to glacier retreat is the formation of proglacial lakes and in some cases the events of glacial lake outburst floods (GLOFs). These can have a devastating effect on important and vulnerable infrastructure downstream such as hydropower stations. Equally important, the operations of hydroelectric power stations will become more complex. With climate change, the complexity and variability of river flow generation will both rise (Renoj et al., 2007) and become increasingly difficult to predict. For example, although the annual average proportion of melt water in river flow has been estimated at 13% for rivers flowing to the Ganges from Nepal, from March through May the monthly average proportion is more than 30% (Chaulagain, 2006). This could have serious implications on river flows and water availability for power plants for about six months per year.

The landmark Qinghai-Tibet Railway, built at a huge cost and associated with important development objectives is partly built on permafrost. Projected widespread permafrost melting on the Tibetan Plateau can threaten future railway services (Chen et al., 2005; Xu et al., 2005).

Likewise, the Yangtze River, China's largest river and a crucial supplier of water to industry, agriculture and 500 million domestic users experienced its lowest upper reaches flow since the 1920's in 2006. With upstream dryland expansion, melting glaciers and aggravated sediment deposits that affect downstream flood discharge capacity (Wang et al., 2005), the Three Gorges Dam, the world's largest hydroelectric installation, is also at risk.

19.10.5   LIVELIHOODS, VULNERABILITY AND ADAPTATION

The term 'livelihood' comprises the capabilities, assets (material and social resources), and activities required for a means of living (Carney, 1998). Sustainable livelihood includes the idea of coping with and recovering from stresses and shocks, and maintaining or enhancing existing capabilities and assets. Climate change has made the future of mountain indigenous people and their livelihoods more vulnerable and uncertain. The available scientific evidence suggests that climate change will place significant stress on the rural livelihoods of mountain people. Efforts to reduce vulnerability and enhance the adaptive capacity of at-risk groups need to take a proactive approach that address the social processes leading to vulnerability and the structural inequalities that are often at the root of social-environmental vulnerabilities.

Adaptation to climate change is both related to vulnerability, which can be defined as the "degree to which individuals and systems are susceptible to or unable to cope with the adverse effects of climate change" (Smit and Pilifosova, 2001) and to future potential impacts, either avoidable or unavoidable. Effective adaptation includes both the establishment of adaptive capacity (awareness, governance, and knowledge) and the adaptation itself (change of behavior, practices, and livelihoods according to new conditions) (Mirza, 2007). Adaptation consists of a multitude of options depending on the scale, context and approach. The scale of adaption may be local, national or regional; the context of the adaptation will determine the type of adaptation (e.g., new farming practices in a rural context or water demand management in an urban context); and the approach to adaptation may focus on general poverty alleviation, enhanced transparency in decision making, or the empowerment of women, among other things. Structural inequalities that make adaptation by poor people more difficult will need to be leveled. It is important to note that poor and marginalized people already face all of the difficulties that we usually associate with climate change. This is nothing new to them. They are already facing poor health, susceptibility to floods and landslides and a lack of adequate shelter, food and water. While they do need climate change adaptation they need poverty alleviation even more.

China and India's rapid economic development which has been moving many tens of thousands of people out of poverty every day, may also provide the best way to handle a changing world. It should be noted that much of the adaptation to climate change will be found outside the sphere of natural sciences. For example, to focus only on flood-safe housing or new types of pest-resistant crops is not enough. The focus must include enhanced capacity to adopt (implying a comprehensive approach) new adaptation strategies. "An adaptation strategy to reduce vulnerability to future climate change needs to be incorporated in regulatory procedures, integrated natural resources management and other development planning procedures" (UNDP-GEF, 2007). As poverty is widespread in the Himalayan region, the empowerment of poor people to adapt to climate change is critical.

Examples of adaptation at different levels may include good governance to mainstream climate change into development and institutional reform (Mirza, 2007), general political reform and associated openness (ibid), health education programs (WHO, 2005) and the development of early warning systems for floods, flash floods and droughts.

19.10.6 LACK OF KNOWLEDGE – UNKNOWN DOWNSTREAM EFFECTS

The impact of climate change on the Himalayan cryosphere is not understood sufficiently to be able to estimate the full scale of the downstream impact of reduced snow and ice coverage. While in-depth studies of glaciers, snow and permafrost have been carried out in some areas they are scattered widely in space and time. Few detailed investigations of the response of snow and ice to climate warming have taken place in the Himalayan and other high ranges. Baseline studies are lacking for most areas, particularly for areas higher than 4,000 amsl and there has been little long-term monitoring of climatic variables, perennial snow and ice, runoff and hydrology in the extraordinary heterogeneity of mountain topography (Messerli et al., 2004; Rees and Collins, 2006). In addition, the one common feature that all mountain areas share with one another complexity caused by topography causes temperature and precipitation to vary over very short distances (Becker and Bugmann, 1997), which in turn makes projections difficult.

Three levels of impact to climate change can be identified: (1) local effects; (2) downstream effects; and (3) global feedback effects. The development of adaptive strategies can be approached from the perspective of each of these three different levels. Firstly, adaptive strategies can be developed at the local level, looking at local effects within the Himalayas and giving priority to local adaptation. Secondly, adaptive strategies can be developed from the perspective of the downstream level, evaluating the downstream effects of climate change and designing adaptive strategies around these effects. Thirdly, adaptive strategies can be on the global level, based on the potential feedback mechanism of the environmental changes in the Himalayas to global warming. All three levels are interlinked and interrelated, but full of uncertainty.

19.10.7   LOCAL EFFECTS

Few model simulations have attempted to address issues related to future climatic change in mountain regions, primarily because the current spatial resolution of models is too crude to adequately represent the topographic and land use details (Beniston, 2003). Most climate models and predictions for high-altitude areas (above 4,000 masl) are dependent on extra potation from hydro-meteorological stations at comparatively low altitudes and upon assumptions based on other, better-studied, parts of the world (Rees and Collins, 2004). The importance of the most widespread cryogenic processes – avalanches, debris flows, rock glaciers, alpine permafrost, and surging glaciers has been recognized and their incidence recorded for certain areas. Yet, almost no basic scientific investigation of these cryogenic processes has taken place in the greater Himalayan region, even though they involve significant hazards, whichmay increase or decrease risk in given areas. The immense diversity of local effects found within the region should be recognized: diversity of climates and topo-climates, hydrology and ecology, andabove all of human cultures and activities. Before effective responses can be developed much work has to be carried out to identify and predict the possible effects of climate change across different systems from glaciers to water resources, from biodiversity to food production, from natural hazards to human health and filtered through diverse contexts. In particular, there has been little engagement with local populations so far to learn from their knowledge and experience in adapting to unique and changing environments and to address their concerns and needs (Xu and Rana, 2005).

19.10.8 DOWNSTREAM EFFECTS

The downstream effects of changing water flow regimes in the large Himalayan rivers are to a great extent, unknown. Few (if any) studies have attempted to model the impact of a 30 to 50% reduction in dry season flow on, for example, downstream economic growth, livelihood conditions and urban water use. It is likely that these changes will have major impacts on downstream societies; however, these impacts are largely unknown. Impacts on water resources will differ depending upon the importance or influence of different sectors (such as tourism, irrigated agriculture, industry and resource extraction), the ecosystems involved and the mitigation measures implemented to reduce water-induced hazards. There are substantial variations within, as well as between these sectors in different countries and valleys in the region.

19.10.9 GLOBAL FEEDBACK EFFECTS

Glaciations at low latitudes have the potential to play an important role in the global radiation budget. A climatic feedback mechanism for Himalayan glaciation shows that a higher glacier free or low albedo surface has a cooling effect over the Himalayas and a warming effect over the Persian Gulf and the Arabian Peninsula (Bush, 2000). The Himalayan region is also an important carbon sink, particularly in terms of carbon storage in the soil of grasslands, wetlands, and forests. Wang et al. (2002) estimate that the organic carbon content of soils subtending grasslands on the Qinghai-Tibet Plateau total 33,500   106 metric tons representing almost one quarter of China's total organic soil carbon and 2.5 percent of the global pool of soil carbon. Climatic variables influence soil carbon stocks through their effects on vegetation and through their influence on the rate of decomposition of soil organic matter. In grassland ecosystems, net ecosystem productivity (that is, the amount of carbon sequestered) is very small compared to the size of fluxes, so there is great potential for changes affecting fluxes to change the net flow of carbon and for grasslands, therefore, to shift from being a CO_2 sink to a CO_2 source (Jones and Donnely, 2004), contributing further to global warming.

19.11 POLICY RECOMMENDATIONS

19.11.1 REDUCING SCIENTIFIC UNCERTAINTY

19.11.1.1 DEVELOP SCIENTIFIC PROGRAMS FOR CLIMATE CHANGE MONITORING

Credible, up-to-date scientific knowledge is essential for the development of a climate change policy, including adaptation and mitigation measures. The current review finds a severe lack of field observations. It is essential to develop a scientific basis, in collaboration with government agencies and academia. Remote sensing allows for regular and repeated monitoring of snow cover, which can be carried out by countries such as China and India, with results shared with those lacking such technological infrastructure. Studies need to include both ground-based and satellite-based monitoring. Well-equipped stations and long-term monitoring, networking and cooperation within and outside the region are essential. Participatory methods of assessing and monitoring climate and environmental change, local perceptions and practices at the local level are also required. Local communities can play a role in determining adaptation practices based on local information and knowledge. School science programs can be developed and introduced in local communities.

19.11.1.2   APPLICATION OF REGIONAL CLIMATE MODELS (RCMS)

The Himalayas are not well represented in global models because of the coarse resolution of such models. Regional climate models, with a higher resolution than global ones, need to be constructed for 'hotspots' and run for shorter periods (20 years or so). The results of RCMs have to be downscaled and applied to impact assessments, particularly for watersheds or subcatchments.

19.11.2   MITIGATION MEASURES

19.11.2.1   BEYOND THE KYOTO PROTOCOL

With rapid regional economic growth, China and India, in particular, should accept equal, albeit differentiated, responsibility to developed countries for controlling increasing carbon emissions. Countries should jointly develop a regional action plan for the control of emissions. Participation of all countries has to be achieved by allowing them to interpret the mandates of international agreements according to their national interests and priorities.

19.11.2.2   LAND-USE MANAGEMENT FOR CARBON SINKS AND REDUCED EMISSIONS

Many countries in the Himalayas have experienced forest recovery (or transition), through policy intervention and the participation of local communities in forest management. Examples include forest conservation in Bhutan, tree plantation in China, community forest user groups in Nepal and joint forest management in India. The forests conserved have contributed significantly to carbon sequestration (Fang et al., 2001).

19.11.2.3   PAYMENT FOR ECOSYSTEM SERVICES (PES)

The mountains of the greater Himalayas provide abundant services to the downstream population in terms of water for household purposes, agriculture, hydropower, tourism, spiritual values and transport. There is a heavy responsibility leaning on the shoulders of upstream land and water managers to ensure reliable provision of good quality water downstream. PES schemes can be developed at different scales, from local to national to regional and involve local communities, governments and the private sector. So far, the opportunities to establish PES schemes in the Himalayas to ensure safe provision of good quality water remain largely unexplored. However, land and water managers, as well as policy and decision makers, should be encouraged to look for win-win solutions in this context.

19.11.2.4   DEVELOPMENT OF ALTERNATIVE TECHNOLOGIES

Novel and affordable technologies and energy resources that do not emit greenhouse gasses are needed. Notable examples in the region include the diffusion of hydropower in Bhutan, solar energy and biogas in China, bio-diesel and wind energy in India, and biogas and microhydropower in Nepal.

19.11.3   ADAPTATION MEASURES:

19.11.3.1   DISASTER RISK REDUCTION AND FLOOD FORECASTING

Floods are the main natural disaster aggravating poverty in the Himalayas and downstream. Technical advances in flood forecasting and management offer an opportunity for regional cooperation in disaster management. Regional cooperation in trans boundary disaster risk management should become a political agenda. Preparedness for disasters is essential (www.disasterpreparedness.icimod.org).

19.11.3.2   SUPPORTING COMMUNITY LED ADAPTATION

One approach to vulnerability and local level adaptation is 'bottom-up' community led processes built on local knowledge, innovations and practices. The focus should be on empowering communities to adapt to a changing climate and environment based on their own decision-making processes and participatory technology development with support from outsiders. For example, Tibetan nomads have already noticed the earlier spring and moved yaks to alpine meadows earlier than previously practiced. Farmers in the flood plains of Bangladesh build houses on stilts, and Nepali farmers store crop seeds for postdisaster recovery. Priority should be given to the most vulnerable groups such as women, the poor and people living in fragile habitats such as along riversides and on steep slopes.

19.11.3.3   NATIONAL ADAPTATION PLANS OF ACTION (NAPAS)

NAPAs are currently being prepared by countries under the initiative of the UN Framework Convention on Climate Change. They are expected (a) to identify the most vulnerable sectors to climate change; and (b) to prioritize activities for adaptation measures in those sectors. NAPAs need to pay more attention to sectors such as water, agriculture, health, disaster reduction and forestry as well as the most vulnerable groups.

19.11.3.4   INTEGRATED WATER RESOURCES MANAGEMENT

Disaster preparedness and risk reduction should be seen as an integral part of water resources management. Integrated water resource management (IWRM) should include future climate change scenario and be scaled up from watershed to river basins. Water allocation for households, agriculture and ecosystems deserves particular attention. Water storage based on local practices should be encouraged in mountain regions.

19.11.4   PUBLIC AWARENESS AND ENGAGEMENT

19.11.4.1   FULL DISCLOSURE AND PRIOR INFORMATION FOR GRASS ROOT SOCIETIES

Indigenous and local communities should be fully informed about the impacts of climate change. They have a right to information and materials in their own languages and ways of communicating.

19.11.4.2 ENGAGEMENT OF THE MEDIA AND ACADEMIA

Awareness and knowledge among stakeholders generally about the impacts of global warming and the threat to the ecosystem, communities and infrastructure are inadequate. The media and academia together can play a significant role in public education, awareness building and trend projection.

19.11.4.3 FACILITATION OF INTERNATIONAL POLICY DIALOGUE AND COOPERATION

Regional and international cooperation needs to advance in order to address the ecological, socioeconomic and cultural implications of climate change in the Himalayas. The international community, including donors, decision-makers and the private and public sectors needs to be involved in regional cooperation ventures. This is of particular importance for achieving sustainable and efficient management of trans boundary rivers.

19.11.4.4 LINKING SCIENCE AND POLICY IN CLIMATE CHANGE

Good science based on credible, salient, legitimate knowledge can often lead to good policies in the context of climate change and mountain specificities, and vice versa (Thompson and Gyawali, 2007). By credible, we mean knowledge that has been derived from field observations and tested by local communities. Salient information is information that is immediately relevant and useful to policy makers. Legitimate information is unbiased in its origins and creation and both fair and reasonably comprehensive in its treatment of opposing views and interests. Policy is a formula for the use of power and application of knowledge. The question then is who has the power and who has the knowledge, scientific or local, or a combination of both. Scientific knowledge is useful, but limited and full of uncertainties on the complex Himalayan scale. So, 'nobody knows best' becomes the model (Lebel et al., 2004). Alternative perspectives carry their own set of values and perceptions about who should be making the rules, where the best knowledge lies to guide decisions and about what other knowledge is needed. Four contrasting perspectives state, market, civil society and the greens and locals merge together in decision-making processes. In such processes, scientists have to generate new knowledge with reduced uncertainty and facilitate dialog with balanced perspectives. In such processes, international cooperation is essential for the transfer of technology from outsiders to locals, to build regional cooperation into global program, and to develop the capacity to downscale important results to the regional Hindu Kush Himalayan scale.

KEYWORDS

- **Climate Change**
- **Food Access**
- **Food Security**
- **Human Health**
- **Kyoto Protocol**
- **Nutritional Value**
- **Public Awareness**

REFERENCES

Alford, D. (1992). *Hydrological Aspects of the Himalayan Region.* Kathmandu: ICIMOD.

Anonymous (2011). Climate Change Agriculture and Food Security. http://ccafs.cgiar.org/news/media-centre/climatehotspots.

Baker, B. B., & Moseley, R. K. (2007). Advancing tree line and retreating glaciers: implications for conservation in Yunnan, China. *Arctic, Antarctic and Alpine Research* 39(2), 200–209.

Ballantyne, C. K., & Benn, D. I. (1994). Paraglacial slope adjustment and resedimentation following recent glacier retreat, Fabergstols dalen, Norway. *Arctic, Antarctic and Alpine Research* 26(3), 255–269.

Barnett, T. P., Adam, J. C., & Lettenmaier, D. P. (2005). Potential impacts of a warming climate on water availability in a snow-dominated region. *Nature* 438(17), 303–309.

Becker, A., & Bugmann, H. (1997). Predicting global change impacts on mountain hydrology and ecology: integrated catchment hydrology/altitudinal gradient studies. Workshop Report 43. Stockholm: International Geo-sphere Biosphere Programme.

Beniston, M. (2003). Climatic change in mountain regions: a review of possible impacts. *Climatic Change* 59, 5–31.

Bush, A. B. G. (2000). A positive climate feedback mechanism for Himalayan glaciations. *Quaternary International* 65–66, 3–13.

Campbell, B. D., & Stafford Smith, B. D. (2000). A synthesis of recent global change research on pasture and rangeland production: reduced uncertainties and their management implications. *Agriculture, Ecosystems and Environment* 82, 39–55.

Carney, D. (1998). Sustainable Rural Livelihoods. London DFID.

Chalise, S. R., & Khanal, N. R. (2001). An introduction to climate, hydrology and landslide hazards in the Hindu Kush-Himalayan region. *In* Tianchi, L., Chalise, S. R., & Upreti, B. N. (eds) *Landslide Hazard Mitigation in the Hindu Kush-Himalayas,* 51–62. Kathmandu: ICIMOD.

Chaulagain, N. P. (2006). Impact of Climate Change on Water Resources of Nepal. Ph. D Thesis. University of Flensburg, Flensburg.

Chen, Y., Ding, Y., & She, Z. (2005). Assessment of Climate and Environment Changes in China (II): Impacts, adaptation and mitigation of climate and environment changes. Beijing: China Science Press.

Dadson, S. J., & Church, M. (2005). Post-glacial topographic evolution of glaciated valleys a stochastic landscape evolution model. *Earth Surface Processes and Landforms* 30(11), 1387–1403.

Dirnbock, T., Dullinger, S., & Grabherr, G. (2003). A regional impact assessment of climate and land-use change on alpine vegetation. *Journal of Biogeography* 30, 401–417.

Dutoit, A., & Ziervogel, G. (2004). Vulnerability and food insecurity: Background concepts for Energy Transition. Available at: www.fao.org/nr/ben/ben_en.htm.

Dyurgerov, M. D., & Meier, M. F. (2005). Glaciers and Changing Earth System: A 2004 Snapshot.

Eriksson, M., Fang, J. & Dekens, J. (2008). How does climate affect human health in the Hindu Kush-Himalaya region, *Regional Health Forum* 12(1), 11–15.

Fang, J. Y., Chen, A. P., Peng, C. H., Zhao, S. Q. & Ci, L. J. (2001). Changes in forest biomass carbon storage in China between 1949 and 1998. *Science* 292, 2320–2322.

Farooq, A. B. & Khan, A. H. (2004). Climate change perspective in Pakistan. Proceedings of Capacity Building APN Workshop on Global Change Research,Islamabad, 39–46.

Gao, X. J., Li, D. L., Zhao, Z. C., & Giorgi, F. (2003). Climate change due to greenhouse effects in Qinghai-Xizang Plateau and along the Qianghai Tibet Railway. *Plateau Meteorol.* 22(5), 458–463.

Hewitt, K. (2005). The Karakoram anomaly: Glacier expansion and the elevation effects of Karakoram Himalaya. *Mountain Research and Development* 25(4), 332–340.

Hofer, T., & Messerli, B. (2006). Floods in Bangladesh: History, Dynamics and Rethinking the Role of the Himalayas. New York: United Nations University Press.

ILO. (2007). Employment by sector. *In*Key indicators of the labour market *(KILM), 5th ed.* www.ilo.org/public/english/employment/strat/kilm/download/kilm04.pdf.

IPCC. (2007a). Climate Change 2007:Impacts, adaptation and vulnerability. Contribution of Working Group II to the Fourth Assessment Report of IPCC. Cambridge. UK. Cambridge University Press.

IPCC. (2007b). Climate Change 2007: mitigation of climate change. Contribution of Working Group III to the Fourth Assessment Report of IPCC. Cambridge. UK. Cambridge University Press.

Ives, J. D., & Messerli, B. (1989). The Himalayan Dilemma: Reconciling Development and Conservation. London: John Wiley and Sons.

Jones, M., & Donnely, A. (2004). Carbon sequestration in temperate grassland ecosystems and the influence of management, climate and elevated CO_2. *New Phytologist* 164, 423–439.

Kattelmann, R. (1987). Uncertainty in assessing Himalayan water resources. *Mountain Research and Development* 7(3), 279–286.

Korner, C. (2004). Mountain biodiversity: its causes and function. *Ambio* 13, 11–17.

Lebel, L., Contreras, A., Pasong, S., & Garden, P. (2004). Nobody knows best: alternative perspectives on forest management and governance in Southeast Asia. *International Environmental Agreements: Politics, Law and Economics,* 4(2), 111–127.

Li, B. L., & Zhou, C. H. (2001). Climatic variation and desertification in West Sandy Land of Northeast China Plain. *Journal of Natural Resources* 16, 234–239.

Liu, S. Y., Ding, Y. J., Li, J., Shangguan, D. H., & Zhang, Y. (2006). Glaciers in response to recent climate warming in Western China. *Quaternary Sciences* 26(5), 762–771.

Ma, Y. S. & Wang, Q. J. (1999). Black soil type of rangeland degradation: an overview and perspective. *Pratacultural Science* 16(2), 5–9.

Macintosh, D. (2005). Asia, eastern, coastal ecology. *In* Schwartz, M. (ed.). *Encyclopaedia of Coastal Science.* 56–67, Dordrecht: Springer.

Mei'e, R., Renzhang, Y., & Haoshend, B. (1985). An outline of China's Physical Geography. Beijing: Foreign Language Press.

Messerli, B., Viviroli, D., & Weingartner, R. (2004). Mountains of the world: vulnerable water towers for the 21st Century. *Ambio,* 13, 29–34.

Millennium Ecosystem Assessment. (2005). Ecosystems and human well-being: Synthesis. Washington DC, Island Press.

Miller, D. J. (2000). Searching for grass and water: Rangeland ecosystem sustainability and herders livelihoods in Western China. Unpublished manuscript submitted to ICIMOD, Kathmandu.

Mirza, M. (2007). Climate change, adaptation and adaptative governance in the water sector in South Asia. Scarborough (Canada): Adaptation and Impacts Research Division (AIRD), Department of Physical and Environmental Sciences, University of Toronto.

New, M., Lister, D., Hulme, M. & Makin, I. (2002). A high resolution data set of surface climate over global land areas. *Climate Research* 21, 1–25.

Nicholls, R. J. (1995). Coastal mega-cities and climate change. *Geo Journal* 37, 369–379.

OECD. (2003). Development and climate change: Focus on water resources and hydropower. Paris: Organization for Economic Co-operation and Development.

Owen, L. A., Finkel, R. C., & Caffee, M. W. (2002). A note on the extent of glaciations throughout the Himalaya during the global last glacial maximum. *Quaternary Science Reviews* 21, 147–157.

Penland, S., & Kulp, M. A. (2005). Deltas. In: Schwartz, M. L. (ed.). *Encyclopedia of Coastal Science,* 362–368. Dordrecht: Springer.

Qin, D. H. (2002). Glacier inventory of China (maps). Xi'an: Xi'an Cartographic Publishing House.

Qiu, G. W., Hao, Y. X., & Wang, S. L. (2001). The impacts of climate change on the interlock area of farming pastoral region and its climatic potential productivity in Northern China. *Arid Zone Research* 18, 23–28.

Ramakrishna, R. N., Keeling, C. D., Hashimoto, H., Jolly, W. M., Piper, S. C., Tuker, C. J., Myneni, R. B. & Running, S. W. (2003). Climate driven increases in global terrestrial net primary production from 1982 to 1999. *Science* 300, 1560–1563.

Rebetez, M., & Dobbertin, M. (2004). Climate change may already threaten scots pine stands in the Swiss Alps. *Theoretical and Applied Climatology* 79(1–2), 1–9.

Rees, G. H. & Collins, D. N. (2004). An Assessment of the Impacts of Deglaciation on the Water Resources of the Himalaya. Wallingford: Centre for Ecology and Hydrology.

Rees, G. H., & Collins, D. N. (2006). Regional differences in response of flow in glacier-fed Himalayan Rivers to climate warming. *Hydrological Processes* 20, 2157–2167.

Ren, J. W., Qin, D. H., Kang, S. C., Hou, S. G., Pu, J. C. & Jin, Z. F. (2003). Glacier variations and climate warming and drying in the central Himalayas. *Chinese Science Bulletin* 48(23), 2478–2482.

Renoj, J., Thayyen, J. T. & Dobhai, D. P. (2007). Role of glaciers and snow cover on headwater river hydrology in monsoon regime Micro-scale study of Din Gad catchment, Garhwal Himalaya, India. *Current Science* 92(3), 376–382.

Ruosteenoja, K., Carter, T. R., Jylha, K. & Tuomenvirta, H. (2003). Future climate in world regions: an inter comparison of model-based projections for the new IPCC emissions scenarios. *The Finnish Environment* 644, 83.

Rupa Kumar, K., Sahai, A. K., Krishna, K., Patwardhan, S. K., Mishra, P. K., Revadkar, J. V., Kamala, K. & Pant, G. B. (2006). High resolution climate change scenario for India for the 21st Century. *Current Science* 90, 334–345.

Sanlaville, P., & Prieur, A. (2005). Asia, Middle East, coastal ecology and geomorphology. In: Schwartz, M. L. (ed.). *Encyclopedia of Coastal Science,* 71–83. Dordrecht Springer.

Sharma, K. P. (1993). Role of Melt water in Major River Systems of Nepal. In: Young, G. J. (ed.). International Symposium on Snow and Glacier Hydrology, Kathmandu, International Association of Hydrological Sciences, Publication No. 218, 113–122.

She, Z. X. (2004). Human-land interaction and socio-economic development, with special reference to the Changjiang Delta. In: *Proceedings of Xiangshan Symposium on Human-land Coupling System of River Delta Regions: Past, Present and Future.* Beijing (in Chinese).

Shrestha, A. B. (2004). Climate change in Nepal and its impact on Himalayan glaciers. In: Hare, W. L., Battaglini, A., Cramer, W., Schaeffer, M., Jaeger, C. (eds). *Climate hotspots: Key vulnerable regions, climate change and limits to warming,* Proceedings of the European Climate Change Forum Symposium. Potsdam: Potsdam Institute for Climate Impact Research

Shrestha, A. B., Wake, C. P., Dibb, J. E. & Mayewski, P. A. (2000). Precipitation fluctuations in the Nepal Himalaya and its vicinity and relationship with some large-scale climatology parameters. *International Journal of Climatology* 20, 317–327.

Shrestha, A. B., Wake, C. P., Mayewski, P. A. & Dibb, J. E. (1999). Maximum temperature trends in the Himalaya and its vicinity: an analysis based on temperature records from Nepal for the period 1971–94. *Journal of Climate* 12, 2775–2787.

Singh, P., & Bengtsson, L. (2004). Hydrological sensitivity of a large Himalayan basin to climate change. *Hydrological Processes* 18, 2363–2385.

Smit, B., & Pilifosova, O. (2001). Adaptation to climate change in the context of sustainable development and equity. In: *Climate Change 2001 Impacts, Adaptation and Vulnerability,* Chapter 18. Cambridge: Cambridge University Press.

Stern, N. (2006). The Economics of Climate Change. London Cambridge University Press.

Tan, L., Cai, Y., Yi, L., An, Z., & Li, L. (2008). Precipitation variations of Longxi, northeast margin of Tibetan Plateau since AD 960 and their relationship with solar activity. *Climate of the Past* 4, 19–28.

Tao, F., Yokozawa, M., & Hayashi, E. (2005). A perspective on water resources in China: interactions between climate change and soil degradation. *Climatic Change* 68, 169–197.

Tarar, R. N. (1982). Water resources investigation in Pakistan with the help of Landsat imagery snow surveys, 1975–1978. In: *Hydrological aspects of alpine and high mountain areas. Proceedings of the Exeter Symposium,* IAHS Publication No 138, 177–190.

Thompson, M., & Gyawali, D. (2007). Introduction: Uncertainty revisited. In: Thompson, M., Warburton, M., & Hatley, T. (eds.). *Uncertainty on a Himalayan Scale.* Kathmandu: Himal Books.

UNDP. (2008). Fighting Climate Change Human Solidarity in a Divided World. New York UNDP.

UNDP-GEF. (2007). Climate change impacts, vulnerability and adaptation in Asia and the Pacific. A background note for the Training Workshop on Environmental Finance and GEF Portfolio Management, 22–25 May 2007, Bangkok, Thailand.

Wang, G. X., Qian, J., Cheng, G. & Lai, Y. (2002). Soil organic carbon pool of grassland soils on the Qinghai-Tibetan Plateau and its global implication. *Science of the Total Environment* 291, 207–217.

Wang, X., Xie, Z. & Feng, Q. (2005). Response of glaciers to climate change in the source region of the Yangtze river. *Journal of Glaciology and Geocryology* 27(4), 498–502.

WHO (2005). Human health impacts from climate variability and climate change in the Hindu Kush Himalaya region. Report from an Inter-Regional Workshop, Mukteshwar, India, October 2005.

Wiggins, S. (2008). Rising Food Prices a Global Crisis. *Briefing paper* No 37. London ODI.

Wilkes, A. (2008). Towards mainstreaming climate change in grassland management policies and practices on the Tibetan PlateauWorking Paper No. 67. Beijing: World Agroforestry Centre, ICRAF China.

Winiger, M., Gumpert, M., & Yamout, H. (2005). Karakoram-Hindu Kush-Western Himalaya: assessing high altitude water resources. *Hydrological Processes* 19(12), 2329–2338.

Woodroffe, C. D., Nicholls, R. J., Saito, Y., Chen, Z. & Goodbred, S. L. (2006). Landscape variability and the response of Asian megadeltas to environmental change. *In* Harvey, N. (ed.). *Global Change and Integrated Coastal Management: The Asia-Pacific Region,* 277–314. New York Springer.

www.agis.agric.za/agisweb/fivims_za.

www.un.org/esa/sustdev/documents/agenda21/english/agenda21chapter9.htm.

Xu, J., & Rana, G. M. (2005). Living in the Mountains'. In: Jeggle, T. (ed.). *Know Risk,* 196–199. Geneva: UN Inter-agency Secretariat of the International Strategy for Disaster Reduction.

Xu, J. C. (2008). The highlands: a shared water tower in a changing climate and changing Asia. Working Paper No. 67. Beijing: World Agro forestry Centre, ICRAF-China.

Xu, J. C., & Wilkes, A. (2004). Biodiversity impact analysis in Northwest Yunnan, Southwest China. *Biodiversity and Conservation* 13(5), 959–983.

Xu, Y., Zhao, Z. C. & Li, D. (2005). Simulations of climate change for the next 50, years over Tibetan Plateau and along the line of Qing Zang Railway. *Plateau Meteorology* 24(5), 698–707.

Xu, Z., Gong, T., & Liu, C. (2007). Detection of decadal trends in precipitation across the Tibetan Plateau'. In: *Methodology in Hydrology.* Proceedings of the Second International Symposium on Methodology in Hydrology held in Nanjing, China, IAHS Publication 311, 271–276.

Yao, T., Duan, K., Xu, B., Wang, N., Guo, X., & Yang, X. (2008). Ice core precipitation record in Central Tibetan plateau since AD 1600. *Climate of the Past Discuss.* 4, 233–248.

Zhao, L., Ping, C. L., Yang, D. Q., Cheng, G. D., Ding, Y. J. & Liu, S. Y. (2004). Change of climate and seasonally frozen ground over the past 30, years in Qinghai-Tibetan plateau, China. *Global and Planetary Change* 43, 19–31.

INDEX